METHODS IN MOLECULAR BIOLOGY

Series Editor
John M. Walker
School of Life and Medical Sciences
University of Hertfordshire
Hatfield, Hertfordshire, UK

For further volumes:
http://www.springer.com/series/7651

For over 35 years, biological scientists have come to rely on the research protocols and methodologies in the critically acclaimed *Methods in Molecular Biology* series. The series was the first to introduce the step-by-step protocols approach that has become the standard in all biomedical protocol publishing. Each protocol is provided in readily-reproducible step-by-step fashion, opening with an introductory overview, a list of the materials and reagents needed to complete the experiment, and followed by a detailed procedure that is supported with a helpful notes section offering tips and tricks of the trade as well as troubleshooting advice. These hallmark features were introduced by series editor Dr. John Walker and constitute the key ingredient in each and every volume of the *Methods in Molecular Biology* series. Tested and trusted, comprehensive and reliable, all protocols from the series are indexed in PubMed.

Intestinal Differentiated Cells

Methods and Protocols

Edited by

Paloma Ordóñez-Morán

Translational Medical Sciences Unit, School of Medicine, Centre for Cancer Sciences, Biodiscovery Institute-3, University Park, University of Nottingham, Nottingham, UK

Humana Press

Editor
Paloma Ordóñez-Morán
Translational Medical Sciences Unit,
School of Medicine, Centre for Cancer
Sciences, Biodiscovery Institute-3
University Park, University of Nottingham
Nottingham, UK

ISSN 1064-3745 ISSN 1940-6029 (electronic)
Methods in Molecular Biology
ISBN 978-1-0716-3075-4 ISBN 978-1-0716-3076-1 (eBook)
https://doi.org/10.1007/978-1-0716-3076-1

This Humana imprint is published by the registered company Springer Science+Business Media, LLC, part of Springer Nature.
The registered company address is: 1 New York Plaza, New York, NY 10004, U.S.A.

Preface

The intestinal epithelium plays an essential role in maintaining the barrier against pathogen invasion as well as in absorbing nutrients. To ensure intestinal homeostasis, the stem cells at the crypt base differentiate into progenitor cells which migrate upward the villus to turn into more specialized cell types. The lineage specification between absorptive and secretory fate at the progenitor cells is tightly regulated. The absorptive cells consist of enterocytes and microfold cells, while the secretory consists of goblet cells, enteroendocrine cells, tuft cells, and paneth cells. All these specialized cells fulfill a function which creates an "ecosystem" that maintains a correct intestinal function.

The goal of this book is to englobe the most up-to-date methods of the intestinal differentiation field. We provide here step-by-step guidance to a variety of techniques for studying intestinal differentiated cells. We aim to provide a comprehensive and easy to follow protocols that are designed to be helpful to both seasoned researchers and newcomers to the field. The protocols included in this volume are separated into four different parts. Part I (Chaps. 1, 2, 3, 4 and 5) describes techniques to characterize intestinal differentiated cell functions by innovative imaging and functional assays. Part II (Chaps. 6, 7 and 8) outlines the powerful of RNAseq and single-cell RNAseq transcriptional profiling methods. These recent years many researchers have better described intestinal stem cells heterogeneity thanks to the development of emerging technologies. Part III (Chaps. 9, 10, 11, 12, 13, 14, 15 and 16) presents protocols for the isolation of intestinal crypts to generate in vitro models to study differentiated cells. Functional analysis of differentiated cells and their environment can currently be performed by using innovative in vitro technologies that allows long-term culture and maintains basic crypt-villus physiology. This method allows a level of tractability that is impossible to achieve in vivo. Finally, Part IV (Chaps. 17, 18, 19 and 20) presents examples of the use of state-of-the-art methods for studying intestinal differentiated cancer cells.

I would like to thank all of the contributors for sharing their expertise and for carefully guiding readers through all the details of their respective techniques. I am very grateful to the series editor, Dr. John Walker, for his help during the editing process and to my family.

Nottingham, UK *Paloma Ordóñez-Morán*

Contents

Contributors

DAVID O. BATES • *Translational Medical Sciences Unit, School of Medicine, Centre for Cancer Sciences, Biodiscovery Institute-3, University Park, University of Nottingham, Nottingham, UK*

ANDREW V. BENEST • *Translational Medical Sciences Unit, School of Medicine, Centre for Cancer Sciences, Biodiscovery Institute-3, University Park, University of Nottingham, Nottingham, UK*

J. BEUMER • *Hubrecht Institute, Royal Netherlands Academy of Arts and Sciences (KNAW) and UMC Utrecht, Utrecht, the Netherlands; Oncode Institute, Hubrecht Institute, Utrecht, the Netherlands*

ANDREA BONILLA-DÍAZ • *Department of Biochemistry and Molecular Biomedicine, Faculty of Biology, Institute of Biomedicine, University of Barcelona, Barcelona, Spain*

HANNAH M. BROWN • *School of Health and Life Science, Teesside University, Middlesbrough, UK; National Horizons Centre, Teesside University, Darlington, UK*

CHIA-WEI CHENG • *Columbia Stem Cell Initiative, Columbia University Irving Medical Center, New York, NY, USA; Koch Institute for Integrative Cancer Research, MIT, Cambridge, MA, USA; Department of Genetics and Development, Columbia University Irving Medical Center, New York, NY, USA*

IRENE CHICOTE • *Stem Cells and Cancer Laboratory, Vall d'Hebron Institute of Oncology (VHIO), CIBERONC, Barcelona, Spain*

H. CLEVERS • *Hubrecht Institute, Royal Netherlands Academy of Arts and Sciences (KNAW) and UMC Utrecht, Utrecht, the Netherlands; Oncode Institute, Hubrecht Institute, Utrecht, the Netherlands; The Princess Maxima Center for Pediatric Oncology, Utrecht, the Netherlands*

WILLIAM DALLEYWATER • *Translational Medical Sciences, School of Medicine, University of Nottingham, Nottingham, UK; Department of Cellular Pathology, Nottingham University Hospitals NHS Trust, Nottingham, UK*

JESSICA R. DIGRAZIA • *Department of Regenerative Medicine and Cell Biology, Medical University of South Carolina, Charleston, SC, USA*

RUSLAN I. DMITRIEV • *Tissue Engineering and Biomaterials Group, Department of Human Structure and Repair, Faculty of Medical and Health Sciences, Ghent University, Ghent, Belgium*

RAYAN EL-KHOLDI • *Columbia Stem Cell Initiative, Columbia University Irving Medical Center, New York, NY, USA; ESPCI Paris, Université PSL, Paris, France, Paris, France*

AMY C. ENGEVIK • *Department of Regenerative Medicine and Cell Biology, Medical University of South Carolina, Charleston, SC, USA*

MELINDA A. ENGEVIK • *Department of Regenerative Medicine and Cell Biology, Medical University of South Carolina, Charleston, SC, USA; Department of Microbiology and Immunology, Medical University of South Carolina, Charleston, SC, USA*

KATHRYN L. FAIR • *Translational Medical Sciences Unit, School of Medicine, Centre for Cancer Sciences, Biodiscovery Institute-3, University Park, University of Nottingham, Nottingham, UK*

KRISHNA R. GAJERA • *Swiss Institute for Experimental Cancer Research (ISREC), École Polytechnique Fédérale de Lausanne-(EPFL-SV), Lausanne, Switzerland; Translational*

Medical Sciences Unit, School of Medicine, Centre for Cancer Sciences, Biodiscovery Institute-3, University Park, University of Nottingham, Nottingham, UK
CHARLYNE GARD • *Aix-Marseille University, INSERM, TAGC, TGML, Marseille, France*
GÉRALDINE GUASCH • *Aix-Marseille University, CNRS, INSERM, Institute Paoli-Calmettes, CRCM, Epithelial Stem Cells and Cancer Team, Marseille, France*
NICHOLAS R. F. HANNAN • *Translational Medical Sciences Unit, School of Medicine, Centre for Cancer Sciences, Biodiscovery Institute-3, University Park, University of Nottingham, Nottingham, UK*
HIKARU HANYU • *Department of Organoid Medicine, Keio University School of Medicine, Tokyo, Japan*
AKIRA HIROTA • *Institute for Chemical Reaction Design and Discovery (WPI-ICReDD), Hokkaido University, Sapporo, Japan*
JOERG HUELSKEN • *Swiss Institute for Experimental Cancer Research (ISREC), École Polytechnique Fédérale de Lausanne-(EPFL-SV), Lausanne, Switzerland*
GEORGINA HYLAND • *Translational Medical Sciences, School of Medicine, University of Nottingham, Nottingham, UK*
MOHAMMAD ILYAS • *Translational Medical Sciences, School of Medicine, University of Nottingham, Nottingham, UK; Department of Cellular Pathology, Nottingham University Hospitals NHS Trust, Nottingham, UK*
MASAMICHI IMAJO • *Institute for Chemical Reaction Design and Discovery (WPI-ICReDD), Hokkaido University, Sapporo, Japan*
M. ANGELES JUANES • *School of Health and Life Science, Teesside University, Middlesbrough, UK; National Horizons Centre, Teesside University, Darlington, UK; Centro de Investigación Príncipe Felipe, Valencia, Spain*
HYUN JUNG KIM • *Department of Inflammation and Immunity, Lerner Research Institute, Cleveland Clinic, Cleveland, OH, USA*
XI LAN • *Koch Institute for Integrative Cancer Research, MIT, Cambridge, MA, USA*
RACHEL LATANICH • *Division of Gastroenterology and Hepatology, Department of Medicine, Johns Hopkins University School of Medicine, Baltimore, MD, USA*
JOSE M. LEMME-DUMIT • *Department of Pediatrics, Center for Vaccine Development and Global Health, University of Maryland School of Medicine, Baltimore, MD, USA*
BÉATRICE LORIOD • *Aix-Marseille University, INSERM, TAGC, TGML, Marseille, France*
MAGDA MARTÍNEZ-ESPUGA • *Department of Environmental and Chemical Engineering, Biodiscovery Institute, University of Nottingham, Nottingham, UK; Translational Medical Sciences Unit, School of Medicine, Centre for Cancer Sciences, Biodiscovery Institute-3, University Park, University of Nottingham, Nottingham, UK*
A. MARTINEZ-SILGADO • *Hubrecht Institute, Royal Netherlands Academy of Arts and Sciences (KNAW) and UMC Utrecht, Utrecht, the Netherlands; Oncode Institute, Hubrecht Institute, Utrecht, the Netherlands*
GURVEER MARVA • *Translational Medical Sciences Unit, School of Medicine, Centre for Cancer Sciences, Biodiscovery Institute-3, University Park, University of Nottingham, Nottingham, UK*
ALVARO MATA • *Department of Environmental and Chemical Engineering, Biodiscovery Institute, University of Nottingham, Nottingham, UK; School of Pharmacy, Biodiscovery Institute, University of Nottingham, Nottingham, UK*
LOUCINÉ MITOYAN • *Aix-Marseille University, CNRS, INSERM, Institute Paoli-Calmettes, CRCM, Epithelial Stem Cells and Cancer Team, Marseille, France*

Gordon W. Moran • *Nottingham Digestive Diseases Centre, University of Nottingham, Nottingham, UK; NIHR Nottingham Biomedical Research Centre, Nottingham University Hospitals NHS Trust, Nottingham, UK*

Sébastien Nin • *Aix-Marseille University, INSERM, TAGC, TGML, Marseille, France*

Maxim Norkin • *Swiss Institute for Experimental Cancer Research (ISREC), École Polytechnique Fédérale de Lausanne-(EPFL-SV), Lausanne, Switzerland*

Irina A. Okkelman • *Tissue Engineering and Biomaterials Group, Department of Human Structure and Repair, Faculty of Medical and Health Sciences, Ghent University, Ghent, Belgium*

Paloma Ordóñez-Morán • *Translational Medical Sciences Unit, School of Medicine, Centre for Cancer Sciences, Biodiscovery Institute-3, University Park, University of Nottingham, Nottingham, UK*

Héctor G. Pálmer • *Stem Cells and Cancer Laboratory, Vall d'Hebron Institute of Oncology (VHIO), CIBERONC, Barcelona, Spain*

Marcela F. Pasetti • *Department of Pediatrics, Center for Vaccine Development and Global Health, University of Maryland School of Medicine, Baltimore, MD, USA*

Sabrina K. Phanor • *Department of Genetics and Development, Columbia University Irving Medical Center, New York, NY, USA*

Isabel Puig • *Stem Cells and Cancer Laboratory, Vall d'Hebron Institute of Oncology (VHIO), CIBERONC, Barcelona, Spain*

Subin Pyo • *Columbia Stem Cell Initiative, Columbia University Irving Medical Center, New York, NY, USA*

Yifan Qiu • *Columbia Stem Cell Initiative, Columbia University Irving Medical Center, New York, NY, USA*

Shahadat Rahman • *Columbia Stem Cell Initiative, Columbia University Irving Medical Center, New York, NY, USA*

Toshiro Sato • *Department of Organoid Medicine, Keio University School of Medicine, Tokyo, Japan; Department of Gastroenterology, Keio University School of Medicine, Tokyo, Japan*

Declan Sculthorpe • *Translational Medical Sciences, School of Medicine, University of Nottingham, Nottingham, UK*

Woojung Shin • *Wyss Institute for Biologically Inspired Engineering at Harvard University, Boston, MA, USA; Department of Bio and Brain Engineering, Korea Advanced Institute of Science and Technology, Daejeon, Republic of Korea*

Janet F. Staab • *Division of Gastroenterology and Hepatology, Department of Medicine, Johns Hopkins University School of Medicine, Baltimore, MD, USA*

Shinya Sugimoto • *Department of Organoid Medicine, Keio University School of Medicine, Tokyo, Japan; Department of Gastroenterology, Keio University School of Medicine, Tokyo, Japan*

Shinya Tanaka • *Institute for Chemical Reaction Design and Discovery (WPI-ICReDD), Hokkaido University, Sapporo, Japan; Department of Cancer Pathology, Faculty of Medicine, Hokkaido University, Sapporo, Japan*

Christopher Terranova • *Genomic Medicine Department, The University of Texas MD Anderson Cancer Center, Houston, TX, USA*

Seyda Ünsal • *Department of Molecular Medicine, Faculty of Health Sciences, University of Southern Denmark, Odense, Denmark*

Francesca Wheat • *Department of Cellular Pathology, University Hospitals of Leicester NHS Trust, Nottingham, UK*

Omer H. Yilmaz • *Koch Institute for Integrative Cancer Research, MIT, Cambridge, MA, USA*

Nicholas C. Zachos • *Division of Gastroenterology and Hepatology, Department of Medicine, Johns Hopkins University School of Medicine, Baltimore, MD, USA*

Peixun Zhou • *School of Health and Life Science, Teesside University, Middlesbrough, UK; National Horizons Centre, Teesside University, Darlington, UK*

Part I

Characterization, Imaging and Functional Assays

Chapter 1

Differentiated Epithelial Cells of the Gut

Andrea Bonilla-Díaz and Paloma Ordóñez-Morán

Abstract

The intestine is a prime example of self-renewal where stem cells give rise to progenitor cells called transit-amplifying cells which differentiate into more specialized cells. There are two intestinal lineages: the absorptive (enterocytes and microfold cells) and the secretory (Paneth cells, enteroendocrine, goblet cells, and tuft cells). Each of these differentiated cell types has a role in creating an "ecosystem" to maintain intestinal homeostasis. Here, we summarize the main roles of each cell type.

Key words Differentiation, Goblet cells, Tuft cells, Microfold cells, Paneth cells, Enterocytes, Enteroendocrine cells, Intestine, Colon

1 Secretory Lineage

1.1 Paneth Cells

Paneth cells are trapezoid-shaped intestinal cells that have short microvilli on their apical surface. These cells are located at the bottom of the small intestinal crypts of Lieberkühn, intercalated within leucine-rich repeat-containing G-protein-coupled receptor 5 (LGR5)$^+$ stem cells [1] (Figs. 1 and 2). In healthy tissue, each crypt contains approximately 5–15 Paneth cells [2]. This cell type has an extensive endoplasmic reticulum and Golgi apparatus with large antimicrobial secretory granules. Due to their secretory function, these cells have a very active lysosomal function [3]. Perhaps one of the most important roles of Paneth cells is the maintenance of gastrointestinal barrier [4]. They prevent virus and bacteria adhesion and absorption without disturbing the intestinal layer of host microbes. During bacterial infections or inflammatory diseases, Paneth cells are increased along the gut and secrete IL-17 that interacts with tumor necrosis factor (TNF)α inflammatory mediators [5]. Paneth cells enhance the release of apical cytoplasmic granules content into the intestinal lumen in the presence of gram-positive and gram-negative bacteria or bacterial products such as lipopolysaccharides [6, 7]. These granules contain Ca^{2+}

Paloma Ordóñez-Morán (ed.), *Intestinal Differentiated Cells: Methods and Protocols*, Methods in Molecular Biology, vol. 2650,
https://doi.org/10.1007/978-1-0716-3076-1_1,

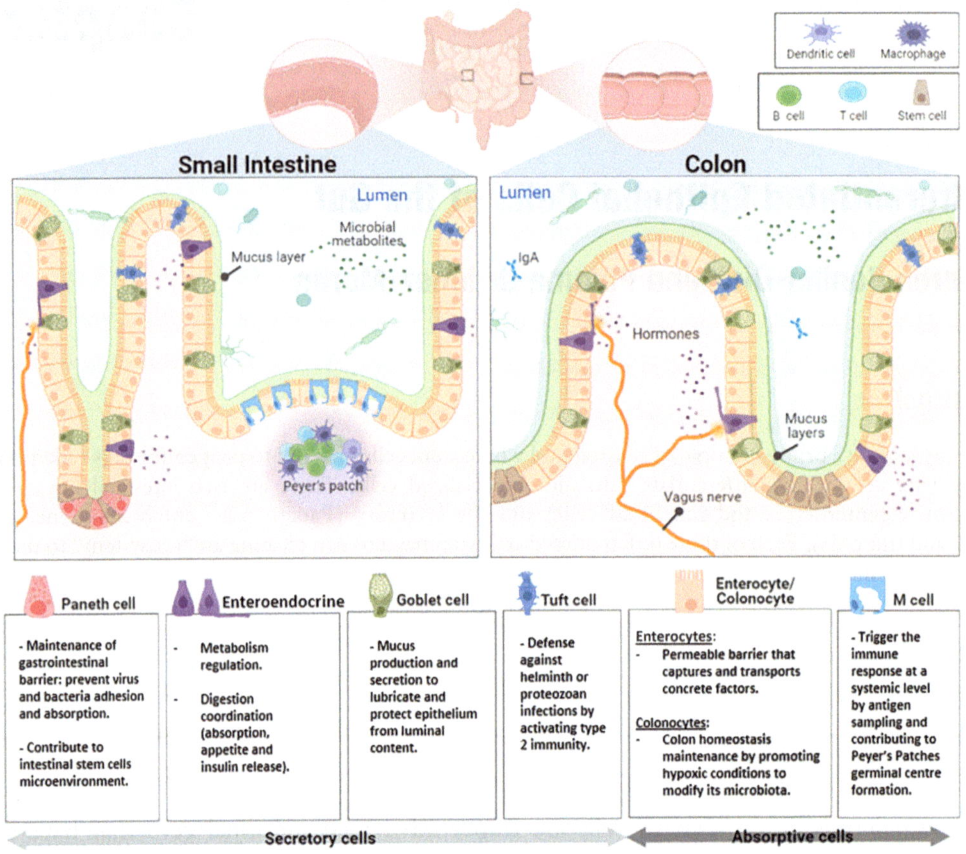

Fig. 1 Scheme of intestinal differentiated cell types (characteristics/functions) of the small intestine and colon. (This schema was created with Biorender.com)

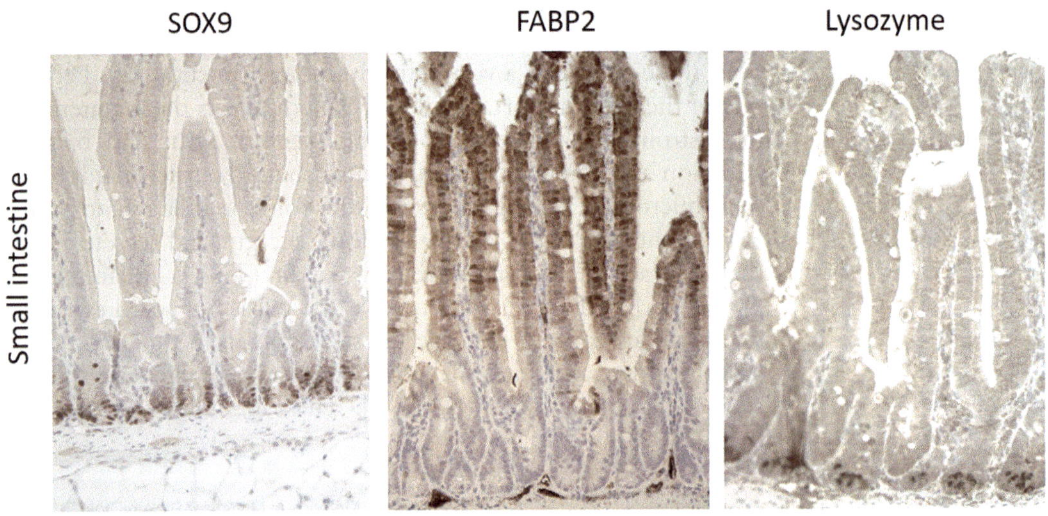

Fig. 2 Immunohistochemistry of the small intestine: SOX9 (progenitors), FABP2 (enterocytes), and lysozyme (Paneth cells)

and different antibacterial proteins such α-defensins, lysozymes, secretory phospholipase A2, angiogenin-4, RegIIIγ, RegIIIβ, and α1-antitrypsin [8]. Polymeric immunoglobulin (Ig) receptor (pIgR) is expressed by Paneth cells, and these proteins form a complex with IgA which is produced by plasma cells. The complexes are secreted into the gut lumen as antigen-specific secretory immunoglobulins (sIgA) [9, 10]. Recent studies have shown that Paneth cells can engulf apoptotic neighbor cells (efferocytosis) to maintain intestinal homeostasis [11].

On the other hand, Paneth cells are part of the intestinal stem cells microenvironment [1, 12]. These cells secrete proteins in an exocrine manner to maintain LGR5^{+} cell's proliferation, stemness, migration, and polarization [13, 14]. Paneth cells activate Wnt pathway by secreting Wnt ligands as Wnt3, Wnt6, and Wnt9b and regulate Notch pathway by secreting Notch ligands as Delta-like ligand (DLL)-1 and DLL-4, which seem crucial for maintaining the stem cell pool. However, studies on Paneth cells ablation in mice have found that these cells are not essential for Lgr5^{+} stem cells survival. When Paneth cells are lost, they are physically substituted by enteroendocrine and tuft cells, which serve as an alternative source of Notch signals [15]. Paneth cells also secrete transforming growth factor (TGF)α and epidermal growth factor (EGF) which have an important role in stemness maintenance [16]. Rodriguez-Colman et al. revealed that there is a metabolic interdependence between Paneth cells and Lgr5^{+} stem cells. Paneth cells provide lactate that supports stem cells oxidative metabolic profile and activity [17]. Furthermore, Paneth cells' location at the bottom of the crypt-villus axis is maintained by the expression of the ephrin type-B (EphB) receptor complex, which is regulated by the Wnt pathway [18]. Interestingly, these cells only move toward the bottom of the crypt, where they remain 3–4 weeks before being phagocytosed [16].

1.2 Enteroendocrine Cells

Enteroendocrine cells (EECs) comprise <1% of the total intestinal epithelial cell population [19] (Fig. 1). In the small intestine, EECs located in the duodenum and jejunum have a lifetime of 3–10 days, while in the ileum and colon, the lifespan is extended up to 60 days (shown by lineage-tracing models) [20–23]. Depending on whether their apical surfaces are in contact with the lumen or not, these cells are called "open EECs" or "closed EECs," respectively [24]. It is not clear how progenitor cells differentiate into EECs, but some studies point out that the loss of EGF turns secretory progenitors into EEC progenitors. One of the key drivers for EECs fate determination is the transcription factor Neurog3, which is activated by the transcription factor Atoh1 to promote early EEC progenitors. Sequentially, Neurog3 overexpression induces NeuroD1 promoting the transition of early EEC progenitors to committed ECC precursors before giving rise to a wide range of mature EECs. Other transcription factors, such as ISL-1, NKX2-2 or PAX4, are activated when ECCs are ready to undergo maturation [25, 26]. Indeed, mice with increased Neurog3 expression in gut epithelium have a higher number of EECs at the expense of goblet cells number [27]. Conversely, mice lacking Neurog3 exhibit an EECs loss throughout the gut [28]. EEC differentiation process follows a crypt-to-villi directionality, where the most differentiated EECs are located near the villus tip. However, a subset of EECs expressing the peptide hormone cholecystokinin (CCK) would possibly migrate downward the crypt base, as suggested by the changes seen in localization along time of ECC-GFP$^+$ in the organoids derived from CCK-GFP$^+$ mice [29].

EECs have a wide range of functions related to metabolism and digestion (absorption, appetite, and insulin release). Proximal EECs secrete about 20 different hormones into the bloodstream in response to meal-related stimuli and express a wide variety of sensory proteins, comprising G-protein-coupled receptors (GPCR), transporters, and ion channels [19, 30]. Thus, several hormones such as glucose-dependent insulinotropic polypeptide (GIP), CCK, and glucagon-like peptides 1 (GLP-1) are promptly secreted after food consumption and have an important role because of their anorexigenic function [30]. Conversely, distal EECs mostly react to gut microbiota metabolites generating hormonal signals related to food ingestion, microbial composition, and epithelial integrity [30].

Many in vitro methods have been developed to study human EECs. Beumer et al. established an organoid-based platform to detect EECs mRNA-expression profile, and the authors identified regulators of human EEC differentiation [31]. It has been largely reviewed that different EEC populations can be distinguished depending on the hormone secreted; however current studies have shown that one cell type can secrete more than one hormone. In fact, different EEC subtypes can harbor a similar transcriptional profile [30, 32–36]. In the proximal intestine, the EECs that secrete secretin are known as S-cells; the cells that secrete ghrelin, somatostatin, and motilin are known as M-cells; the cells that secrete GIP are known as K-cells; and the ones secreting CCK are known as I-cells. In the distal intestine, the aforementioned cells are progressively substituted by EECs that secrete neurotensin known as N-cells and by cells that secrete GLP-1 and GLP-2 known as L-cells. In the colon, L-cells also secrete a rising gradient of peptide YY (PYY) and insulin-like peptide 5 [37–40]. Recent studies using single-molecule fluorescence in situ hybridization found that at fetal stages some K- and L-cells secrete insulin and express β-cells-like genes. This EEC subpopulation would possibly be involved in systemic effects on fetal glucose homeostasis, intestinal growth, and vessel formation. Understanding better the mechanisms behind the loss of insulin on K- and L-cells in the postnatal period would help to develop new therapies for diabetes [41]. Remarkably, there is a subset of EECs known as enterochromaffin cells (PYY^+ and CCK^+) that have neuropods in their basolateral part and that participate in gastrointestinal function via the enteric nervous system [20]. These cells secrete peptides that can rapidly activate the local terminations of vagal afferent neurons via GPCR. Enterochromaffin cells also secrete other components such as serotonin, which activate ligand-dependent ion channels to elicit membrane depolarization [30].

1.3 Goblet Cells

Goblet cells are the most abundant secretory cells in the intestine and have a lifespan of 3–7 days. They are displayed along the mucosal epithelia and progressively increase in numbers from the proximal small intestine to the distal colon [42] (Fig. 1). In mouse organoids, Wnt and Notch inhibition dramatically increase goblet cells [43]. Loss of Notch increases the zinc-finger transcription factor Klf4 expression, leading to a rise in goblet cells [44], and genetic deletion of Klf4 in the colon drastically reduces the number of goblet cells [45]. Likewise, loss of SHP2-mediated MAPK signaling also downregulates goblet cells while increases Wnt/β-catenin signaling, having a positive impact in the stem cell niche [46]. Furthermore, the transcription factor SAM pointed domain-containing ETS transcription factor (SPDEF) downstream effector of ATOH1 promotes progenitors' differentiation toward goblet cells in transgenic overexpression and knockout mice models [47, 48].

Goblet cells' main role consists of producing and secreting mucus throughout the small intestine and colon to lubricate and protect the mucosal epithelium. Therefore, these cells accumulate vesicles on the apical zone of the cytoplasm with densely packed proteins such as mucin 2 (MUC2), resistin-like molecule β, and Fc-γ-binding protein [49–52]. Mucins can be released to the extracellular compartment via two mechanisms, by exocytosis or via single-vesicle secretion fused to the plasma membrane [53]. During exocytosis, the interplay between the inflammasome, endocytosis, autophagy, and ROS generation allows goblet cells vesicles aggregation [54, 55]. On the other hand, the single-vesicle secretion process is controlled by SNAPS, Vamp, syntaxins, and Munc18 proteins [56], a mechanism well-described in airway cells that needs to be better studied in the intestine [57].

The small intestinal goblet cells protect epithelial cells by creating a nonattached layer [42]. Goblet cells located at the villi are faster-producing mucus than those located at the crypt, where mucus secretions are released every 2 h [58]. The colonic goblet cells produce two mucus layers. Whereas the inner layer carries out an antibacterial function, the outer layer holds the gut microbiota [42]. Colonic goblet cells at the crypt secrete low amount of mucus (crypt mucus plume), while only the colonic goblet cells located at the upper part of the crypt secrete stored bulks with a different spatial organization that makes small molecules, like ions, more penetrable (intercrypt mucus) [59, 60]. Intriguingly, intercrypt mucus malfunction is related to ulcerative colitis disease [60].

The immune system can regulate goblet cells' function [61]. In response to worm infections, T helper type 2 (Th2) cytokines are induced to promote worm expulsion which has been associated with mucus production of upregulated goblet cells [62–64]. These cells may be involved in intestinal immune regulation as SPDEF regulates its differentiation in the lung and gut, which may be mastering Th2 in a normal inflammatory response [65–67]. Furthermore, small intestinal goblet cells can retain antigens of the lumen to present them to $CD103^{+}$ dendritic cells situated in the lamina propria [68, 69].

1.4 Tuft Cells

Tuft cells are rare cells (less than 1% of the epithelium) and are mainly found at the villus tip. They show long and thick microvilli at their narrow apical membrane (Fig. 1). In addition, they contain a turbovesicular network holding a bulk of empty vesicles in the apical zone. Another remarkable feature of this cell type is the absence of rough endoplasmic reticulum and secretory vesicles [70]. Interestingly, tuft cells can still act as enteroendocrine signalers interconnecting their neighboring cells [70, 71]. Their microvilli are similar to the luminal surface of taste buds and to the lateral microvilli mechanosensory Merkel skin cells [72]. Transcriptional analysis and antibody labeling revealed that tuft cells express elements of the canonical bitter and umami (the fifth taste stimulated by glutamate ion and some nucleotides). They also express the transient receptor potential cation channel subfamily M member 5 (TRPM5), a protein that plays a central role in taste transduction, and the choline acetyltransferase, an enzyme that contributes to the generation of the neurotransmitter acetylcholine. All together indicates that tuft cells may have a chemosensing role in the intestine [73–77]. It has been hypothesized that the homeostatic physiologic function of gastrointestinal tuft cells is linked to the sensing of dietary ligands. Thus, loss of chemosensing could trigger an increased expression of catecholamines to boost lipolysis and thermogenesis [78].

Tuft cells also play a central role in defense against helminth or protozoan infections by activating type 2 immunity. Once helminths invade the small intestine of the host, tuft cells secrete IL-25. As a result, IL-25 triggers IL-13 production by Th2 and type 2 innate lymphoid cell (ILC2) in the lamina propria, activating the "weep and sweep" response [79–81]. This effect promotes mucus-secreting goblet cell hyperplasia and/or smooth muscle hypercontractility [79]. Interestingly, tuft cell hyperplasia disappears in $Trpm5^{-/-}$ and G protein subunit alpha transducin 3 ($Gnat3^{-/-}$) mice during *Trichomonas muris* infection, meaning that a chemosensing function may play a role in type 2 immunity [81].

Several studies revealed the existence of a doublecortin-like kinase 1 (DCLK1)$^+$ tuft cells subset that harbor stem cell abilities as they are long lived and quiescent even after oncogenic mutations. After tissue damage events, DCLK1$^+$ cells become active and participate in intestinal regeneration [82, 83]. Although tuft cells arise from stem cells, mechanisms driving tuft cells' generation are not entirely defined. While in other secretory lineages ATOH1 is a master regulator, in mature tuft cells ATOH1 expression is lost [82, 84, 85]. This could indicate that they undergo an alternative differentiation pathway by a secretory progenitor-independent manner. Some master factors are being proposed, such as transcriptional repressor growth factor independent 1B (GFI1B), which is expressed in tuft progenitor cells and mature cells, or the transcription factor POU domain class 2 transcription factor 3 (POU2F3), a tuft-cell-specific regulator that is also expressed in taste receptor cells [78, 85, 86].

2 Absorptive Lineage

2.1 Enterocytes

Enterocytes are the most abundant epithelial cell type found in the intestine and represent about 70–80% of intestinal cells. They are distributed along the villi, especially at the tip, and their lifetime comprises only 1–5 days to avoid their damage [87]. Enterocytes are hyperpolarized as their apical membrane contains microvilli with filamentous glycocalyx anchored at the top, best-known as brush border domains. These microvilli structures help to better capture lumen nutrients and to be protected from luminal bile salts [88] (Figs. 1 and 2). Colonic enterocytes have approximately three times shorter microvilli than small intestinal enterocytes. The apical basolateral membranes of enterocytes present deep invaginations with tight junctions that separate apical basolateral membranes from the apical part, preventing permeation of extracellular molecules but allowing the diffusion of small hydrophilic molecules and ions [89].

Enterocytes' fate determination starts when stem cells receive Notch signals by close DLL$^+$ cells, which allow the Notch-responsive hairy and enhancer of split 1 (HES1) gene expression. Then, HES1 inhibits ATOH1 expression which represses the secretory lineage commitment factors. Progressively Wnt signaling reduction plus incoming Notch signals allows stem cells to become absorptive progenitors. Deletion of Atoh1 in mice causes differentiation toward the absorptive lineage, and forced expression of Atoh1 in mice transforms absorptive transit-amplifying cells into postmitotic secretory cells [90, 91]. Interestingly, the enterocytes can harbor different gene signatures depending on their differentiation state and proximal or distal location [92].

One of enterocytes' main tasks is capturing and transporting concrete molecules. Their brush border domain hydrolyzes nutrients that are absorbed by ion channels, transporters, or receptors on the apical membrane along the secretory pathway. This allows vesicular transport from the lumen to the lamina propria cells and to the bloodstream. Related to its absorptive function, enterocytes have a protective role since they act as a permeable barrier against microorganisms, toxins, and antigens. Indeed, the apical membrane of enterocytes captures antigens that are later processed by the endosomal pathway [88, 93]. Recent studies suggest that enterocytes could participate in SARS-CoV-2 infection as the angiotensin-converting enzyme 2 (ACE2) receptor is present at the enterocytes cell membrane and their expression would allow virions attachment and posterior endocytosis [94]. Enterocytes located at the follicular-associated epithelium (FAE) present low levels of antimicrobial effects and a reduced absorption program. This could be caused by the influence of different mesenchymal niche signals [95]. Interestingly, enterocytes maintain colon homeostasis by modifying the gut microbiota. In a healthy scenario, they maintain anaerobic conditions enhancing the obligate anaerobic bacteria survival, while during dysbiosis their metabolism changes and promotes epithelium oxygenation increasing the number of facultative aerobic bacteria. Therefore, colonic enterocytes adopt two different metabolic states throughout homeostasis or dysbiosis. This would represent an interesting therapy approach for a wide range of colon diseases [96].

2.2 Microfold Cells

Microfold (M) cells reside in gut-associated lymphoid tissue such as the follicle-associated epithelium covering the Peyer's patches (PP) of the small intestine [97, 98] (Fig. 1). M cells have an approximated lifespan of 5 days, and they have microfolds (short microvilli) on the apical membrane and a thin glycocalyx. In addition, M cells show basolateral invaginations containing B and T cells and some myeloid cells [99]. M cells act as an antigen sampling mechanism as PP have no afferent lymphatic vessels to drain lymph [100]. Therefore, M cells sense antigens through membrane receptors on the apical surface [101]. Once antigens are bound to the membrane, M cells engulf and transport them from the intestine lumen to its basolateral membrane by a process named transcytosis. Consequently, mononuclear phagocytes and B cells located at the subepithelial dome of PP can recognize the antigens and prompt the cellular response producing S-IgA [102]. S-IgA establishes mutual relationships within the gut microbiome and protects it from pathogens. M cells allow a germinal center formation and maturation in PP which is indispensable for somatic hypermutation and affinity maturation of B cells. Mice lacking M cells have lower B cell follicles and subsequent lower levels of IgA than wild-type mice [103, 104]. Thus, M cells have a key role in originating an immune response not only at the intestinal mucosa level but also at the systemic level. In addition to this immunosurveillance role, some pathogens such as *Salmonella* can use M cells as an infection gate; however the mechanism still remains unclear [102].

Differentiation of progenitor cells toward M cells differs from other cell types as it does not follow crypt-to-villi migration. Instead, subepithelial stromal cells in the PP present receptor activator of nuclear factor κ-B ligand (RANKL) which induces SPIB expression. This pathway activation promotes enterocytes and progenitors to become M cells [103]. Indeed, Spib knockout mice results in an absence of M cells expression [105, 106].

References

1. Sato T, van Es JH, Snippert HJ et al (2011) Paneth cells constitute the niche for Lgr5 stem cells in intestinal crypts. Nature 469:415–418. https://doi.org/10.1038/NATURE09637
2. Elphick DA, Mahida YR (2005) Paneth cells: their role in innate immunity and inflammatory disease. Gut 54:1802–1809. https://doi.org/10.1136/GUT.2005.068601
3. Clevers HC, Bevins CL (2013) Paneth cells: maestros of the small intestinal crypts. Annu Rev Physiol 75:289–311. https://doi.org/10.1146/ANNUREV-PHYSIOL-030212-183744
4. Nita HS, Kuiechun H, Dipica H et al (2010) Enteric defensins are essential regulators of intestinal microbial ecology. Nat Immunol 11:76–83. https://doi.org/10.1038/NI.1825
5. Takahashi N, Vanlaere I, de Rycke R et al (2008) IL-17 produced by Paneth cells drives TNF-induced shock. J Exp Med 205:1755. https://doi.org/10.1084/JEM.20080588
6. Yohichi S, Yoshiaki H, Kazuyuki O, Tomio K (1995) Carbamylcholine- and catecholamine-induced intracellular calcium dynamics of epithelial cells in mouse ileal crypts. Gastroenterology 108:1345–1356. https://doi.org/10.1016/0016-5085(95)90681-9
7. Tokiyoshi A, Donald PS, Carole LW et al (2000) Secretion of microbicidal alpha-defensins by intestinal Paneth cells in response to bacteria. Nat Immunol 1:113–118. https://doi.org/10.1038/77783
8. Dupont A, Heinbockel L, Brandenburg K, Hornef MW (2015) Antimicrobial peptides and the enteric mucus layer act in concert to protect the intestinal mucosa. Gut Microbes 5:761–765. https://doi.org/10.4161/19490976.2014.972238
9. Satoh Y, Ishikawa K, Tanaka H, Ono K (1986) Immunohistochemical observations of immunoglobulin a in the Paneth cells of germ-free and formerly-germ-free rats. Histochemistry 85(3):197–201. https://doi.org/10.1007/BF00494804
10. Tang Q-J, Wang L-M, Tao K-Z et al (2006) Expression of polymeric immunoglobulin receptor mRNA and protein in human Paneth cells: Paneth cells participate in acquired immunity. Am J Gastroenterol 101(7):1625–1632. https://doi.org/10.1111/j.1572-0241.2006.00605.x
11. Shankman LS, Fleury ST, Evans WB et al (2021) Efferocytosis by Paneth cells within the intestine. Curr Biol 31:2469–2476.e5. https://doi.org/10.1016/j.cub.2021.03.055
12. Kiminori N, Naoya S, Akiko T, Tokiyoshi A (2016) Paneth cell α-defensins and enteric microbiota in health and disease. Biosci Microbiota Food Health 35:57–67. https://doi.org/10.12938/BMFH.2015-019
13. Matthias S, Andrea S, Mark S, Riccardo F (2017) The organoid reconstitution assay (ORA) for the functional analysis of intestinal stem and niche cells. J Vis Exp. https://doi.org/10.3791/56329
14. Roel N, Hans C (2017) Wnt/β-catenin signaling, disease, and emerging therapeutic modalities. Cell 169:985–999. https://doi.org/10.1016/J.CELL.2017.05.016
15. van Es JH, Wiebrands K, López-Iglesias C et al (2019) Enteroendocrine and tuft cells support Lgr5 stem cells on Paneth cell depletion. Proc Natl Acad Sci U S A 116:26599–26605. https://doi.org/10.1073/PNAS.1801888117
16. Nikolenko VN, Oganesyan MV, Sankova MV et al (2021) Paneth cells: maintaining

dynamic microbiome-host homeostasis, protecting against inflammation and cancer. BioEssays 43:1–8. https://doi.org/10.1002/bies.202000180

17. Rodríguez-Colman MJ, Schewe M, Meerlo M et al (2017) Interplay between metabolic identities in the intestinal crypt supports stem cell function. Nature 543:424–427. https://doi.org/10.1038/NATURE21673
18. Donghai W, Karina P, Elena D et al (2011) Paneth cell marker expression in intestinal villi and colon crypts characterizes dietary induced risk for mouse sporadic intestinal cancer. Proc Natl Acad Sci U S A 108:10272–10277. https://doi.org/10.1073/PNAS.1017668108
19. Gribble FM, Reimann F (2016) Enteroendocrine cells: chemosensors in the intestinal epithelium. Annu Rev Physiol 78:277–299. https://doi.org/10.1146/annurev-physiol-021115-105439
20. Diego VB, Rafiq AS, Alan E et al (2015) Neuroepithelial circuit formed by innervation of sensory enteroendocrine cells. J Clin Invest 125:782–786. https://doi.org/10.1172/JCI78361
21. Cheng H, Leblond CP (1974) Origin, differentiation and renewal of the four main epithelial cell types in the mouse small intestine I. Columnar cell. Am J Anat 141:461–479. https://doi.org/10.1002/aja.1001410403
22. Thompson EM, Price YE, Wright NA (1990) Kinetics of enteroendocrine cells with implications for their origin: a study of the cholecystokinin and gastrin subpopulations combining tritiated thymidine labelling with immunocytochemistry in the mouse. Gut 31: 406–411. https://doi.org/10.1136/GUT.31.4.406
23. Tsubouchi S, Leblond CP (1979) Migration and turnover of entero-endocrine and caveolated cells in the epithelium of the descending colon, as shown by radioautography after continuous infusion of 3H-thymidine into mice. Am J Anat 156:431–451. https://doi.org/10.1002/AJA.1001560403
24. Sakata I, Sakai T (2010) Ghrelin cells in the gastrointestinal tract. Int J Pept 2010:1–7. https://doi.org/10.1155/2010/945056
25. Li HJ, Ray SK, Singh NK et al (2011) Basic helix-loop-helix transcription factors and enteroendocrine cell differentiation. Diabetes Obes Metab 13(Suppl 1):5–12. https://doi.org/10.1111/J.1463-1326.2011.01438.X
26. May CL, Kaestner KH (2010) Gut endocrine cell development. Mol Cell Endocrinol 323: 70–75. https://doi.org/10.1016/J.MCE.2009.12.009
27. López-Díaz L, Renu NJ, Keeley TM et al (2007) Intestinal neurogenin 3 directs differentiation of a bipotential secretory progenitor to endocrine cell rather than goblet cell fate. Dev Biol 309:298–305. https://doi.org/10.1016/J.YDBIO.2007.07.015
28. Jenny M, Uhl C, Roche C et al (2002) Neurogenin3 is differentially required for endocrine cell fate specification in the intestinal and gastric epithelium. EMBO J 21:6338–6347. https://doi.org/10.1093/EMBOJ/CDF649
29. Sei Y, Lu X, Liou A et al (2011) A stem cell marker-expressing subset of enteroendocrine cells resides at the crypt base in the small intestine. Am J Physiol Gastrointest Liver Physiol 300:G345–G356. https://doi.org/10.1152/AJPGI.00278.2010
30. Gribble FM, Reimann F (2019) Function and mechanisms of enteroendocrine cells and gut hormones in metabolism. Nat Rev Endocrinol 15:226–237. https://doi.org/10.1038/S41574-019-0168-8
31. Beumer J, Puschhof J, Bauzá-Martinez J et al (2020) High-resolution mRNA and secretome atlas of human enteroendocrine cells. Cell 181:1291–1306. https://doi.org/10.1016/J.CELL.2020.04.036/ATTACHMENT/9E0FD085-4E63-4AF7-877C-99F80ED0513C/MMC1.XLSX
32. Abdella MH, Paul R, Lynne SC et al (2012) Overlap of endocrine hormone expression in the mouse intestine revealed by transcriptional profiling and flow cytometry. Endocrinology 153:3054–3065. https://doi.org/10.1210/EN.2011-2170
33. Egerod KL, Engelstoft MS, Grunddal KV et al (2012) A major lineage of enteroendocrine cells coexpress CCK, secretin, GIP, GLP-1, PYY, and neurotensin but not somatostatin. Endocrinology 153:5782–5795. https://doi.org/10.1210/EN.2012-1595
34. Svendsen B, Pedersen J, Albrechtsen NJW et al (2015) An analysis of cosecretion and coexpression of gut hormones from male rat proximal and distal small intestine. Endocrinology 156:847–857. https://doi.org/10.1210/EN.2014-1710
35. Cho H-J, Kosari S, Hunne B et al (2014) Differences in hormone localisation patterns of K and L type enteroendocrine cells in the mouse and pig small intestine and colon. Cell Tissue Res 359(2):693–698. https://doi.org/10.1007/S00441-014-2033-3

36. Theodorakis MJ, Carlson O, Michopoulos S et al (2006) Human duodenal enteroendocrine cells: source of both incretin peptides, GLP-1 and GIP. Am J Physiol Endocrinol Metab 290:E550–E559. https://doi.org/10.1152/AJPENDO.00326.2004
37. Sjölund K, Sandén G, Håkanson R, Sundler F (1983) Endocrine cells in human intestine: an immunocytochemical study. Gastroenterology 85:1120–1130. https://doi.org/10.1016/S0016-5085(83)80080-8
38. Andreas S, Yvette T (2009) Regulation of food intake: the gastric X/A-like endocrine cell in the spotlight. Curr Gastroenterol Rep 11:448–454. https://doi.org/10.1007/S11894-009-0069-4
39. Itoh Z (1997) Motilin and clinical application. Peptides (NY) 18:593–608. https://doi.org/10.1016/S0196-9781(96)00333-6
40. Grosse J, Heffron H, Burling K et al (2014) Insulin-like peptide 5 is an orexigenic gastrointestinal hormone. Proc Natl Acad Sci U S A 111:11133–11138. https://doi.org/10.1073/PNAS.1411413111
41. Egozi A, Llivichuzhca-Loja D, McCourt BT et al (2021) Insulin is expressed by enteroendocrine cells during human fetal development. Nat Med 27:2104–2107. https://doi.org/10.1038/s41591-021-01586-1
42. McCauley HA, Guasch G (2015) Three cheers for the goblet cell: maintaining homeostasis in mucosal epithelia. Trends Mol Med 21:492–503. https://doi.org/10.1016/j.molmed.2015.06.003
43. Xiaolei Y, Henner FF, van Es JH et al (2014) Niche-independent high-purity cultures of Lgr5+ intestinal stem cells and their progeny. Nat Methods 11:106–112. https://doi.org/10.1038/NMETH.2737
44. Amr MG, Gaurav A, Agnieszka BB et al (2008) Notch inhibits expression of the Krüppel-like factor 4 tumor suppressor in the intestinal epithelium. Mol Cancer Res 6:1920–1927. https://doi.org/10.1158/1541-7786.MCR-08-0224
45. Katz JP, Perreault N, Goldstein BG et al (2002) The zinc-finger transcription factor Klf4 is required for terminal differentiation of goblet cells in the colon. Development 129:2619–2628. https://doi.org/10.1242/dev.129.11.2619
46. Heuberger J, Kosel F, Qi J et al (2014) Shp2/MAPK signaling controls goblet/paneth cell fate decisions in the intestine. Proc Natl Acad Sci 111:3472–3477. https://doi.org/10.1073/PNAS.1309342111
47. Noah TK, Kazanjian A, Whitsett J, Shroyer NF (2010) SAM pointed domain ETS factor (SPDEF) regulates terminal differentiation and maturation of intestinal goblet cells. Exp Cell Res 316:452–465. https://doi.org/10.1016/J.YEXCR.2009.09.020
48. Gregorieff A, Stange DE, Kujala P et al (2009) The Ets-domain transcription factor Spdef promotes maturation of goblet and Paneth cells in the intestinal epithelium. Gastroenterology 137:1333–1345. https://doi.org/10.1053/J.GASTRO.2009.06.044
49. Kim YS, Ho SB (2010) Intestinal goblet cells and mucins in health and disease: recent insights and progress. Curr Gastroenterol Rep 12:319–330. https://doi.org/10.1007/S11894-010-0131-2
50. Birchenough GMH, Nyström EEL, Johansson MEV, Hansson GC (2016) A sentinel goblet cell guards the colonic crypt by triggering Nlrp6-dependent Muc2 secretion. Science (1979) 352:1535–1542. https://doi.org/10.1126/SCIENCE.AAF7419
51. Morampudi V, Dalwadi U, Bhinder G et al (2016) The goblet cell-derived mediator RELM-β drives spontaneous colitis in Muc2-deficient mice by promoting commensal microbial dysbiosis. Mucosal Immunol 9:1218–1233. https://doi.org/10.1038/MI.2015.140
52. He W, Wang ML, Jiang HQ et al (2003) Bacterial colonization leads to the colonic secretion of RELMβ/FIZZ2, a novel goblet cell-specific protein. Gastroenterology 125:1388–1397. https://doi.org/10.1016/J.GASTRO.2003.07.009
53. Herath M, Hosie S, Bornstein JC et al (2020) The role of the gastrointestinal mucus system in intestinal homeostasis: implications for neurological disorders. Front Cell Infect Microbiol 10:248
54. Patel KK, Miyoshi H, Beatty WL et al (2013) Autophagy proteins control goblet cell function by potentiating reactive oxygen species production. EMBO J 32:3130–3144. https://doi.org/10.1038/EMBOJ.2013.233
55. Martinon F, Burns K, Tschopp J (2002) The inflammasome: a molecular platform triggering activation of inflammatory caspases and processing of proIL-beta. Mol Cell 10:417–426. https://doi.org/10.1016/S1097-2765(02)00599-3
56. Specian RD, Neutra MR (1980) Mechanism of rapid mucus secretion in goblet cells stimulated by acetylcholine. J Cell Biol 85:626–640. https://doi.org/10.1083/jcb.85.3.626
57. Adler KB, Tuvim MJ, Dickey BF (2013) Regulated mucin secretion from airway epithelial cells. Front Endocrinol (Lausanne) 4:

1–9. https://doi.org/10.3389/FENDO.2013.00129

58. Johansson MEV (2012) Fast renewal of the distal colonic mucus layers by the surface goblet cells as measured by in vivo labeling of mucin glycoproteins. PLoS One 7:e41009. https://doi.org/10.1371/JOURNAL.PONE.0041009

59. Specian RD, Neutra MR (1982) Regulation of intestinal goblet cell secretion. I. Role of parasympathetic stimulation. Am J Physiol 5: G370–G379. https://doi.org/10.1152/ajpgi.1982.242.4.g370

60. Nyström EEL, Martinez-Abad B, Arike L et al (2021) An intercrypt subpopulation of goblet cells is essential for colonic mucus barrier function. Science (1979):372–eabb1590. https://doi.org/10.1126/SCIENCE.ABB1590

61. Turner J-E, Stockinger B, Helmby H (2013) IL-22 mediates goblet cell hyperplasia and worm expulsion in intestinal helminth infection. PLoS Pathog 9:e1003698. https://doi.org/10.1371/JOURNAL.PPAT.1003698

62. Oeser K, Schwartz C, Voehringer D (2015) Conditional IL-4/IL-13-deficient mice reveal a critical role of innate immune cells for protective immunity against gastrointestinal helminths. Mucosal Immunol 8:672–682. https://doi.org/10.1038/MI.2014.101

63. Marillier RG, Michels C, Smith EM et al (2008) IL-4/IL-13 independent goblet cell hyperplasia in experimental helminth infections. BMC Immunol 9. https://doi.org/10.1186/1471-2172-9-11

64. Finkelman FD, Shea-Donohue T, Morris SC et al (2004) Interleukin-4- and interleukin-13-mediated host protection against intestinal nematode parasites. Immunol Rev 201:139–155. https://doi.org/10.1111/J.0105-2896.2004.00192.X

65. Whitsett JA, Alenghat T (2014) Respiratory epithelial cells orchestrate pulmonary innate immunity. Nat Immunol 2015 16(1):27–35. https://doi.org/10.1038/ni.3045

66. Chen G, Krfhagen TR, Xu Y et al (2009) SPDEF is required for mouse pulmonary goblet cell differentiation and regulates a network of genes associated with mucus production. J Clin Invest 119:2914–2924. https://doi.org/10.1172/JCI39731

67. Johansson MEV, Hansson GC (2014) Is the intestinal goblet cell a major immune cell? Cell Host Microbe 15:251–252. https://doi.org/10.1016/J.CHOM.2014.02.014

68. McDole JR, Wheeler LW, McDonald KG et al (2012) Goblet cells deliver luminal antigen to CD103+ dendritic cells in the small intestine. Nature 483:345–349. https://doi.org/10.1038/NATURE10863

69. Knoop KA, McDonald KG, McCrate S et al (2015) Microbial sensing by goblet cells controls immune surveillance of luminal antigens in the colon. Mucosal Immunol 8:198–210. https://doi.org/10.1038/MI.2014.58

70. Trier JS, Allan CH, Marcial MA, Madara JL (1987) Structural features of the apical and tubulovesicular membranes of rodent small intestinal tuft cells. Anat Rec 219:69–77. https://doi.org/10.1002/AR.1092190112

71. Gebert A, Al-Samir K, Werner K et al (2000) The apical membrane of intestinal brush cells possesses a specialised, but species-specific, composition of glycoconjugates-on-section and in vivo lectin labelling in rats, guinea-pigs and mice. Histochem Cell Biol 113: 389–399. https://doi.org/10.1007/s004180000148

72. Luciano L, Reale E (1979) A new morphological aspect of the brush cells of the mouse gallbladder epithelium. Cell Tissue Res 201: 37–44. https://doi.org/10.1007/BF00238045

73. Kaske S, Krasteva G, König P et al (2007) TRPM5, a taste-signaling transient receptor potential ion-channel, is a ubiquitous signaling component in chemosensory cells. BMC Neurosci 8(1):1–12. https://doi.org/10.1186/1471-2202-8-49

74. Gautron L, Rutkowski JM, Burton MD et al (2013) Neuronal and nonneuronal cholinergic structures in the mouse gastrointestinal tract and spleen. J Comp Neurol 521:3741–3767. https://doi.org/10.1002/CNE.23376

75. Schütz B, Jurastow I, Bader S et al (2015) Chemical coding and chemosensory properties of cholinergic brush cells in the mouse gastrointestinal and biliary tract. Front Physiol 6:1–14. https://doi.org/10.3389/FPHYS.2015.00087

76. Bezençon C, le Coutre J, Damak S (2007) Taste-signaling proteins are coexpressed in solitary intestinal epithelial cells. Chem Senses 32:41–49. https://doi.org/10.1093/CHEMSE/BJL034

77. Kinnamon SC (2009) Umami taste transduction mechanisms. Am J Clin Nutr 90:753S–755S. https://doi.org/10.3945/AJCN.2009.27462K

78. Ushiama S, Ishimaru Y, Narukawa M et al (2016) Catecholamines facilitate fuel expenditure and protect against obesity via a novel network of the gut-brain axis in transcription

factor Skn-1-deficient mice. EBioMedicine 8: 60–71. https://doi.org/10.1016/j.ebiom.2016.04.031

79. Gerbe F, Sidot E, Smyth DJ et al (2016) Intestinal epithelial tuft cells initiate type 2 mucosal immunity to helminth parasites. Nature 529:226–230. https://doi.org/10.1038/NATURE16527
80. Liang HE, Leinhardt RL, Bando JK et al (2011) Divergent expression patterns of IL-4 and IL-13 define unique functions in allergic immunity. Nat Immunol 13:58–66. https://doi.org/10.1038/NI.2182
81. Howitt MR, Lavoie S, Michaud M et al (2016) Tuft cells, taste-chemosensory cells, orchestrate parasite type 2 immunity in the gut. Science 351:1329–1333. https://doi.org/10.1126/SCIENCE.AAF1648
82. Westphalen CB, Asfaha S, Hayakawa Y et al (2014) Long-lived intestinal tuft cells serve as colon cancer-initiating cells. J Clin Invest 124: 1283–1295. https://doi.org/10.1172/JCI73434
83. Nakanishi Y, Seno H, Fukuoka A et al (2013) Dclk1 distinguishes between tumor and normal stem cells in the intestine. Nat Genet 45: 98–103. https://doi.org/10.1038/NG.2481
84. Gerbe F, van Es JH, Makrini L et al (2011) Distinct ATOH1 and Neurog3 requirements define tuft cells as a new secretory cell type in the intestinal epithelium. J Cell Biol 192:767–780. https://doi.org/10.1083/JCB.201010127
85. Bjerknes M, Khandanpour C, Möröy T et al (2012) Origin of the brush cell lineage in the mouse intestinal epithelium. Dev Biol 362: 194–218. https://doi.org/10.1016/J.YDBIO.2011.12.009
86. Matsumoto I, Ohmoto M, Narukawa M et al (2011) Skn-1a (Pou2f3) specifies taste receptor cell lineage. Nat Neurosci 14:685–687. https://doi.org/10.1038/NN.2820
87. de Santa BP, van den Brink GR, Roberts DJ (2003) Development and differentiation of the intestinal epithelium. Cell Mol Life Sci 60:1322–1332
88. Massey-Harroche D (2000) Epithelial cell polarity as reflected in enterocytes. Microsc Res Tech 49:353–362. https://doi.org/10.1002/(SICI)1097-0029(20000515)49:4<353::AID-JEMT4>3.0.CO;2-8
89. Steed E, Balda MS, Matter K (2010) Dynamics and functions of tight junctions. Trends Cell Biol 20:142–149. https://doi.org/10.1016/J.TCB.2009.12.002
90. Gehart H, Clevers H (2019) Tales from the crypt: new insights into intestinal stem cells. Nat Rev Gastroenterol Hepatol 16:19–34. https://doi.org/10.1038/s41575-018-0081-y
91. VanDussen KL, Samuelson LC (2010) Mouse atonal homolog 1 directs intestinal progenitors to secretory cell rather than absorptive cell fate. Dev Biol 346:215–223. https://doi.org/10.1016/J.YDBIO.2010.07.026
92. Haber AL, Biton M, Rogel N et al (2017) A single-cell survey of the small intestinal epithelium. Nature 551:333–339. https://doi.org/10.1038/nature24489
93. Miron N, Cristea V (2012) Enterocytes: active cells in tolerance to food and microbial antigens in the gut. Clin Exp Immunol 167: 405–412
94. Mönkemüller K, Fry LC, Rickes S (2020) COVID-19, coronavirus, SARS-CoV-2 and the small bowel. Rev Esp Enferm Dig 112: 383–388. https://doi.org/10.17235/reed.2020.7137/2020
95. Cohen NI, Massalha HI, Ben-Moshe SI et al (2021) Spatial gene expression maps of the intestinal lymphoid follicle and associated epithelium identify zonated expression programs. PLoS Biol 19:1–16. https://doi.org/10.1371/journal.pbio.3001214
96. Litvak Y, Byndloss MX, Bäumler AJ (2018) Colonocyte metabolism shapes the gut microbiota. Science (1979) 362:1–6. https://doi.org/10.1126/science.aat9076
97. Mabbott NA, Donaldson DS, Ohno H et al (2013) Microfold (M) cells: important immunosurveillance posts in the intestinal epithelium. Mucosal Immunol 6:666–677. https://doi.org/10.1038/MI.2013.30
98. Kobayashi N, Takahashi D, Takano S et al (2019) The roles of Peyer's patches and microfold cells in the gut immune system: relevance to autoimmune diseases. Front Immunol 10:1–15. https://doi.org/10.3389/fimmu.2019.02345
99. Corr SC, Gahan CC, Hill C (2008) M-cells: origin, morphology and role in mucosal immunity and microbial pathogenesis. FEMS Immunol Med Microbiol 52:2–12. https://doi.org/10.1111/J.1574-695X.2007.00359.X
100. Andrea R, Jason GC (2016) Peyer's patches: organizing B-cell responses at the intestinal frontier. Immunol Rev 271:230–245. https://doi.org/10.1111/IMR.12400
101. Mabbott NADDSOHwIRmA (2013) Microfold (M) cells: important immunosurveillance posts in the intestinal epithelium. Mucosal

A key function of the small intestine is the absorption of nutrients and water from the intestinal lumen to the vascular system. Thus, it is unsurprising that absorptive enterocytes are the most numerous cells in the small intestinal epithelium [3, 4]. Enterocytes have an elaborate apical membrane that is covered with actin-rich protrusions known as microvilli, which facilitates absorption by increasing the cell surface area. The apical membrane, or brush border, of mature enterocytes contains enzymes and transporters that promote intestinal absorption of amino acids and sodium. The sodium gradient, established by ion transporters such as the sodium hydrogen exchangers (NHE3 and NHE2), the sodium glucose co-transporter (SGLT1), downregulated in adenoma (DRA), and putative anion transporter (PAT1), is the driving force for intestinal water absorption resulting in hydration [5–9]. Intestinal epithelial cells also function as a mucosal barrier through the formation of tight junction complexes between cells. The intestinal epithelial barrier regulates the paracellular transport of antigens and prevents the flow of luminal contents into the blood and lymphatic system while still allowing passive permeability of water and small ions [7].

In the small intestine, enterocytes are interspersed with mucin-secreting goblet cells, which provides a protective mucus layer over the epithelium preventing encroachment of luminal microbes and noxious luminal contents [10, 11]. Goblet cells also secrete chemokines, cytokines, antimicrobial proteins, and wound-healing peptides, further contributing to mucosal immunity. Similarly, Paneth cells are also secretory epithelial cells that secrete antimicrobial peptides. Paneth cells reside in the crypts of the small intestine and play a role in host defense and maintenance of stem cells [12, 13].

Enteroendocrine cells make up less than 1% of the intestinal epithelium, yet enteroendocrine cells exert broad influence over the gastrointestinal tract by regulating gut motility, secretion, insulin release, appetite, food digestion, and absorption through secretion of gut hormones [14–16]. Enteroendocrine cells detect nutrients and sense microbial metabolites, allowing them to respond to changes in the luminal environment [17]. Another chemosensory cell type in the gastrointestinal tract is tuft cells. Tuft cells have a unique morphological structure characterized by long microvilli that appear as an apical tuft. Intestinal tuft cells represent a small population of the epithelium (approximately 0.4%) and are identified by the presence of long actin filaments and marked by doublecortin-like kinase 1 (DCLK1 or DCAMKL) [18, 19]. Additionally, tuft cells express taste receptors including transient receptor potential cation channel, subfamily M, member 5 (TRPM5), T1R1, T1R3, and α-gustducin [19–23]. Alterations in the function of differentiated cell types in the small intestine result in an array of intestinal pathophysiologies.

In the 1940s, Coons pioneered the use of fluorescent derivatives of antibodies to label and locate antigens [24]. Since that time the field of immunofluorescence labeling and imaging has rapidly developed, allowing scientists to visualize cellular components that were previously elusive. Multilabeled indirect immunofluorescence techniques and advanced microscopy allow researchers to examine several target molecules within the same tissue in three-dimensional images, allowing for the reconstruction of cellular structures and greater understanding of cellular organization and function [25].

Here, we present a detailed procedure for immunohistochemical identification and analysis of differentiated intestinal cells in paraffin-embedded tissue. We have applied this procedure to paraffin-embedded mouse small intestinal tissue using widefield and confocal microscopy. Importantly, this procedure also works well for immunostaining human and pig tissue. We provide notes to address pitfalls and areas for optimization of the protocol to accommodate different antibodies and reagents. We also include a list of commercially available antibodies that we have used to label the differentiated intestinal cell lineages (Tables 1 and 2). The method herein provides researchers with a means of studying the major differentiated cell lineages of the small intestine and securing reliable and reproducible immunofluorescence identification of individual cell types of interest (*see* **Note 1**).

Table 1
Primary antibodies

Primary antibodies	Company	Catalog #	Dilution	Species	Conjugated
Chromogranin A	Proteintech	23342-1-AP	1:400	Rabbit	Unconjugated
Claudin-4	Thermo Fisher	36-4800	1:200	Rabbit	Unconjugated
DCLK1	Abcam	ab109029	1:2000	Rabbit	Unconjugated
DPPIV	R&D Systems	AF954	1:200	Goat	Unconjugated
E-cadherin	R&D Systems	AF748	1:200	Goat	Unconjugated
Gamma actin	Santa Cruz Biotechnology	Sc-65638	1:100	Mouse	Conjugated AF647
Lysozyme	Abcam	ab108508	1:200	Rabbit	Unconjugated
Muc2	Genetex	GTX100664	1:200	Rabbit	Unconjugated
Villin	Santa Cruz Biotechnology	Sc-58897	1:50	Mouse	Conjugated AF488
ZO-1	Sigma Aldrich	MABT11	1:200	Rat	Unconjugated

Table 2
Secondary antibodies

Secondary antibodies	Company	Catalog #	Dilution	Species	Conjugated
Donkey anti-rabbit IgG	Jackson ImmunoResearch	711-545-152	1:200	Donkey anti-rabbit	AF488
Donkey anti-rabbit IgG	Jackson ImmunoResearch	711-165-152	1:200	Donkey anti-rabbit	Cy3
Donkey anti-mouse IgG	Jackson ImmunoResearch	711-545-150	1:200	Donkey anti-mouse	AF488
Donkey anti-mouse IgG	Jackson ImmunoResearch	711-605-150	1:200	Donkey anti-mouse	AF647
Donkey anti-rat IgG	Jackson ImmunoResearch	712-165-153	1:200	Donkey anti-rat	Cy3
Donkey anti-goat IgG	Jackson ImmunoResearch	705-545-003	1:200	Donkey anti-goat	AF488
Donkey anti-goat IgG	Jackson ImmunoResearch	705-605-003	1:200	Donkey anti-goat	AF647

2 Materials

2.1 Deparaffinization of Tissue Sections

1. Slide warmer.
2. Tissue-Tek slide rack.
3. Tissue-Tek staining dish.
4. HistoClear II.
5. 100% ethanol.
6. 95% ethanol.
7. 70% ethanol.
8. 50% ethanol.
9. Deionized water.

2.2 Antigen Retrieval

1. Cuisinart 6 qt electric pressure cooker.
2. Citrate buffer: 2.94 g of trisodium citrate (dihydrate), 1 L distilled water, mix to dissolve, adjust pH to 6.0, add 0.5 mL Tween 20, and mix well. Store at 4 °C.
3. Tissue-Tek staining dish.
4. Rubber bands.
5. Deionized water.
6. 1× PBS.

2.3 Protein Block

1. Pap or hydrophobic barrier pen.
2. Dako protein block serum-free.
3. Humidified chamber.

2.4 Mouse-on-Mouse Block

1. Vector Laboratories' mouse-on-mouse blocking reagent.
2. 1× PBS.

2.5 Primary Antibodies (See Notes 2 and 3)

1. Dako antibody diluent, background reducing.
2. Primary antibodies.

2.6 Wash Steps

1. Tissue-Tek staining dish.
2. 1× PBS.

2.7 Secondary Antibodies

1. Dako antibody diluent.
2. Secondary antibodies: donkey anti-mouse Alexa Fluor 488, donkey anti-mouse Alexa Fluor 647, donkey anti-goat Alexa Fluor 488, donkey anti-goat Alexa Fluor 647, donkey anti-rat Alexa Fluor Cy3, donkey anti-rabbit Alexa Fluor 488, donkey anti-rabbit Cy3, gamma actin anti-mouse Alexa Fluor 647, villin anti-mouse Alexa Fluor 488.

2.8 Nuclei Counterstain

1. Hoechst or DAPI.
   - 1.1 Hoechst No. 33342 trihydrochloride trihydrate 25 mg.
   - 1.2 To the 25 mg of Hoechst, add 2.5 mL DI H_20.
   - 1.3 Mix with the pipette multiple times to ensure all powder is dissolved.
   - 1.4 Aliquot in labeled tubes, and label 10 mg/mL (16.2 mM) in H_2O stock solution.
2. 1× PBS.

2.9 Wash Steps

1. Tissue-Tek staining dish.
2. 1× PBS.

2.10 Mounting Slides

1. ProLong Gold Antifade.
2. #1.5 coverslips.

2.11 Imaging and Processing

1. Microscope.
2. Adobe Photoshop.

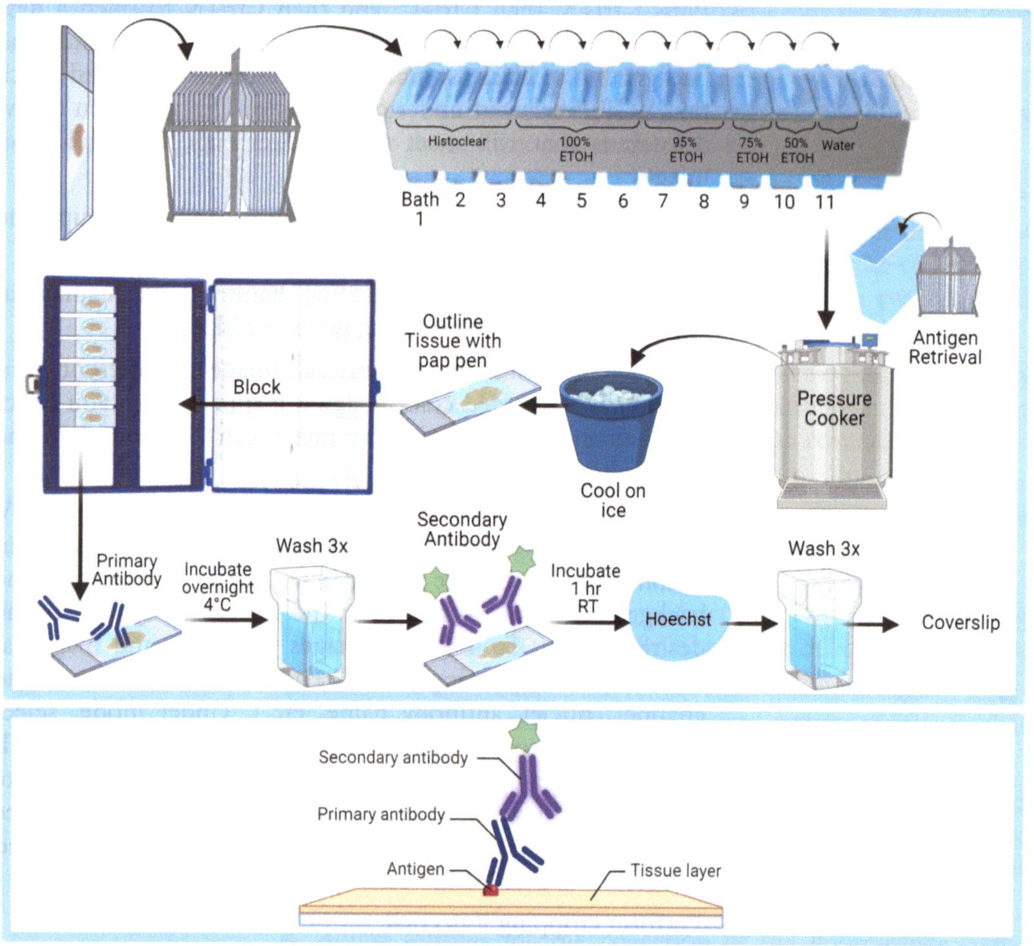

Fig. 1 Graphical depiction of the staining process for paraffin-embedded tissue sections. Intestinal tissue sections are deparaffinized and rehydrated and heat-mediated antigen retrieval performed. Tissue sections are then blocked and primary antibodies incubated overnight at 4° C. After washing in PBS, tissue sections are incubated with fluorescently conjugated secondary antibodies and then counterstained with a nuclear stain. Sections are washed and then mounted before image acquisition

illustrates the major steps in the immunostaining process. Figure 2 shows immunostaining delineating goblet cells (Muc2), Paneth cells (lysozyme), and enterocytes (DPPIV and villin). Figure 3 shows representative widefield and confocal images of enteroendocrine markers (CgA and claudin 4) and tuft cells (DCLK1 and gamma actin) in adult mouse small intestine. All antibodies used are listed in Tables 1 and 2. Both widefield and confocal microscopes can provide micrographs suitable for publication. With high-quality immunostaining, widefield and confocal microscopes can both yield meaningful data and provide the user with flexibility in choosing the optimal system based on their resources and data requirements.

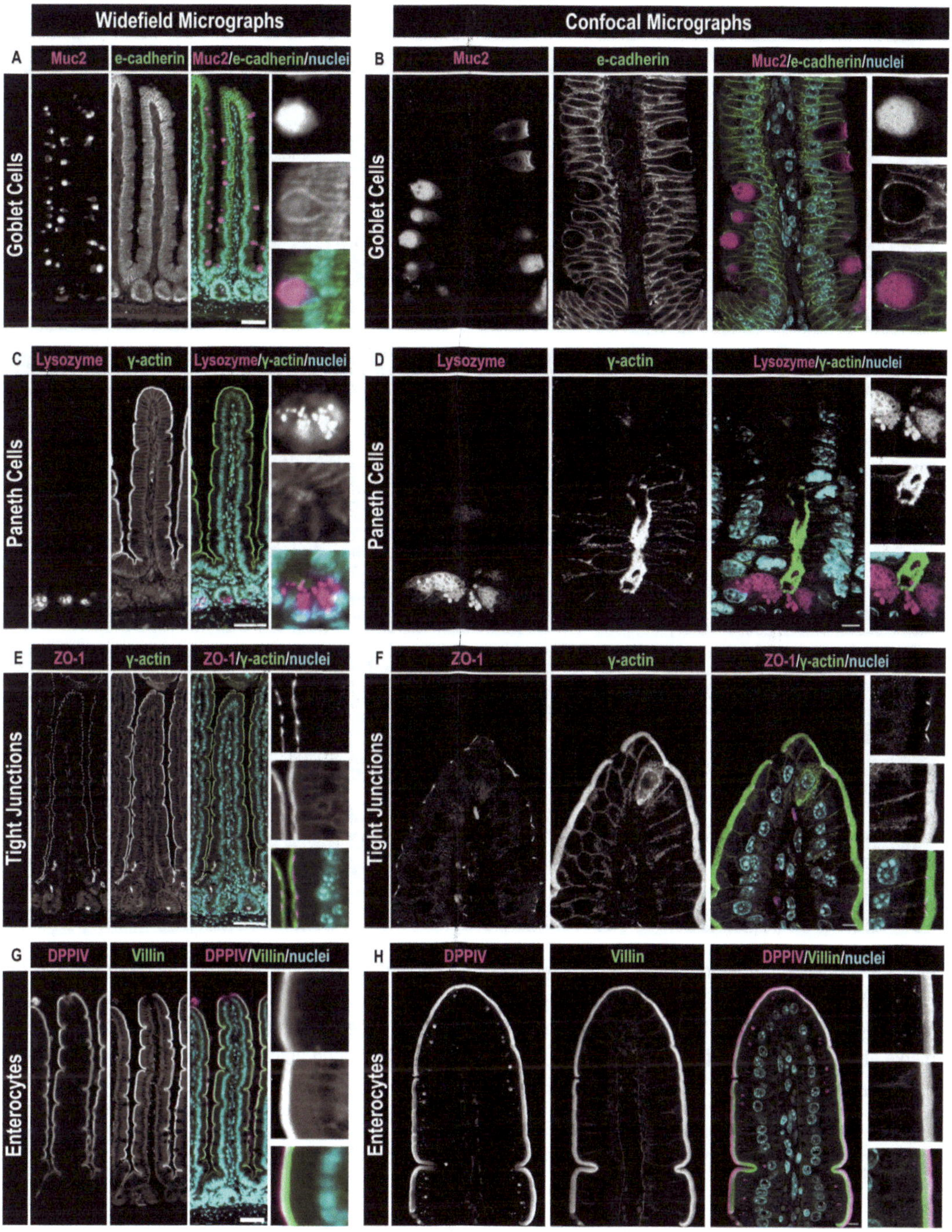

Fig. 2 Representative immunofluorescence images of the small intestine from adult mice fixed in 10% neutral buffered formalin acquired with a widefield and confocal microscope. (A&B) Muc2 immunostaining demonstrating small intestinal goblet cells and the cell adhesion protein e-cadherin. (C&D) Representative images of lysozyme and gamma actin staining in the small intestine of adult mice. (E&F) Tight junctions (identified here by ZO-1) form part of the intestinal epithelial barrier restricting the access of harmful luminal contents. ZO-1 localizes to the apical domain and is part of the apical junctional complex. (G&H) Enterocytes are absorptive epithelial cells and are the most common cell type in the small intestine. Villin and DPPIV are present in the brush border of mature enterocytes

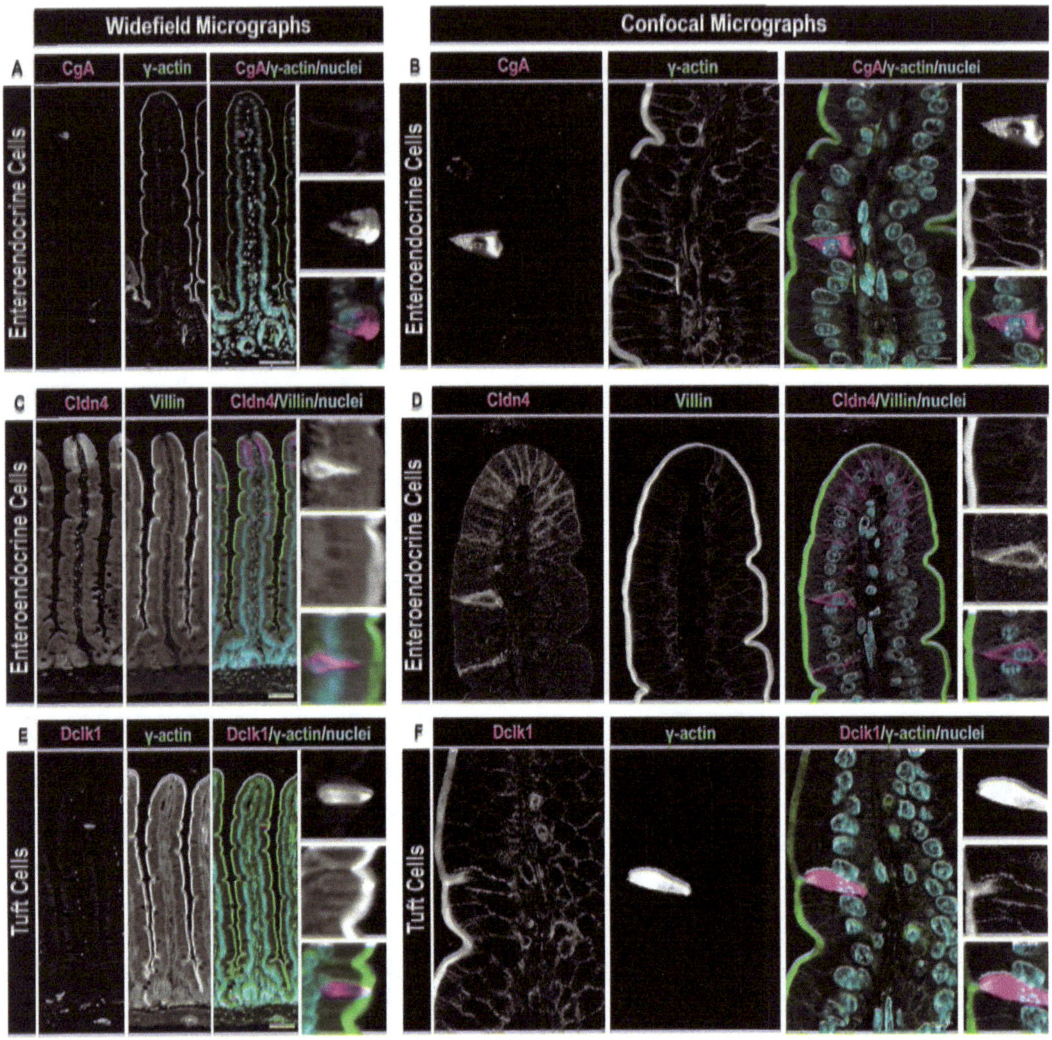

Fig. 3 Examples of small intestinal tissue staining using antibodies that label enteroendocrine cells and tuft cells. (A&B) Widefield and confocal micrographs depicting chromogranin A (CgA)-positive enteroendocrine cells. (C&D) Claudin-4 is a tight junction membrane protein that labels enteroendocrine cells [28], while villin marks absorptive enterocytes. (E&F) Tuft cells are identified by their long apical microvilli that are positive for gamma actin [29] and by DCLK1 immunostaining

4 Notes

1. Three things are essential for good immunostaining. The first is proper tissue fixation and processing, the second is effective antigen retrieval, and the third requirement is quality antibodies that are highly specific.
2. To identify high-quality commercially available antibodies, we first query the human protein atlas or the available literature to determine which cells express the protein of interest and where the protein is normally localized within cells. If the literature

provides high-quality staining and the antibody catalog number and antibody concentration, then we order the recommended antibody and test it to verify whether our staining coincides with the literature. If no suitable antibody is found in the literature, we next search for primary antibodies and observe the immunostaining that the companies have featured to identify whether the staining matches other sources and whether the antibody has been tested for immunohistochemistry in our target tissue and species. Many companies do not have high-quality fluorescence images but still have good antibodies. Often, multiple antibodies need to be tried to find an appropriate antibody for experiments. Companies that we generally use for high-quality antibodies include Abcam, Novus Biologicals, Genetex, and Cell Signaling Technology.

3. If using mouse tissue expressing a fluorescent reporter, the fluorescence is quenched by the tissue processing and paraffin-embedding procedure. We recommend using antibodies against the fluorescent reporter (i.e., GFP, mCherry, etc.) if using paraffin-embedded sections. Alternatively, tissue can be frozen embedded to maintain the endogenous fluorescence.
4. Incomplete removal of paraffin can cause poor staining of sections. To determine whether deparaffinization was not performed correctly, look at your tissue section under a microscope with a 20× objective, and see whether the staining across the whole tissue is consistent (as opposed to patchy staining). It is helpful to look at nuclei staining to determine if paraffin was efficiently removed.
5. For deparaffinization using xylene, incubate slides for 5 min per bath and follow safety protocols (3 × 5 min).
6. The deparaffinization steps are not time sensitive, and the amount of time in HistoClear can vary with longer incubations if necessary. However, the rehydration times (the time the slides are in the different ethanol concentrations) need to be closely followed.
7. The HistoClear and ethanol solutions can be used multiple times before replacing the solutions. Solutions need to be changed if they become cloudy or contain particulates. Check to make sure that you have an adequate amount of ethanol (which evaporates quickly) in each Tissue-Tek staining dish before the rehydration steps commence.
8. Antigen retrieval is required because epitopes become masked by protein cross-linking during most fixation methods. Epitope retrieval is necessary to reverse protein cross-links to unmask antigenic sites for antibody accessibility [27].

9. A variety of antigen retrievals are commercially available and can be used for certain antibodies. Alternative antigen retrieval buffers may be required for specific antibodies such as a Tris/EDTA antigen retrieval buffer pH 9.
10. If using less than eight slides, a Coplin jar can be used, add approximately 50 mL of antigen retrieval to the Coplin jar, transfer slides to Coplin jar containing antigen retrieval, and place the lid on but do not completely tighten the lid. Ensure the lid is loose to prevent pressure from building up inside of the Coplin jar.
11. The trivet or metal rack in the Cuisinart electric pressure cooker is needed to prevent the Tissue-Tek dish or Coplin jar from melting or becoming deformed in the pressure cooker.
12. Heat-induced antigen retrieval will take longer than the set 15 min because of the time required for the correct pressure to be reached. The red float on the top will rise as pressure builds, and when the high-pressure level is reached, the light in the lower right-hand corner of the LED screen will stop flashing and remain lit, and the timer will begin to count down.
13. Do not let your tissue sections dry out, be sure to outline the tissue with a pap pen, and add protein block to each slide individually, leaving the other slides submerged in PBS.
14. Alternative protein blocks can be used for immunostaining. These include PBS with 5% fetal calf serum or 1% bovine serum albumin typically with 5% normal donkey serum or goat serum. Additionally, 0.5% fish gelatin can be used as a blocking buffer with Triton X-100.
15. For primary antibody diluents, 1% normal serum (donkey or goat) in PBS can be substituted for Dako's antibody diluent, background reducing.
16. Primary antibodies that are fluorescently conjugated (such as villin and gamma actin) can be added to tissue sections with secondary antibodies for 1 h at room temperature rather than incubated at 4 °C overnight.
17. Secondary antibodies can also be diluted in PBS.
18. Save original files separately to ensure that you have the raw image files for scientific journals that require submission of original images without any image processing.

References

1. Barker N, van Es JH, Kuipers J, Kujala P, van den Born M, Cozijnsen M et al (2007) Identification of stem cells in small intestine and colon by marker gene Lgr5. Nature 449(7165):1003–1007. https://doi.org/10.1038/nature06196
2. Leblond CP, Stevens CE (1948) The constant renewal of the intestinal epithelium in the albino rat. Anat Rec 100(3):357–377. https://doi.org/10.1002/ar.1091000306
3. Hooper LV (2015) Epithelial cell contributions to intestinal immunity. Adv Immunol 126:129–172. https://doi.org/10.1016/bs.ai.2014.11.003
4. de Santa BP, van den Brink GR, Roberts DJ (2003) Development and differentiation of the intestinal epithelium. Cell Mol Life Sci 60(7):1322–1332. https://doi.org/10.1007/s00018-003-2289-3
5. Loo DD, Zeuthen T, Chandy G, Wright EM (1996) Cotransport of water by the Na+/glucose cotransporter. Proc Natl Acad Sci U S A 93(23):13367–13370. https://doi.org/10.1073/pnas.93.23.13367
6. Schultheis PJ, Clarke LL, Meneton P, Miller ML, Soleimani M, Gawenis LR et al (1998) Renal and intestinal absorptive defects in mice lacking the NHE3 Na+/H+ exchanger. Nat Genet 19(3):282–285. https://doi.org/10.1038/969
7. Field M (2003) Intestinal ion transport and the pathophysiology of diarrhea. J Clin Invest 111(7):931–943. https://doi.org/10.1172/JCI18326
8. Kato A, Romero MF (2011) Regulation of electroneutral NaCl absorption by the small intestine. Annu Rev Physiol 73:261–281. https://doi.org/10.1146/annurev-physiol-012110-142244
9. Engevik AC, Engevik MA (2021) Exploring the impact of intestinal ion transport on the gut microbiota. Comput Struct Biotechnol J 19:134–144. https://doi.org/10.1016/j.csbj.2020.12.008
10. Johansson ME, Phillipson M, Petersson J, Velcich A, Holm L, Hansson GC (2008) The inner of the two Muc2 mucin-dependent mucus layers in colon is devoid of bacteria. Proc Natl Acad Sci U S A 105(39):15064–15069. https://doi.org/10.1073/pnas.0803124105
11. Knoop KA, Newberry RD (2018) Goblet cells: multifaceted players in immunity at mucosal surfaces. Mucosal Immunol 11(6):1551–1557. https://doi.org/10.1038/s41385-018-0039-y
12. Bevins CL, Salzman NH (2011) Paneth cells, antimicrobial peptides and maintenance of intestinal homeostasis. Nat Rev Microbiol 9(5):356–368. https://doi.org/10.1038/nrmicro2546
13. Sato T, van Es JH, Snippert HJ, Stange DE, Vries RG, van den Born M et al (2011) Paneth cells constitute the niche for Lgr5 stem cells in intestinal crypts. Nature 469(7330):415–418. https://doi.org/10.1038/nature09637
14. Sternini C, Anselmi L, Rozengurt E (2008) Enteroendocrine cells: a site of 'taste' in gastrointestinal chemosensing. Curr Opin Endocrinol Diabetes Obes 15(1):73–78. https://doi.org/10.1097/MED.0b013e3282f43a73
15. Yu Y, Yang W, Li Y, Cong Y (2020) Enteroendocrine cells: sensing gut microbiota and regulating inflammatory bowel diseases. Inflamm Bowel Dis 26(1):11–20. https://doi.org/10.1093/ibd/izz217
16. Gribble FM, Reimann F (2016) Enteroendocrine cells: Chemosensors in the intestinal epithelium. Annu Rev Physiol 78:277–299. https://doi.org/10.1146/annurev-physiol-021115-105439
17. Worthington JJ, Reimann F, Gribble FM (2018) Enteroendocrine cells-sensory sentinels of the intestinal environment and orchestrators of mucosal immunity. Mucosal Immunol 11(1):3–20. https://doi.org/10.1038/mi.2017.73
18. Gerbe F, Brulin B, Makrini L, Legraverend C, Jay P (2009) DCAMKL-1 expression identifies tuft cells rather than stem cells in the adult mouse intestinal epithelium. Gastroenterology 137(6):2179–2180; author reply 80-1. https://doi.org/10.1053/j.gastro.2009.06.072
19. Gerbe F, Legraverend C, Jay P (2012) The intestinal epithelium tuft cells: specification and function. Cell Mol Life Sci 69(17):2907–2917. https://doi.org/10.1007/s00018-012-0984-7
20. Hofer D, Puschel B, Drenckhahn D (1996) Taste receptor-like cells in the rat gut identified by expression of alpha-gustducin. Proc Natl Acad Sci U S A 93(13):6631–6634. https://doi.org/10.1073/pnas.93.13.6631
21. Perez CA, Huang L, Rong M, Kozak JA, Preuss AK, Zhang H et al (2002) A transient

receptor potential channel expressed in taste receptor cells. Nat Neurosci 5(11): 1169–1176. https://doi.org/10.1038/nn952

22. Bezencon C, le Coutre J, Damak S (2007) Taste-signaling proteins are coexpressed in solitary intestinal epithelial cells. Chem Senses 32(1):41–49. https://doi.org/10.1093/chemse/bjl034
23. Kaske S, Krasteva G, Konig P, Kummer W, Hofmann T, Gudermann T et al (2007) TRPM5, a taste-signaling transient receptor potential ion-channel, is a ubiquitous signaling component in chemosensory cells. BMC Neurosci 8:49. https://doi.org/10.1186/1471-2202-8-49
24. Coons AH, Creech HJ, Jones RN (1941) Immunological properties of an antibody containing a fluorescent group. Proc Soc Exp Biol Med 47(2):200–202
25. Liu E, Vega S, Treiser M, Sung H, Moghe P (2011) Fluorescence imaging of cell-biomaterial interactions. In: Comprehensive biomaterials. Elsevier, Amsterdam, pp 291–303
26. Goding JW (1986) Immunofluorescence. In: Monoclonal antibodies: principles and practice, vol 255. Elsevier, Amsterdam
27. Bogen SA, Vani K, Sompuram SR (2009) Molecular mechanisms of antigen retrieval: antigen retrieval reverses steric interference caused by formalin-induced cross-links. Biotech Histochem 84(5):207–215. https://doi.org/10.3109/10520290903039078
28. Nagatake T, Fujita H, Minato N, Hamazaki Y (2014) Enteroendocrine cells are specifically marked by cell surface expression of claudin-4 in mouse small intestine. PLoS One 9(6): e90638. https://doi.org/10.1371/journal.pone.0090638
29. Huh WJ, Roland JT, Asai M, Kaji I (2020) Distribution of duodenal tuft cells is altered in pediatric patients with acute and chronic enteropathy. Biomed Res 41(2):113–118. https://doi.org/10.2220/biomedres.41.113

Chapter 3

Novel Approach to Measure Transepithelial Electrical Resistance in Intestinal Cells

Gurveer Marva, Seyda Ünsal, Andrew V. Benest, David O. Bates, and Paloma Ordóñez-Morán

Abstract

The technique electric cell-substrate impedance sensing (ECIS) can be used to detect and monitor the behavior of intestinal cells. The methodology presented was designed to achieve results within a short time frame, and it was tailored to use a colonic cancer cell line. Differentiation of intestinal cancer cells has previously been reported to be regulated by retinoic acid (RA). Here, colonic cancer cells were cultured in the ECIS array before being treated with RA, and any changes in response to RA were monitored after treatment. The ECIS recorded changes in impedance in response to the treatment and vehicle. This methodology poses as a novel way to record the behavior of colonic cells and opens new avenues for in vitro research.

Key words Intestine, Colon, Caco-2 cells, Differentiation, ECIS, TEER

1 Introduction

Morphological changes are often a consequence of cellular differentiation. The new technology electrical cell-substrate impedance sensing (ECIS) can detect in real time changes in morphology and membrane barrier function through changes in impedance. Once different parameters are established, it can be used to monitor movement and spreading [1–5]. Impedance is measured by passing an alternating current between two electrodes (where cells are cultured), and the potential difference across the electrodes is recorded. This can be translated into other values such as resistance and capacitance using Ohm's law. The cells growing on the electrodes act as insulators so as they attach to, spread, and proliferate on the electrode surface, and they interfere with the current and subsequently the resistance. Therefore, any changes in the impedance are in response to phenotypic and behavioral changes of the

Paloma Ordóñez-Morán (ed.), *Intestinal Differentiated Cells: Methods and Protocols*, Methods in Molecular Biology, vol. 2650, https://doi.org/10.1007/978-1-0716-3076-1_3,

cells which could include new junctions' formation, migration, and morphological changes, so the ECIS can be used to detect differentiation of cells [1–5].

Caco-2 human colon cancer cells are a commonly used model to study membrane barrier function, as they are able to form a monolayer of cells which can differentiate to a phenotype similar to epithelial enterocytes [6–9]. These properties make them favorable to be studied in a novel method that involves measuring the impedance across a monolayer of cells. This technology, initially designed by Giaever and Keese [2, 3], was used in our experiments to detect changes in impedance, resistance, and capacitance when Caco-2 cells were treated with different compounds to identify differentiation and changes in membrane permeability. This was an opportunity to apply the new ECIS technology designed by Applied BioPhysics and to optimize a method that would assess cell barrier function and differentiation real time.

The intestine has a hierarchical organization of tissues, and the intestinal stem cells are maintained by Wnt signaling [10, 11]. Suppression of this pathway has previously been described to inhibit the maintenance of intestinal stem cell renewal, and it has been shown to prevent terminal differentiation by inhibiting the expression of HOXA5 in these cells through a negative feedback mechanism [12]. The gene for HOXA5 contains a retinoic acid response element (RARE) which means it can be regulated by retinoic acid (RA). Retinoids have been found to induce differentiation in colorectal cancer cells by increasing the expression of HOXA5; moreover, stable expression of HOXA5 resulted in terminal differentiation of cancer cells [12]. Signal transduction of RA involves it binding to the nuclear RA receptor which, in turn, binds to the RARE which is a series of tandem repeats within the DNA sequence of a gene [13]. Binding to the RARE attracts other transcription factors which upregulate the expression of the gene. We describe here the methodology testing RA only on short-term studies by using ECIS with Caco-2 cells. Long-term time courses experiments would be also used for testing cell differentiation.

2 Materials

2.1 Cells Preparation

1. Caco-2 cells (ATCC) (Fig. 1) (*see* **Note 1**).
2. DMEM.
3. 10% FBS.
4. Pen/strep.
5. 10 mM L-cysteine (stock).
6. Gelatine 10% (stock).
7. Drug/compound to test (all-trans retinoic acid was used in our experiments) (*see* **Note 2**).

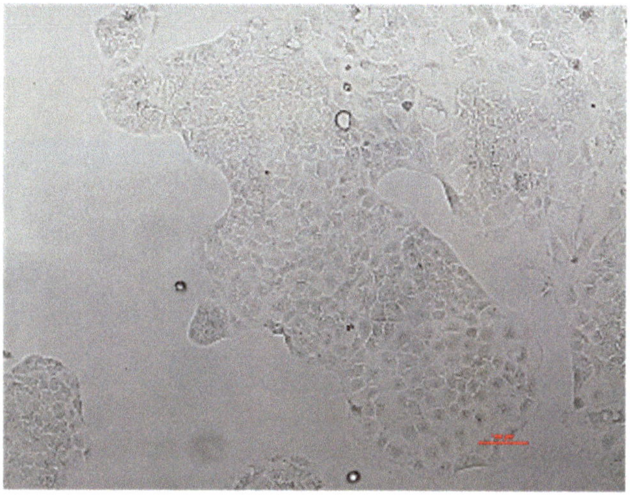

Fig. 1 Caco-2 cells imaged after 168 h by microscopy (phase contrast). Scale bar, 100μm

2.2 ECIS Measurements

1. ECIS (Applied BioPhysics); station controller and an array station, which was kept inside the incubator at 37 °C with 5% CO_2 and ECIS software (Fig. 2) (www.biophysics.com).
2. Eight-well arrays (8W10E) (Fig. 3).

3 Methodology

3.1 Preparation of the Array (Use Class II Laminar Flow Hood)

1. Stabilize the arrays by coating each well with 200μL of 10 mM L-cysteine.
2. Leave for 15 min at room temperature.
3. Aspirate out the cysteine and wash with distilled water three times.
4. Add 200μL of 1% gelatine to each well.
5. Incubate for 30 min at 37 °C with 5% CO_2.
6. Aspirate out the gelatine and wash with distilled water three times.
7. Air dry and either store or use.
8. Fill each well with 400μL cell-free culture medium.

3.2 Cell Seeding

1. Cells were cultured in 150 cm^2 flasks in DMEM with 10% FBS + pen/strep and regularly monitored under microscope to assess confluency.
2. When 70% confluency was observed, splitting was performed by removing media and washing with enough phosphate-buffered saline, PBS (Gibco), to sufficiently cover the flask surface area.

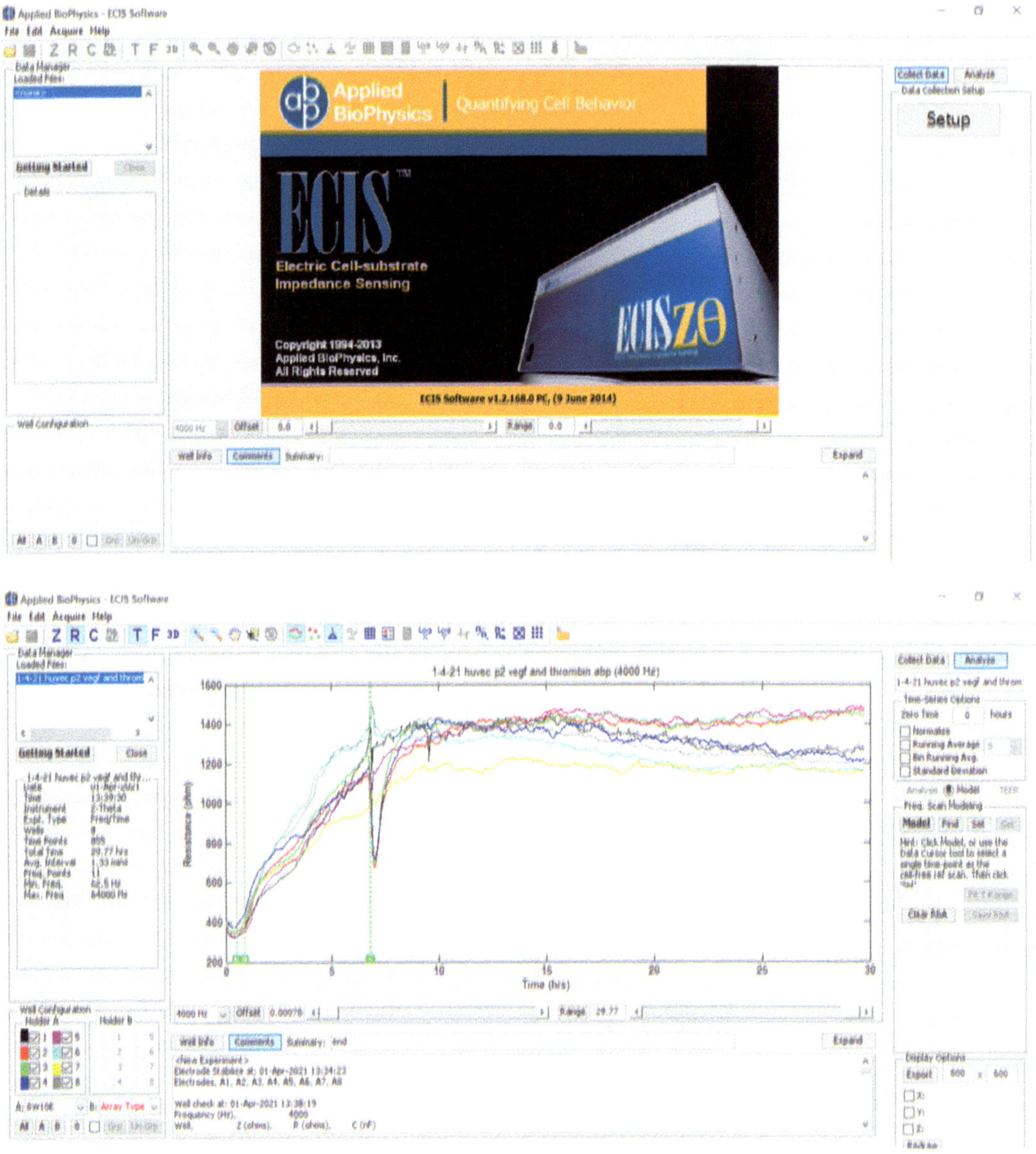

Fig. 2 ECIS (Applied BioPhysics) software, screenshots of the computer screen

3. The cells were then incubated with trypsin (TrypLE™ Express, Gibco) at 37 °C/7 min.
4. Cell dissociation was observed under the microscope, and then the trypsin was deactivated using complete media.
5. The media was then extracted and centrifuged at 0.5 rcf/5 min.
6. The supernatant was disposed, and the cell pellet was resuspended in 1 mL of media.
7. 10μL of the resuspension was aspirated out and used in a hemocytometer.

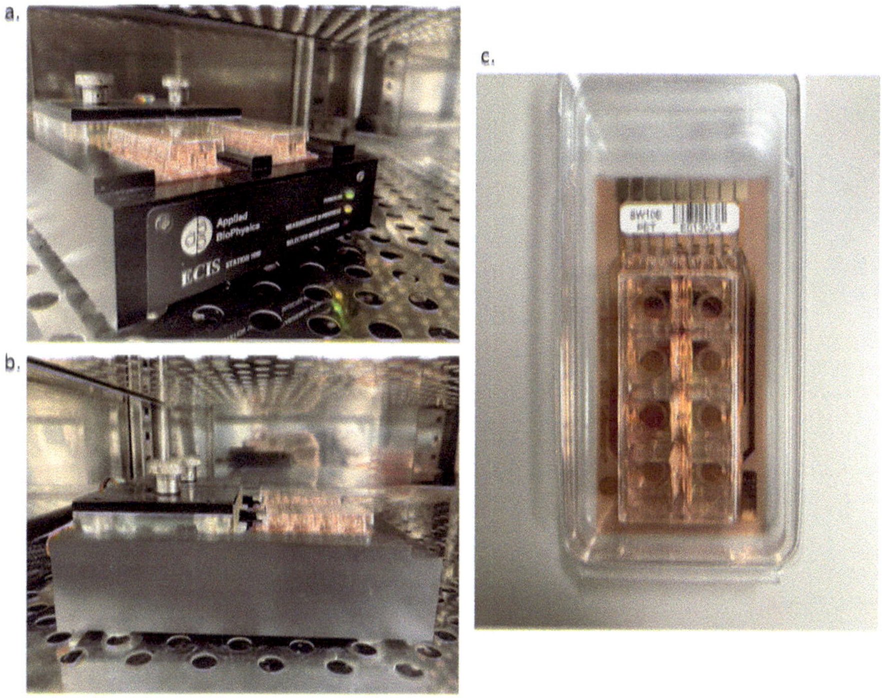

Fig. 3 ECIS (Applied BioPhysics): (**a**, **b**) Station controller and an array station, which was kept inside the incubator at 37 °C with 5% CO_2 (**c**). Eight-well arrays (8W10E)

8. Cells were counted and the formula below was used to determine the number of cells per milliliter:

$$\text{Number of cells per millilitrer} = (\text{Number of cells counted}/4) \times 10^4 \qquad (1)$$

9. A minimum number of 160,000 cells per well is required for this method.
10. Each well can contain 400μL of media and so Eq. 1 needs to be multiplied by 0.4 to account for this.

3.3 ECIS Measurements

1. Place the array into the array station for data collection (Fig. 3).
2. System software setup: The data collection software is then set up as instructed using the manual (*see* **Note 3** and Fig. 2):
   (i) Start by clicking "Setup."
   (ii) Click "Check" to ensure the wells are all recording correctly (this is confirmed by a green mark appearing next to each well number).
   (iii) Then data collection can be started by pressing "Start."

(iv) Clicking "Pause" will stop the recording so that the array can be removed for media change or drug treatments and inspection.

(v) Once the array has been put back in the station, click "Continue" to proceed with the recording.

(vi) To end the recording, click "Stop."

(vii) The data will automatically save under the name and location setup when the recording was started.

3. Start a multifrequency data collection that should last a minimum of 15 min.
4. Pause the recording, and remove the array.
5. Seed the cells at a minimum density of 160,000 cells per well (*see* **Note 4**).
6. Begin again the recording.
7. Leave the ECIS array inside the incubator.
8. A peak and plateau in resistance should be observed at approx. 30 h (*see* **Note 5**) (Fig. 4).
9. Drug treatments can then be administered depending on the aim of the project (*see* **Note 6**).

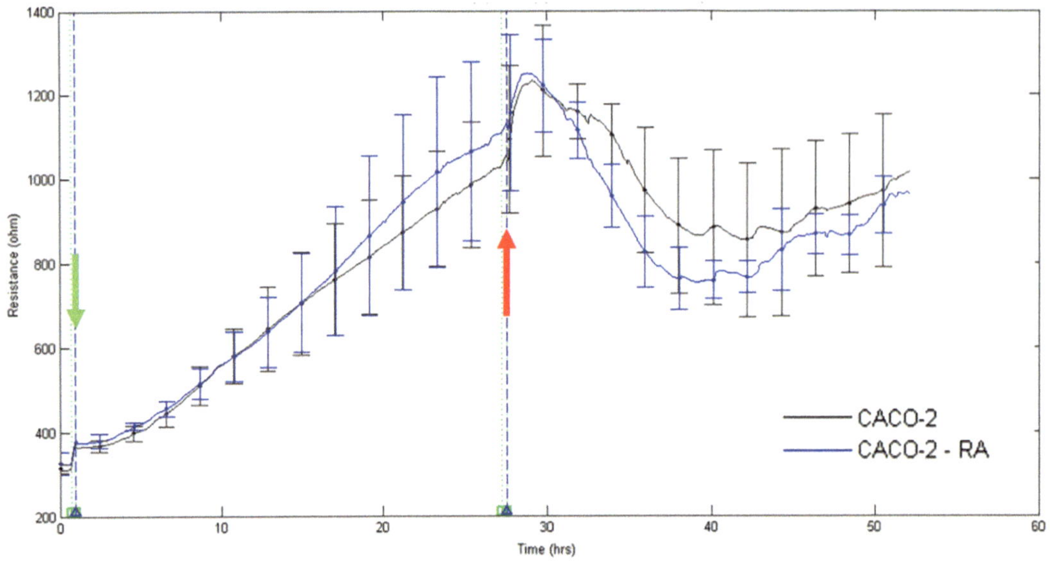

Fig. 4 Experiment which was set up with Caco-2 cells in all eight wells, four wells for control and four wells for treatment with RA. The cells were seeded at a density of 160,000 cells per well, and error bars are indicative of the standard deviation about the mean from the four replicates. The green arrow indicates seeding, and the red arrow indicates change of media. This figure shows that the cell seeding density enabled Caco-2 cells to proliferate and cause a peak in resistance at approximately 30 h

4 Notes

1. Caco-2 cells should be ideally passaged every 2 or 3 days. These cells differentiate in culture so we need to be cautious about the number of passages done before the experiment and we recommend culturing the cells at 40–50% confluency. By using a microscope, the phenotype can be checked visually.
2. We have used in our experiments the compound all-trans retinoic acid (RA) to test differentiation in vitro of our colon cells [12]. The compound can be chosen dependent on your objectives. Differentiation by the compound of interest can be previously validated by RT-qPCR experiments testing several specific markers such as the enterocyte marker *ANPEP*.
3. The machine has the capacity to collect data at various frequencies, and it is strongly recommended to record across "multiple frequencies" to ensure sufficient data collection. More information can be found here: https://www.biophysics.com/whatIsECIS.php.
4. We have tested different seeding numbers for our Caco-2 cells experiments. We recommend using 160,000 cells per array, but this number will change dependent on the type of cell used.
5. The peak and plateau in resistance is indicative of the formation of a confluent layer in the well.
6. This experiment was designed to have a shorter timeframe; moreover the Caco-2 cells did not survive well at low seeding densities which was detrimental to the experiment.

References

1. Szulcek R, Bogaard HJ, van Nieuw Amerongen GP (2014) Electric cell-substrate impedance sensing for the quantification of endothelial proliferation, barrier function, and motility. J Vis Exp. https://doi.org/10.3791/51300
2. Giaever I, Keese CR (1986) Use of electric fields to monitor the dynamical aspect of cell behavior in tissue culture. IEEE Trans Biomed Eng BME-33:242–247
3. Giaever I, Keese CR (1984) Monitoring fibroblast behavior in tissue culture with an applied electric field. Proc Natl Acad Sci 81:3761–3764
4. Bagnaninchi PO, Drummond N (2011) Real-time label-free monitoring of adipose-derived stem cell differentiation with electric cell-substrate impedance sensing. Proc Natl Acad Sci U S A 108(16):6462–6467
5. Marziano M, Tonello S, Cantù E, Abate G, Vezzoli M, Rungratanawanich W, Serpelloni M, Lopomo NF, Memo M, Sardini E, Uberti D (2019) Monitoring Caco-2 to enterocyte-like cells differentiation by means of electric impedance analysis on printed sensors. Biochim Biophys Acta Gen Subj 1863:893–902
6. Yamashita S, Furubayashi T, Kataoka M, Sakane T, Sezaki H, Tokuda H (2000) Optimized conditions for prediction of intestinal drug permeability using Caco-2 cells. Eur J Pharm Sci 10:195–204
7. Sun H, Chow EC, Liu S, Du Y, Pang KS (2008) The Caco-2 cell monolayer: usefulness and limitations. Expert Opin Drug Metab Toxicol 4:395–411
8. Artursson P, Palm K, Luthman K (2001) Caco-2 monolayers in experimental and theoretical

predictions of drug transport. Adv Drug Deliv Rev 46:27–43

9. Kerns EH, Di L, Petusky S, Farris M, Ley R, Jupp P (2004) Combined application of parallel artificial membrane permeability assay and Caco-2 permeability assays in drug discovery. J Pharm Sci 93:1440–1453
10. Katoh M, Katoh M (2007) WNT signaling pathway and stem cell signaling network. Clin Cancer Res 13:4042
11. Tian H, Biehs B, Chiu C, Siebel CW, Wu Y, Costa M, de Sauvage FJ, Klein OD (2015) Opposing activities of notch and Wnt signaling regulate intestinal stem cells and gut homeostasis. Cell Rep 11:33–42
12. Ordóñez-Morán P, Dafflon C, Imajo M, Nishida E, Huelsken J (2015) HOXA5 counteracts stem cell traits by inhibiting Wnt signaling in colorectal cancer. Cancer Cell 28:815–829
13. Mangelsdorf DJ, Umesono K, Kliewer SA, Borgmeyer U, Ong ES, Evans RM (1991) A direct repeat in the cellular retinol-binding protein type II gene confers differential regulation by RXR and RAR. Cell 66:555–561

Chapter 4

In Vivo Model for Isolating Epithelial Cells of the Anorectal Transition Zone

Loucinė Mitoyan, Charlyne Gard, Sébastien Nin, Béatrice Loriod, and Géraldine Guasch

Abstract

Different epithelia line the body and organs and form a continuous lining of cells. The junction of two different types of epithelia represents a special region called transition zone (TZ). TZ are small areas found in numerous places in the body such as between the esophagus and the stomach, in the cervix, in the eye, and between the anal canal and the rectum. These zones are associated with diverse pathologies such as cancers; however, the cellular and molecular mechanisms involved in tumor progression are poorly investigated. We recently characterized the role of anorectal TZ cells during homeostasis and after injury using an in vivo (lineage tracing) approach. To follow TZ cells, we previously developed a mouse model of lineage tracing using cytokeratin 17 (Krt17) as a promoter and GFP as a reporter. Krt17 is expressed by TZ but also by anal glands located below the TZ in the stroma that can interfere with TZ cell population isolation and analysis afterward. In this chapter, we provide a new dissection method to remove specifically anal glands without affecting anorectal TZ cells. This protocol allows the specific dissection and isolation of anal canal, TZ, and rectum epithelia.

Key words Anorectal transition zone, Epithelial cells, Cellular heterogeneity, Intestine, Rectum

1 Introduction

Our body and our organs are lined with epithelium constituted with different kinds of cells arranged in a single or multiple layers of cells. Epithelia form a continuous lining through the body, and in numerous places, the meeting of different types of epithelia constitutes special regions called transition zones (TZ). TZ are conserved among species and are found throughout the body, at the beginning of the gastrointestinal (GI) tract, between the esophagus and the stomach, and at the end of the GI tract between the anus and the rectum [1–3]. In the cervix, TZ is located between the

Authors Loucinė Mitoyan and Charlyne Gard have equally contributed to this chapter.

Paloma Ordóñez-Morán (ed.), *Intestinal Differentiated Cells: Methods and Protocols*, Methods in Molecular Biology, vol. 2650,
https://doi.org/10.1007/978-1-0716-3076-1_4,

endo- and the ectocervix [4]. It is also found in the eye in a region called the limbus located between the conjunctiva and the cornea [5]. Interestingly, TZ are frequently associated with pathologies such as ulcers and cancers.

Molecularly, TZ constitutively express cytokeratin 17 (Krt17), a protein usually expressed in injured tissues such as skin. Other wound-associated proteins are expressed in the TZ as well such as β6-integrin and tenascin-C suggesting a pre-lesional state of TZ [6]. A study from 2010 showed that anorectal TZ contains a unique small cell population with stem cell features. These cells are slow-cycling and express a common stem cell marker CD34 [7]. Together, these data suggest anorectal TZ as a niche prone to pathologies development. We hypothesized that TZ stem cells may contribute to maintaining surrounding epithelia during homeostasis and upon injury. Using a mouse model of lineage tracing, we recently showed that TZ stem cells maintain anal canal epithelium during homeostasis. After an injury, a small TZ population identified as hybrid cells that express specific markers of both squamous and glandular epithelia contributes to repairing the glandular epithelium of the rectum. Most of the time, different cell types can coexist within the same population, forming subpopulations with distinct roles [8–11]. Therefore, to study the role of these cells under different conditions and unveil the heterogeneous population of anorectal TZ, it was necessary to properly isolate the tissue and extract anorectal cells. In this chapter, we will focus on how to dissect the anorectal TZ from mice, extract cells from the tissue, and isolate various populations of cells using flow cytometry.

2 Materials

2.1 Mice

All animals used are in the C57/BL6J background. Both male and female mice are used in all the experiments. In vivo experiments were performed in accordance with the European Community guidelines for the care and use of laboratory animals. All experiments were approved by the European and national regulation (protocols #4572) ethical committee CE14. All animal manipulations were performed by qualified personnel having sufficient experience with the described procedures to minimize any possible pain and discomfort to the animals. The housing conditions of all the mice followed strictly the ethical regulation. Mice are housed in individually ventilated cages (IVC, sealsafe plus TECHNIPLAST) according to SPF FELASA standards, and food and water were given at libitum. For social enrichment, six mice per cages are generally housed with sterile nesting materials such as cotton or compressed wood chips for nidification. The room temperature ranged from 20 to 25 °C. The relative ambient humidity was 55%

+/−15. Seminatural light cycle of 12:12 was used. Transgenic mouse lines K17CreERT2xRosa26-lsl-ZsGreen1 (Jackson Laboratories, stock no. 007906) were induced with tamoxifen for 2 days and sacrificed 2 days later for FACS-cells isolation.

2.2 Tissue Anorectal Isolation and Anal Glands Removal

1. Falcon® clarified polypropylene 50 mL conical bottom tube, 352070.
2. Sterile scissors F.S.T. 14060-11 stainless germany 12s5.
3. Sterile dissecting forceps F.S.T. 11008-13 stainless germany 15s13.
4. Falcon® 60 × 15 mm TC-treated Easy-Grip Style Cell Culture Dish, 353004.
5. Gibco® DPBS 1×, 14190-094.
6. Leica® stereomicroscope MZ6.
7. Sterile small scissors F.S.T. 15000-00 stainless germany 17s9.
8. Sterile dissecting forceps F.S.T. by DUMONT Switzerland 11255-20 Inox Biology or DERRON 14-88 stainless germany 10322/12.
9. Trimmer MOSER 1556 AKKU.

2.3 Sample Preparation for Flow Cytometry (FACS)

1. Sterile scissors F.S.T. 14060-11 stainless germany 12s5.
2. Dissecting forceps F.S.T. 11008-13 stainless germany 15s13.
3. Swann-Morton disposable sterile scalpels, 0510.
4. Life Technologies® Hank's Balanced Salt Solution (HBSS) 1×, 14170-112.
5. Sigma® collagenase 20% from *Clostridium histolyticum*, C2674.
6. Sigma® DNase I (10 mg/mL), DN25.
7. Fetal bovine serum (FBS) chelated [12].
8. Falcon® polypropylene 5 mL round-bottom tube, 352054.
9. Gibco® cold DPBS 1×, 14190-094.
10. Falcon® clarified polypropylene 50 mL conical bottom tube, 352070.
11. Falcon® 60 × 15 mm TC-treated Easy-Grip Style Cell Culture Dish, 353004.
12. Sarstedt® serological pipette 10 mL (86.1254.001) and 25 mL (86.1685.001).
13. Falcon® sterile nylon cell strainers, 70 μm (352350) and 40 μm (352340).
14. Thermo Scientific® Centrifuge Heraeus Megafuge 40R.
15. Life Technologies® 0.25% trypsin-EDTA, 25200056.

Table 1
Antibodies used during FACS including markers for dead cells, microenvironment cells (immune cells, macrophages, fibroblasts, and endothelial cells), and epithelial cells

Population	Antibodies	Species	Supplier	Catalog-number	Dilution
Dead cells	eFLUOR		Thermo Fisher	65-0895	1/1000
Immune cells	PE-Cy7-CD45	Rat	eBioscience	25-0451	1/200
Macrophages	PE-Cy7-CD11b	Rat	BD Pharmingen	561098	1/200
Fibroblasts	PE-Cy7-PDGFRα (CD140α)	Mouse	Biolegend (Ozyme)	BLE323507	1/50
Endothelial cells	PE-Cy7-CD31	Rat	BD Pharmingen	561410	1/100
Epithelial cells	PE-EpCam	Rat	Biolegend	118206	1/100
TZ subpopulation	Biotin-CD34	Rat	eBioscience	13-0341	1/50
	APC-Streptavidin		BD Pharmingen	554607	1/200

16. Staining buffer (DPBS 1× + 2% FBS).
17. Epithelial cell culture medium without calcium [12].
18. Epithelial cell culture medium without calcium with 10% FBS.
19. Reagiergelab® 1.5 mL tubes, 72.690.
20. Parafilm® "M" p7793-1EA.
21. Antibodies used in sorting strategy (Table 1).

3 Methods

3.1 Anorectal Tissue Isolation and Anal Glands Removal

Lineage tracing allows the following of labeled cells to understand their role and behaviors in vivo. To follow specifically TZ cells, we previously developed a mouse model of lineage tracing using Krt17 as a promoter [3]. Krt17 is overexpressed in transition zones. However, Krt17 is also expressed by anal glands which can interfere with the TZ cell population afterward. To overcome this issue and isolate the purest TZ sample possible, we developed a new method to specifically remove anal glands without damaging the TZ:

1. Sacrifice mice by cervical dislocation according to standard protocol.
2. Shave the anorectal region using a trimmer as shown in Fig. 1a.
3. Put the mouse on its back and cut the tail off (Fig. 1b, c).
4. Lift the tissue with forceps and cut around the anus with scissors, keeping the rectum intact (Fig. 1d–g).

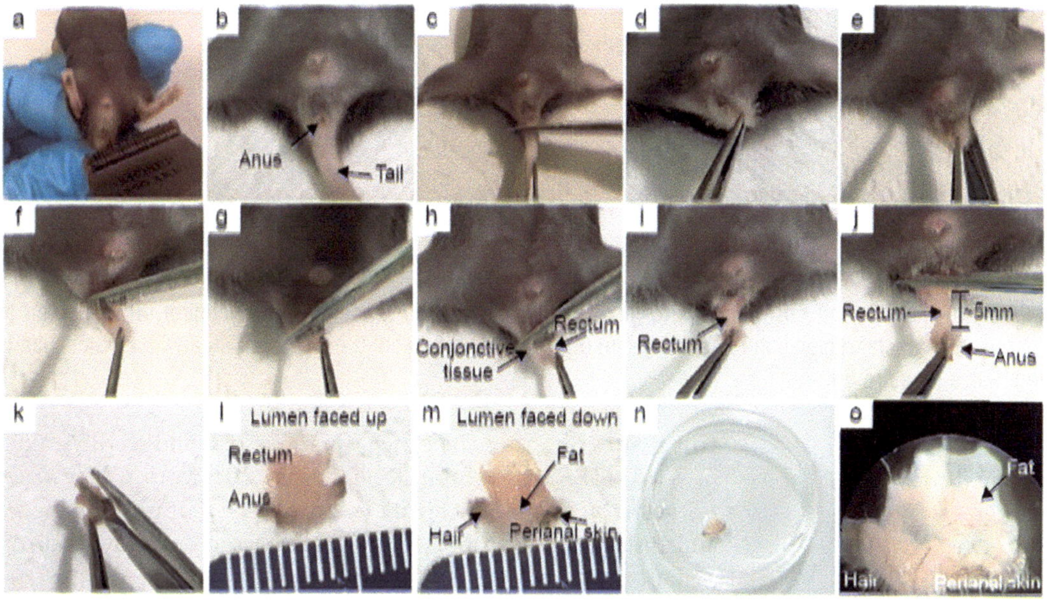

Fig. 1 Anorectal TZ isolation. TZ isolation starts by shaving the anal region to ease following dissection (**a**). (**b**, **c**) Cut the tail off. Next, cut carefully around the anus (**d**–**g**). Remove any conjunctive tissues left on the rectum (**h**, **i**). Cut approximately 5 mm away from the anus (**j**). Open the tissue to (**k**). Flip the tissue lumen faced down (**l**, **m**). Place the tissue in a 60 mm plate with DPBS 1× (**n**). Microdissect fat, hair, and perianal skin under the stereomicroscope (**o**)

5. Cut remaining conjunctive tissue that may be stuck to the rectum (Fig. 1h).
6. Anorectal transition zone and rectum are clearly visible; cut sharply at minimum 5 mm from the anus (Fig. 1i, j).
7. Next, to cut the anorectal region in half, insert scissors directly in the anal orifice and cut sharply. Lumen epithelium is now visible (Fig. 1k, l).
8. Flip the tissue and place the tissue lumen faced down (Fig. 1m) (*see* **Note 1**) in the cell culture dish. 9. Put 500 μL of DPBS 1× in a 60 mm cell culture dish (Fig. 1n). DPBS 1× will prevent the tissue from dehydration.
9. Under the stereomicroscope and using forceps and small scissors, start removing fat tissues to reveal anal glands (Fig. 1o). Hair follicles and skin have to be removed as well to avoid contaminations during further single-cell analysis.
10. Anal glands are easily visible under the stereomicroscope (*see* **Note 2** and Fig. 2). Gentle pressure with forceps on anal glands will further highlight them and ease removal. Cut the glands off. Anal glands are then discarded.
11. Put 10 mL of epithelial cell culture medium in a 50 mL Falcon. Pool all anorectal tissues together in the Falcon, process to 3.2, or keep it at 4 °C overnight (*see* **Note 3**).

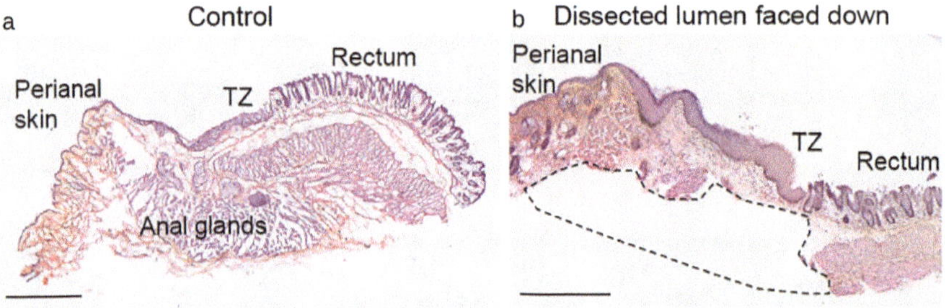

Fig. 2 Hematoxylin and eosin staining of anal glands. Untouched anorectal tissue control (**a**) compared to dissected anorectal region (lumen faced down) (**b**). Dot lines represent the dissected area with no anal glands left. Scale bars are 500 μm

3.2 Sample Preparation for FACS

1. Cut tissues one after the other using scissors, and sterile scalpel in 4 mL HBSS 1× + 50 μL collagenase 20% from *Clostridium histolyticum* in a 60 mm plate (*see* **Note 4**). Cut in small pieces and put Parafilm around the plate to avoid overflows during agitation.
2. Agitate under 280 rpm during 45 min at 37 °C.
3. Add 4.2 μL DNase I (10 mg/mL) and agitate for 10 more min. DNase will prevent any tissue stickiness. During this time, fill 5 mL Falcon tubes with 5 mL of FBS to prevent cells from sticking to the tube. This tube will be used later in **step 12**.
4. Resuspend the mix with 12 mL DPBS 1× cold (*see* **Note 5**) with a 25 mL pipette, and mix inside the pipet to further dissociate all the tissue pieces. Add to a 50 mL Falcon tube.
5. Fill the tube until 50 mL with DPBS 1× cold and centrifuge for 10 min at 1000 rpm at 4 °C. Discard the supernatant carefully.
6. Place a 70 μm strainer on a 50 mL Falcon. Resuspend the pellet with 20 mL DPBS 1× cold (use 10 mL pipette to further dissociate), and filter through 70 μm strainer. Wet strainer first with DPBS 1×. Add 200 μL of FBS to promote cell survival.
7. Move the 70 μm strainer to a 60 mm plate; put 1 mL of 0.25% trypsin-EDTA on the top to dissociate leftover tissues (*see* **Note 6**).
8. Incubate for 5 min at 37 °C.
9. Block trypsin activity with 3 mL of epithelial cell culture medium without calcium, and mix it with the rest of the filtered cells.
10. Place a 40 μm strainer on a 50 mL Falcon and wet with epithelial cell culture medium. Filter everything.
11. Centrifuge for 10 min at 1000 rpm at 4 °C. Discard the supernatant carefully, leaving approximately 250 μL.

12. Discard the FBS from 5 mL tubes used earlier in **step 3** and rinse twice with DPBS 1×.
13. Resuspend the pellet in staining buffer and transfer in a 5 mL Falcon tube.
14. Incubate with primary antibodies (30–45 min on ice). At the same time, add 1/1000 of FVD780 to mark dead cells. Cover with foil and tap the tube every 5 min (Table 1).
15. Fill to top with staining buffer to wash and centrifuge for 5 min at 1000 rpm at 4 °C.
16. Discard supernatant by inverting the tube. There will be about 200 μL left in the tube. Add a maximum of 300 μL of staining buffer.
17. Add secondary antibodies for 20 min at 4 °C (Table 1).
18. Fill to top with staining buffer to wash.
19. Centrifuge for 5 min at 1000 rpm at 4 °C.
20. Discard supernatant by inverting the tube, and resuspend the sample in 250 μL of staining buffer to obtain a final volume of 500 μL (*see* **Note 7**).

    Cells are ready to be sorted (*see* **Note 8**).

3.3 FACS Sorting

In order to decipher TZ population heterogeneity and understand the hierarchical relationships within TZ cells and with its surrounding epithelia that are the anal canal and rectum, we sorted these three populations:

1. Under the PSM, prepare three sterile 1.5 mL tubes (*see* **Note 9**), one for each population to sort.
2. Cells are sorted using FACS Aria II equipment (BD Biosciences) in a collecting medium that is epithelial cell culture medium with no calcium and enriched with 10% chelated FBS (*see* **Note 10**).
3. Put 200–300 μL of collecting medium in each tube.
4. Sort cells in corresponding tubes according to the strategy already published [3].
5. Record FACS-cell counts. According to our previous observations [3], we consider that the real count of alive cells after sorting is half the count given by the FACS.
6. Next, all three samples corresponding to the three populations are pooled together (*see* **Note 11**):
   - TZ cells are the main population of interest, so we use all cells from this population.
   - Anal canal population is entirely used as well and mixed with TZ cells.

- Usually, the rectum population is 10 times more than TZ and anal canal populations. We complete the sample containing TZ and anal canal cells with rectum cells to reach the desired total number of cells.

  Optimal cell concentration ranges from 700 to 1200 cells/μL and is best at 1000 cells/μL according to the manufacturer recommendations.

7. Centrifuge during 10 min at 1000 rpm at 4 °C.
8. Remove the supernatant carefully with a pipette to let only enough volume to have 1000 cells/μL.

(Example: if we sort 5000 TZ cells, 3000 anal canal cells, and 100,000 rectum cells, then all TZ and anal canal cells will be pooled together along with 12,000 rectum cells for a total of 20,000 cells in a final volume of 20 μL.)

4 Notes

1. Be careful to flip the tissue lumen faced down. Otherwise, if anal glands dissection is done lumen faced up, anorectal epithelia may come off with anal glands (Fig. 3).
2. Anal glands are white-colored structures (Fig. 4), and their presence and size are age- and mice-dependent.
3. It is possible to dissect tissues the day before cell sorting and keep overnight at 4° C.
4. To ensure effective enzymatic dissociation, do not pool more than four anorectal tissues in one 60 mm cell culture dish.
5. During FACS sample preparation, use cold DPBS 1× and all media cold to ensure cell viability.
6. Trypsinization of the 70 μm strainer is a crucial step to properly dissociate the tissue. Cells positive for CD34 may remain on the strainer if the enzymatic activity of the trypsin is not efficient.

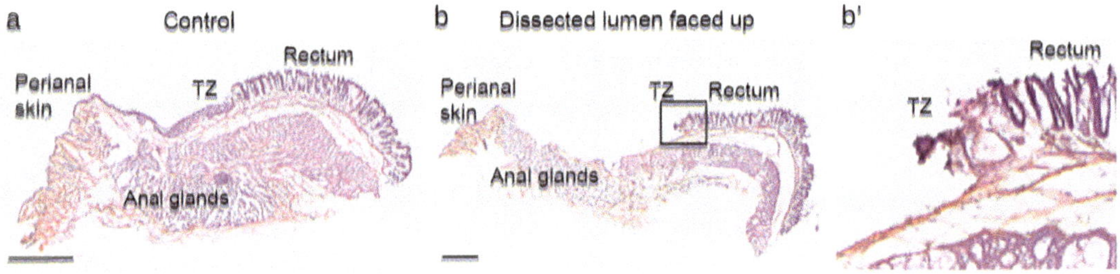

Fig. 3 Anal glands dissection with the lumen faced up. Undissected anorectal tissue (**a**) compared to anal glands removal with the lumen faced up (**b**). Zoom in the remaining of transition zone after dissection (**b′**). Scale bars are 500 μm

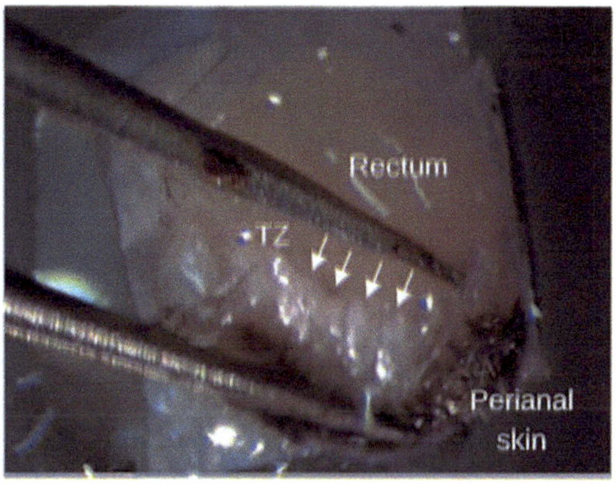

Fig. 4 Anal glands are easily visible. Anal glands, denoted with white arrows, are white structures located above the perianal skin

7. At the end of FACS sample preparation, the resuspension volume depends on the estimated concentration of cells. If it is too concentrated, aggregates may appear. In that case, make sure to filter the sample before sorting cells. Otherwise, aggregates may plug the cell sorter.
8. Do not wait to sort cells after sample preparation. Anorectal epithelial cells are very fragile and die readily.
9. Use 1.5 mL tube to collect cells after cell sorting as it is easier to dump supernatant compared to 2 mL tubes.
10. Collecting medium is enriched with 10% FBS to ensure cell viability as collecting cells in DPBS alone decreases viability.
11. FACS from one single anorectal tissue results in a small sorted TZ population that is not sufficient for analysis whence the need to pool multiple tissues together; at least five tissues need to be pooled.

Acknowledgments

This work was supported by ANR grant #ANR20-CE13-0009-01 (G.G.) and partly supported by research funding from the Canceropôle Provence-Alpes-Côte d'Azur, Institut National du Cancer and Région Sud, grants from the Excellence Initiative of Aix-Marseille University A∗Midex, "Investissement d'avenir" (CapoStromEx) and Inserm Plan Cancer AAP single cell (G.G.). L.M. is a recipient of the French ministerial research fellowship and the Ligue Nationale Contre le Cancer fellowship. Processing of the slides for histology was performed by Emilie Agavnian in the experimental histopathology (ICEP) core facility at the Institut Paoli Calmettes, and the cell sorting by FACS was performed by Manon Richaud and Françoise Mallet in the flow cytometry facility at the CRCM. Mouse housing and caring was performed by Jean-Christophe

Orsoni, Arnaud Capel, and Jean-Charles Graziano in the mouse facility at the CRCM.

The authors would like to thank Lisa Bargier and Dr. Véronique Chevrier for reading and corrections.

References

1. Wang X et al (2011) Residual embryonic cells as precursors of a Barrett's-like metaplasia. Cell 145:1023–1035
2. Guasch G et al (2007) Loss of TGFβ signaling destabilizes homeostasis and promotes squamous cell carcinomas in stratified epithelia. Cancer Cell 12:313–327
3. Mitoyan L et al (2021) A stem cell population at the anorectal junction maintains homeostasis and participates in tissue regeneration. Nat Commun 12:2761
4. Herfs M et al (2012) A discrete population of squamocolumnar junction cells implicated in the pathogenesis of cervical cancer. Proc Natl Acad Sci 109:10516–10521
5. Amitai-Lange A et al (2015) Lineage tracing of stem and progenitor cells of the murine corneal epithelium: lineage tracing of limbal and corneal epithelial cells. Stem Cells 33:230–239
6. Mcnairn AJ, Guasch G (2011) Epithelial transition zones: merging microenvironments, niches, and cellular transformation. Eur J Dermatol 21:21–28
7. Runck LA, Kramer M, Ciraolo G, Lewis AG, Guasch G (2010) Identification of epithelial label-retaining cells at the transition between the anal canal and the rectum in mice. Cell Cycle 9:3111–3117
8. Tasic B (2018) Single cell transcriptomics in neuroscience: cell classification and beyond. Curr Opin Neurobiol 50:242–249
9. Navin NE (2014) Cancer genomics: one cell at a time. Genome Biol 15:452
10. Dalerba P et al (2011) Single-cell dissection of transcriptional heterogeneity in human colon tumors. Nat Biotechnol 29:1120–1127
11. Treutlein B et al (2014) Reconstructing lineage hierarchies of the distal lung epithelium using single-cell RNA-seq. Nature 509:371–375
12. Nowak J. Fuchs E (2009) Isolation and culture of epithelial stem cells. Methods Mol. Bio 482, 215–232

Chapter 5

In Vitro and in Vivo Assays for Testing Retinoids Effect on Intestinal Progenitors' Lineage Commitments

Krishna R. Gajera, Kathryn L. Fair, Gordon W. Moran, Nicholas R. F. Hannan, Joerg Huelsken, and Paloma Ordóñez-Morán

Abstract

The intestine consists of epithelial cells surrounded by a complex environment as mesenchymal cells and the gut microbiota. With its impressive stem cell regeneration capability, the intestine is able to constantly replenish cells lost through apoptosis or abrasion by food passing through. Over the past decade, researchers have identified signaling pathways involved in stem cell homeostasis such as retinoids pathway. Retinoids are also involved in cell differentiation of healthy and cancer cells. In this study, we describe several approaches in vitro and in vivo to further investigate the effect of retinoids on stem cells, progenitors, and differentiated intestinal cells.

Key words Intestine, Retinoids, Differentiation, Stem cells, Secretory, Absorptive, Organoids

1 Introduction

Our intestine is constantly exposed to a harsh and dynamic environment due to food intake and the gut microbiome. The high renewal rate of the intestinal epithelium, therefore, plays an essential role in intestinal homeostasis [1]. There are several factors and signaling pathways that regulate stem cells and their regenerative capability [2]. For instance, surrounding mesenchymal cells and Paneth cells secrete Wnt creating a gradient of high Wnt expression at the crypt bottom [3]. On the other hand, an opposing gradient of BMP (bone morphogenic protein) is noted at the villus ends which functions to regulate proliferation to a limit [4]. In addition, Notch signaling comes into play for maintaining stem cells in their pluripotent state and preventing their differentiation into secretory lineages [5]. During injury and disease states, the Hippo pathway plays an important role in increasing proliferation of intestinal stem cell population to aid in the regeneration process [6]. Maintaining

Paloma Ordóñez-Morán (ed.), *Intestinal Differentiated Cells: Methods and Protocols*, Methods in Molecular Biology, vol. 2650,
https://doi.org/10.1007/978-1-0716-3076-1_5,

the gradients and balance between these various pathways is crucial for intestinal stem cell homeostasis.

Previous research illustrated the role of retinoic acid in stem cell differentiation in inflammation and preventing regeneration; however not much is known about its cross talk with stem cells. Several studies conducted on animal models, such as mice or zebrafish, support the relevant role of retinoids in development and homeostasis but present certain limitations in modeling the exact human physiology [7, 8]. Retinol, a metabolite of vitamin A, is taken up by the intestinal epithelial cells and catabolized into all-trans retinoic acid (atRA) by the enzyme aldehyde dehydrogenase 1A1 (ALDH1A1). The effects of atRA exerted upon binding to retinoic acid receptor (RAR) include reduced proliferation and enhanced differentiation to goblet cells and enterocytes which support secretion of mucus to protect the epithelial barrier [8, 9]. Furthermore, research has shown that atRA treatment leads to increased expression of tight junctions in the epithelium which is linked to enhanced barrier functionality of the intestinal epithelium [10]. Moreover, atRA has been linked to reduced regeneration during inflammation [11]. A more recent study has shown that atRA treatment leads to a reduced regenerative phenotype of organoids and increased enterocyte maturation [12]. We have previously shown that retinoids induce differentiation in colorectal cancer cells by increasing the expression of the transcription factor HOXA5 [13]. Hence, we describe here several in vitro and in vivo approaches that can be used to test retinoic acid effects on intestinal progenitors' lineage commitments.

2 Materials

2.1 In Vitro Assays

2.1.1 2D Experiments

1. SW480 cells (ATCC) (*see* **Note 1**) (Fig. 1).
2. DMEM.
3. PBS 1X.
4. 10% FBS.
5. Pen/strep.
6. Compound to test: atRA, 9-cis retinoic, and/or retinal (*see* **Note 2**).

2.1.2 3D Experiments

Biological samples were obtained from patients undergoing colonoscopy at the Queen's Medical Centre Hospital (Nottingham, UK) under informed consent and approval according to UK ethical regulations. The study followed the guidelines of the Declaration of Helsinki, and patients' identity remained anonymous:

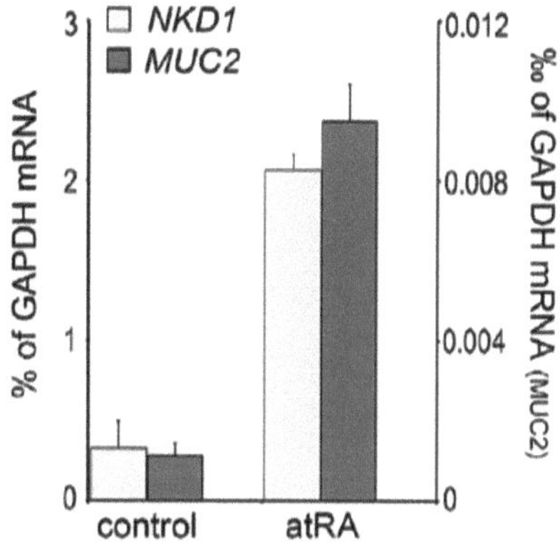

Fig. 1 atRA treatment upregulates *MUC2* (marker of goblet cells) and *NKD1* in human colon cancer cell lines. Bar chart showing gene expression of *MUC2* and *NKD1* for atRA treatment compared to control. (Permission from Cancer Cell to publish in this book; part of this figure was included in [13])

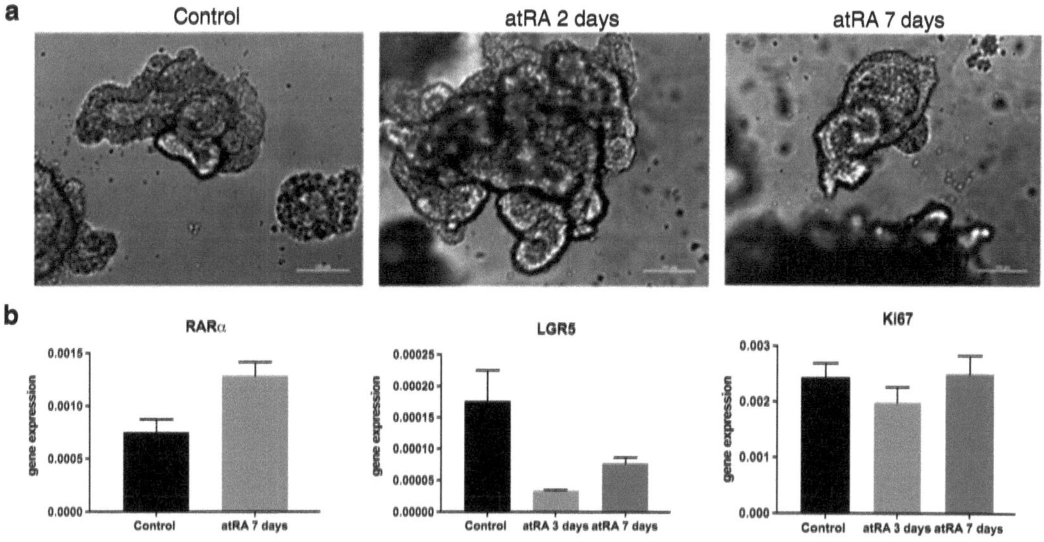

Fig. 2 atRA treatment effects in human organoids. (**a**) Representative images of organoid morphology of control and upon treatment with atRA for 2 and 7 days. Images taken at 10× magnification. (**b**) Bar chart showing gene expression of *RARα*, *LGR5*, and *MKI67* after atRA treatment for 3 and 7 days compared to control. Data represents two independent experiments with triplicate repeats

1. Human intestinal organoids (*see* **Note 3**) (Fig. 2).
2. AdvDMEM/F12.
3. Pen/strep.
4. PBS 1X.
5. Human intestinal organoid media (Table 1).

Table 1
Intestinal organoid media composition. mRspondin-1-Fc and Noggin-His recombinant proteins were produced in our lab [14]

Human intestinal media		Manufacturer	Stock concentration	Final concentration	Dissolved in
Basal intestinal media					
Advanced DMEM		Gibco	1×	1×	
GlutaMax	Source of L-glutamine	Gibco	100×	1×	
N2	Supplement	Gibco	100×	1×	
B27	Supplement	Gibco	50×	1×	
Pen/Strep	Prevent infections	Gibco	100×	1×	
N-Acetyl-l-cysteine	Antioxidant	Sigma-Aldrich	81.5 mg/mL	1 mM	Sterile water
Nicotinamide	Amide of nicotinic acid	Sigma-Aldrich	3.42 M	10 mM	Sterile water
Gastrin	Cholecystokinin activator	Sigma-Aldrich	480 μM	10 nM	950 μL water +50 μL DMSO
A83-01	ALK5 inhibitor	Sigma-Aldrich	11 mM	500 nM	DMSO
SB202190	p38 inhibitor	Sigma-Aldrich	1000×	1×	DMSO
CHIR99021	GSK3β inhibitor / Wnt activator	Sigma-Aldrich	44 mM	3 μM	DMSO
Growth factors					
Noggin-his	BMP antagonist	Produced in-house [14]		100 ng/mL	Sterile PBS
EGF	Activator of EGF signalling	Gibco		50 ng/mL	Sterile water
mRspondin-Fc	Wnt activator	Produced in-house		100 ng/mL	Sterile PBS

6. Compound to test: atRA, 9-cis retinoic, and/or retinal (*see* **Note 2**).
7. Matrigel.

2.2 In Vivo Assay

Biological samples were obtained from patients treated at the University Hospital of Lausanne (Lausanne, Switzerland) under informed consent and approval of the local tumor bank according to Swiss ethical regulations. The study followed the guidelines of

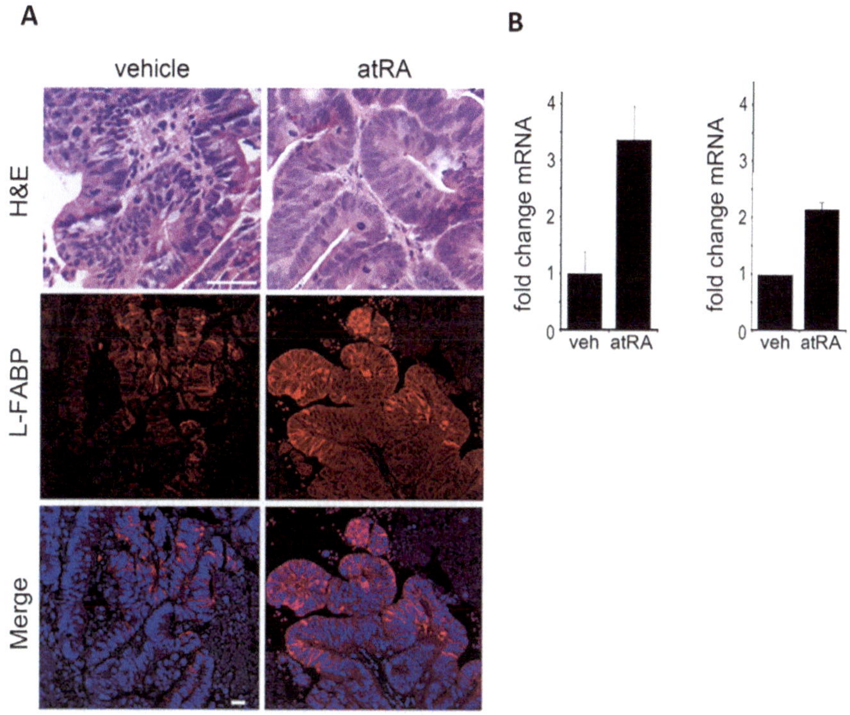

Fig. 3 atRA treatment induces differentiation in colorectal cancer human organoids. (**a**) Representative images of xenografts morphology (H&E) and L-FABP (enterocytes marker) staining when treated with atRA. (**b**) Bar chart showing gene expression of *L-FABP* (left) and *MUC2* (right, goblet cells marker) after atRA or vehicle treatment. *GAPDH* housekeeping gene was used (Permission from Cancer Cell to publish the data in this book; part of this figure was included in [13])

the Declaration of Helsinki, and patients' identity remained anonymous:

1. Intestinal organoids derived from colorectal cancer (CRC) patients (Fig. 3) [13].
2. AdvDMEM/F12.
3. Pen/strep.
4. PBS 1×.
5. Matrigel.
6. Compound to test: atRA, 9-cis retinoic, and retinal (*see* **Note 2**).

3 Methods

3.1 In Vitro Assays

3.1.1 2D Experiments

1. Culture SW480 cells in 150 cm^2 flasks in DMEM with 10% FBS + pen/strep, and regularly monitor under microscope to assess confluency.

2. Wash with PBS 1×.
3. Trypsinize colon cancer cells (SW480), 5 min at 37 °C, 5% CO_2.
4. Spin down at 200 g during 5 min at 4 °C.
5. Wash with PBS 1×.
6. Spin down at 200 g during 5 min at 4 °C.
7. Resuspend the cells in media and seed in tissue-culture-treated plates.
8. Leave the cells O/N inside the incubator at 37 °C, 5% CO_2.
9. Treat the cells with atRA/compound or placebo (*see* **Note 2**).
10. Incubate for 2 days at 37 °C with 5% CO_2.
11. Images were taken using Nikon Eclipse widefield fluorescent microscope to observe any changes in organoid morphology. Image editing can be done using NIS Elements software.
12. Perform further studies by gene expression analysis. The procedure to obtain cDNA and do the qRT-PCRs is described in Chap. 18 of this book (Fig. 1) (*see* **Note 3**).

3.1.2 3D Experiments

1. Embed the intestinal organoids in Matrigel, and keep them in the incubator at 37 °C and 5% CO_2 with human intestinal organoid media (Table 1). For these experiements, B27 without vitamin A can be used. Cells were cultured in 24 well plates (tissue-culture-treated plastic plates) (*see* **Note 4**).
2. Wash with PBS 1×.
3. Mechanically disaggregate the cells and collect in an Eppendorf.
4. Add 20 mL of medium to resuspend the cells (collect them in a 50 mL polypropylene conical tube).
5. Centrifuge (5 min, 4 °C, 200 × g) and discard the supernatant.
6. Resuspend the pellet in 40 mL of cold PBS 1×.
7. Centrifuge (5 min, 4 °C, 200 × g) and discard the supernatant.
8. Treat the cells with 1.7×10^{-4} M of atRA (other concentrations can be used). This concentration was used to treat human organoids over 3-day and 7-day time period. Control wells containing only the media were used to ensure effects observed were due to the atRA treatment and not any other factor.
9. Images were taken using Nikon Eclipse Ti2 widefield fluorescent microscope to observe any changes in organoid morphology. Image editing was done using NIS Elements software (Fig. 2).

10. Perform further studies by gene expression analysis. The procedure to obtain cDNA and do the qRT-PCR is described in Chap. 18 of this book (Fig. 2).

3.2 In Vivo Assays

3.2.1 Preparation of Cells Before Surgery

1. Use intestinal organoids that are plated in p24 well plates (*see* **Note 5**).
2. Aspirate medium.
3. Wash twice with PBS 1×.
4. Mechanically disaggregate the cells with a blue tip and collect the organoids in an Eppendorf.
5. Add 20 mL of medium to dilute the Matrigel, and resuspend the cells (collect them in a 50 mL polypropylene conical tube).
6. Centrifuge (5 min, 4 °C, 200 g) and discard the supernatant.
7. Wash with PBS 1×, and centrifuge.
8. Dissociate the cells with trypsin-EDTA for 5 min at 37 °C.
9. Mechanically disaggregate the cluster of cells into single cells with a micropipette.
10. Count the cells and leave on ice.

3.2.2 Generation of Patient-Derived Xenografts

1. Prepare the heating pad (disinfect with EtOH) and the tools inside the hood.
2. Anesthetize mouse with isoflurane.
3. Once the mouse is asleep, place the mouse on the heating pad, and apply ophthalmic gel to both eyes to prevent dryness.
4. Do the earmarks to the mouse if needed.
5. Flush ethanol 70% on the left flank of the mouse.
6. Make a small incision (0.5 cm).
7. Inject the cells subcutaneously into both flanks of athymic nude mice. 50,000 cells/sample are embedded in 50 μL. Inject the 50 μL in each flank.
8. Slowly remove the needle and apply a cotton wool for a few seconds.
9. Sew the body wall with the sewing needle and the string (tight knot).
10. The skin incision is closed with staples, which are removed within 10 days.
11. Use antiseptic solution to clean the wound.
12. Leave the mouse on the heat pad until awake.
13. Dissolve the analgesic in the drinking water.

14. The mice are injected every other day intraperitoneally with 2.5 mg/kg body weight of atRA or placebo dissolved in corn oil. Treatment with atRA starts 1 day after engraftment.
15. Monitor every day if there is weight loss or any sign of toxicity during the experiment.
16. After 12 weeks the animals treated with atRA or placebo are euthanized.
17. Primary tumors can be carefully isolated and fixed with formalin O/N or can be frozen for RNA analyses. Samples can be sectioned and H&E stained (Fig. 3).

4 Notes

1. SW480 cells should be ideally passaged every 2 or 3 days. The cells may change their morphology if cultured at a high confluence. By using a microscope, check visually the phenotype.
2. We have used in our experiments the compound all-trans retinoic acid (atRA) to test differentiation in vitro of our intestinal cells. This compound can be chosen depending on the study objectives. We have also detected effect on intestinal differentiation with retinal and 9-cis retinoic acid [13].
3. For 3D experiments in vitro, the human intestinal organoids were derived from inflammatory bowel disease non-inflamed tissue (Rec reference: 17/EM/0126, Queen's Medical Centre Hospital, Nottingham, UK). Briefly, healthy human intestinal organoids were cultured at 37 °C with 5% CO_2 in 250 μL intestinal media with growth factors as detailed in Table 1 in a 24- or 48-well plate. The organoids were detached from the wells using a 1 mL pipette tip, resuspended in 7 mL advanced DMEM media, and then centrifuged twice at 200 g for 1 min to separate the crypts from the Matrigel. The pellet was resuspended in 5 mL media using a 200 μL pipette tip to disaggregate the organoids and centrifuged again. The pellet was finally resuspended in 70% Matrigel and 30% intestinal media with growth factors. This was then added to the center of each well of a 48-well plate and incubated at 37 °C, 5% CO_2, for 5 min followed by addition of 250 μL of media with growth factors to each well. The organoids were maintained and split on a 1:2 ratio every 5–6 days. Stocks were frozen for future use.
4. Stem cell pool and cell differentiation process done by the compound of interest can be further validated by RT-qPCR experiments selecting the specific markers (Figs. 1 and 2) [13].
5. The protocol of human intestinal organoids was used also to maintain the CRC patient-derived organoids [13]. These can be cultured without Rspondin1 as are derived from colorectal tumors that usually have Wnt pathway constitutively active.

Acknowledgments

This work was supported by MRC-NIRG (MR/V009559/1), GUTS UK and CICRA UK charities.

References

1. Stevens CE, Leblond CP (1947) Rate of renewal of the cells of the intestinal epithelium in the rat. Anat Rec 97(3):373
2. Takahashi T, Shiraishi A (2020) Stem cell signaling pathways in the small intestine. Int J Mol Sci 21(6):2032
3. Andreu P, Colnot S, Godard C, Gad S, Chafey P, Niwa-Kawakita M, Laurent-Puig P, Kahn A, Robine S, Perret C, Romagnolo B et al (2005) Crypt-restricted proliferation and commitment to the Paneth cell lineage following Apc loss in the mouse intestine. Development 132(6):1443–1451
4. Qi Z, Li Y, Zhao B, Xu C, Liu Y, Li H et al (2017) BMP restricts stemness of intestinal Lgr5+ stem cells by directly suppressing their signature genes. Nat Commun 8(1):13824
5. Tian H, Biehs B, Chiu C, Siebel CW, Wu Y, Costa M et al (2015) Opposing activities of Notch and Wnt signaling regulate intestinal stem cells and gut homeostasis. Cell Rep 11(1):33–42
6. Mo JS, Park HW, Guan KL (2014) The hippo signaling pathway in stem cell biology and cancer. EMBO Rep 15(6):642–656
7. Martin BL, Kimelman D (2010) Brachyury establishes the embryonic mesodermal progenitor niche. Genes Dev 24:2778–2783
8. Chanchevalap S, Nandan MO, Merlin D, Yang VW (2004) All-trans retinoic acid inhibits proliferation of intestinal epithelial cells by inhibiting expression of the gene encoding Kruppel-like factor 5. FEBS Lett 578(1–2):99–105
9. Snyder LM, Arora J, Kennett MJ, Weaver V, Cantorna MT (2020) Retinoid signaling in intestinal epithelial cells is essential for early survival from gastrointestinal infection. Front Immunol 11(2479)
10. Yamada S, Kanda Y (2019) Retinoic acid promotes barrier functions in human iPSC-derived intestinal epithelial monolayers. J Pharmacol Sci 140(4):337–344
11. Bhatlekar S, Viswanathan V, Fields JZ, Boman BM (2018) Overexpression of HOXA4 and HOXA9 genes promotes self-renewal and contributes to colon cancer stem cell overpopulation. J Cell Physiol 233(2):727–735
12. Lukonin I, Serra D, Challet Meylan L, Volkmann K, Baaten J, Zhao R et al (2020) Phenotypic landscape of intestinal organoid regeneration. Nature 586(7828):275–280
13. Ordóñez-Morán P, Dafflon C, Imajo M, Nishida E, Huelsken J (2015) HOXA5 counteracts stem cell traits by inhibiting Wnt signaling in colorectal cancer. Cancer Cell 28:815–829
14. Hacker DL, Ordóñez-Morán P (2020) Large-scale production of recombinant noggin and R-Spondin1 proteins required for the maintenance of stem cells in intestinal organoid cultures. In: Ordóñez-Morán P (ed) Intestinal stem cells: methods and protocols. Springer, New York, pp 171–184

Part II

Transcriptional Profiling of Intestinal Cells

Chapter 6

TORNADO-seq: A Protocol for High-Throughput Targeted RNA-seq-Based Drug Screening in Organoids

Maxim Norkin and Joerg Huelsken

Abstract

Organoids are 3D ex vivo cell aggregates derived from primary tissue and shown to closely recapitulate tissue homeostasis. Organoids deliver certain advantages compared to 2D cell lines and mouse models, especially in drug-screening studies and translational research projects. The application of organoids in the research field is fast-emerging and new techniques for organoid manipulation are constantly developing. Despite recent advances, RNA-seq-based drug-screening platforms in organoids are not yet established. Here, we provide a detailed protocol for performing TORNADO-seq, a targeted RNA-seq-based drug-screening method in organoids. Analyzing complex phenotypes with a large number of carefully selected read-outs allows to directly classify and group drugs even without structural similarity or overlapping mode of actions from prior knowledge. Our assay principle combines cost-effectiveness and sensitive detection of multiple cell identities, signaling pathways, and key drivers of cellular phenotypes and can be applied to many systems where this new form of high-content screening can provide information not obtainable otherwise.

Key words Targeted RNA-seq organoids, High-throughput screening, High-content screening, NGS-based screening in organoids

1 Introduction

The development of extracellular matrix-based gels in the 1990s opened the window for performing experiments with cells in 3D [1–4]. Further, Toshiro Sato from the Clevers group generated a functional 3D ex vivo model of intestinal epithelium [5]. He created a long-term culturing method of $Lgr5^+$ stem cells utilizing a cocktail of growth factors that allowed to grow 3D intestinal organoids. This milestone achievement boosted research in the organoid field. Nowadays, organoids are widely accepted as a valuable research tool.

A variety of organoid-based drug-screening platforms were established in recent years. These platforms utilize many wt and cancer models of different organs. Intestinal and colorectal cancer

Paloma Ordóñez-Morán (ed.), *Intestinal Differentiated Cells: Methods and Protocols*, Methods in Molecular Biology, vol. 2650, https://doi.org/10.1007/978-1-0716-3076-1_6,

organoid models have been studied the most extensively. Organoids were used in high-throughput drug screenings [6, 7], CRISPR screenings [8, 9], microscopy-based high-content drug screenings [10, 11], and translational studies [12–15].

While most drug screens are based on a few parameters such as the activity of signaling pathways or the measurement of cell viability, they lack the rich data required for the analysis of complex phenotypes in heterogeneous organoid models and in most cases are insufficient in predicting the mechanism of action of drugs. The platforms which rely on the detection of multiple parameters would allow to overcome these limitations. NGS-based drug screening where the analysis of a treated sample is based on the expression of hundreds of genes is an excellent example of such a strategy. Unfortunately, currently, there are no RNA-seq-based platforms for performing high-throughput screenings in organoids due to technical difficulties (mainly organoids being trapped in solidifying drops of Matrigel) and the high cost of RNA sequencing. Several groups established bulk RNA-seq-based screening methods for cell lines [16, 17], but the high cost of whole transcriptome sequencing is not compatible with high-throughput assays on a large collection of samples. Several targeted approaches were suggested [18, 19] but were also not compatible with high-throughput drug screening in organoids or suffer from PCR amplification bias and difficulties in detecting weakly expressed targets. Thus, there is currently a need for NGS-based drug-screening platforms in organoids that would quantify and evaluate complex cellular phenotypes and alterations upon drug treatment.

Here we present a detailed protocol of the recently established TORNADO-seq method [20]. TORNADO-seq is a targeted RNA-seq-based drug-screening platform in organoids that allows analysis of complex phenotypes and predicts the mechanism of action of drugs. TORNADO-seq compares favorably with other high-content methods with respect to availability, cost, and applicability to organoid-based screens. TORNADO-seq shows high reproducibility, requires only 6 h to perform, is cost-efficient (5 USD per sample including organoid culture, library preparation, and sequencing cost), and doesn't require any specialized equipment. Due to the implemented UMI counting and sequencing of amplicons, TORNADO-seq lacks PCR- and ligation-based bias and misdetections.

2 Materials

2.1 mRNA Isolation

1. The New England Biolabs (NEB) Magnetic mRNA Isolation Kit (S1550S).
2. 1.2 ml low-profile 96-well plates (AB-1127, Thermo Fisher Scientific).

3. Aluminum sealing (6570, Corning).
4. Orbital shaker.
5. 96-well plate magnetic stand for 1.2 ml profile plates (AM10027, Thermo Fisher Scientific).
6. 96-well plate side magnet (12331D, Thermo Fisher Scientific, DynaMag).
7. Non-skirted PCR plate (AB0600, Thermo Fisher Scientific).
8. V-bottom 96-well plate (6,018,323, 96-well micro test plates, V-bottom, PP Ratiolab).
9. Multichannel pipette.
10. A thermocycler compatible with 96-well plate format.

2.2 RT Reaction

1. Tth DNA polymerase (11,480,022,001, Roche).
2. Non-skirted PCR plate (AB0600, Thermo Fisher Scientific).
3. Mixture of designed rev gene-specific primers.
4. 96-well plate cooling rack.
5. A thermocycler compatible with 96-well plate format.
6. Agencourt AMPure magnetic beads (A63881, Beckman Coulter Inc.).

2.3 Library Amplification (First- and Second-Stage PCR)

1. GoTaq G2 HS Polymerase (M7423, Promega).
2. First-stage common primer.
3. Mixture of designed fw gene-specific primers.
4. Fw and Rev. second-stage indexing primers.
5. A thermocycler compatible with 96-well plate format.
6. Agencourt AMPure magnetic beads (A63881, Beckman Coulter Inc.).
7. QIAquick Gel Extraction Kit (28,704, Qiagen).
8. Agarose and the equipment for running agarose gel.

3 Methods

3.1 Designing the Panel of Genes of Interest

Designing a panel of genes is one of the most important steps for a successful outcome in the experiment and implies a good knowledge of the model system of choice. Use available gene expression profiles for the chosen model system to extract the genes of interest depending on the scope of the study. For example, use available single-cell RNA-seq data to select marker genes of different cell populations present in the system of choice. Another approach for choosing the genes of interest is to perform a bulk RNA sequencing on organoids treated with a carefully selected number of drugs.

This approach is useful in the case of designing a platform for NGS-based drug screening and allows to identify the most differentially expressed genes upon several treatment conditions.

3.2 Designing the Primers for Targeted RNA-Seq for a Selected Number of Genes

Design primers following the next general primer design rules. Primers are designed with a melting temperature of 59–61 °C corresponding to the gene-specific part of the primer and 75–76 ° C corresponding to the full primer sequence. Primers should be designed in a way to exclude secondary structures (stable above 54 °C) in the full primer sequence. Melting temperatures can be determined using OligoAnalyzer (https://eu.idtdna.com/calc/analyzer/). In order to avoid unspecific annealing, primers should be chosen to have unique 3′ ends (last 5 nt), at least 7 °C difference in annealing temperature to the most similar nontarget sequence and lack of expression of the most similar genome sequence. PCR products between 150 and 170 bp in length should be selected to span at least two exons with average intron sizes of over 1500 nt. Primer dimers can be evaluated with the program Primer Pooler v1.41 [21] at the reaction temperature of 58 °C: the primer dimers having free energy values of less than −3 kcal/mol should be further analyzed, and those which form complementary structures at their 3′ ends should be removed, and new primers for corresponding genes should be designed (*see* **Note 1**):

1. Design gene-specific part (GSP) of fw and rev primers for genes of interest obtaining a PCR product between 150 and 170 bp in length for each amplicon and satisfying the rules mentioned above (e.g., fw GSP primer, 5’-GACTCCTCCTGCTTTCTGTC-3′; rev GSP primer, 5’- GTTATAATTGCTCTCATGCTGAGC - 3′).
2. Design elongated fw and rev primers which should have the next structure:
   (a) Fw primer:

   5′-GCGGAAGCAGTGGTATCAACGCAGAGTGACTCCTCCTGCTTTCTGTC-3′,

   where GCGGAAGCAGTGGTATCAACGCAGAGT is common part of the fw primer, a part of SMART-seq primer sequence, and GACTCCTCCTGCTTTCTGTC is gene-specific part of fw primer.

   (b) Rev. primer:

   5’CGTCGGCAGCGTCAGATGTGTATAAGAGACAGNNNNNNNNNNGTTATAATTGCTCTCATGCTGAGC-3′,

   where CGTCGGCAGCGTCAGATGTGTATAAGAGACAG is common part of the rev primer, a part of Read1 sequencing primer, NNNNNNNNNN is 10 bp UMI, and GTTATAATTGCTCTCATGCTGAGC is gene-specific part of rev primer.

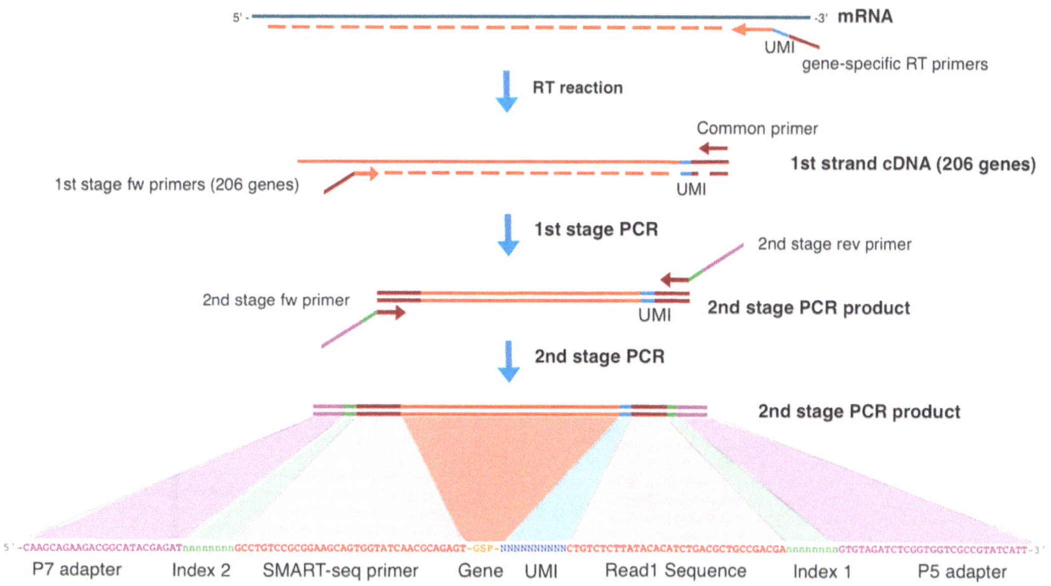

Fig. 1 Schematic representation of library preparation for targeted RNA sequencing. cDNA is synthesized from polyA mRNA in an RT reaction using a mixture of 206 gene-specific reverse primers (GSP) carrying a UMI and a common sequence (part of Illumina Read1 sequence) needed for further amplification steps. The obtained cDNAs are further amplified in a first-stage PCR with one common primer and the mixture of 206 gene-specific forward primers (carrying a second common part derived from SMART-seq). The obtained products are barcoded in a second-stage PCR using a set of forward and reverse primers (carrying index1 and index2 and P5 and P7 adapters for Illumina sequencing)

3. You will need to order next extra primers:

   Common reverse primer: 5′- TCGTCGGCAGCGTCAG-3′ (for first-stage PCR amplification step)

   Set of second-stage amplification primers:

   P7_adapter_2nd_stage_primer:

   5’CAAGCAGAAGACGGCATACGAGATTCGCCTTAGCCTGTCCGCGGAAGCAGTGGTATCAACGCAGAGT-3’

   P5_adapter_2nd_stage_primer:

   5’-AATGATACGGCGACCACCGAGATCTACACTATCCTCTTCGTCGGCAGCGTC-3’

   The full library structure is represented in Fig. 1.

3.3 Targeted RNA-seq Library Construction

3.3.1 mRNA Isolation

All library preparation steps are performed in 96-well plates. The protocol workflow is depicted in Fig. 2. mRNA is isolated with The New England Biolabs (NEB) Magnetic mRNA Isolation Kit (S1550S), and the isolation is performed according to the manufacturer’s protocol with alight adjustments:

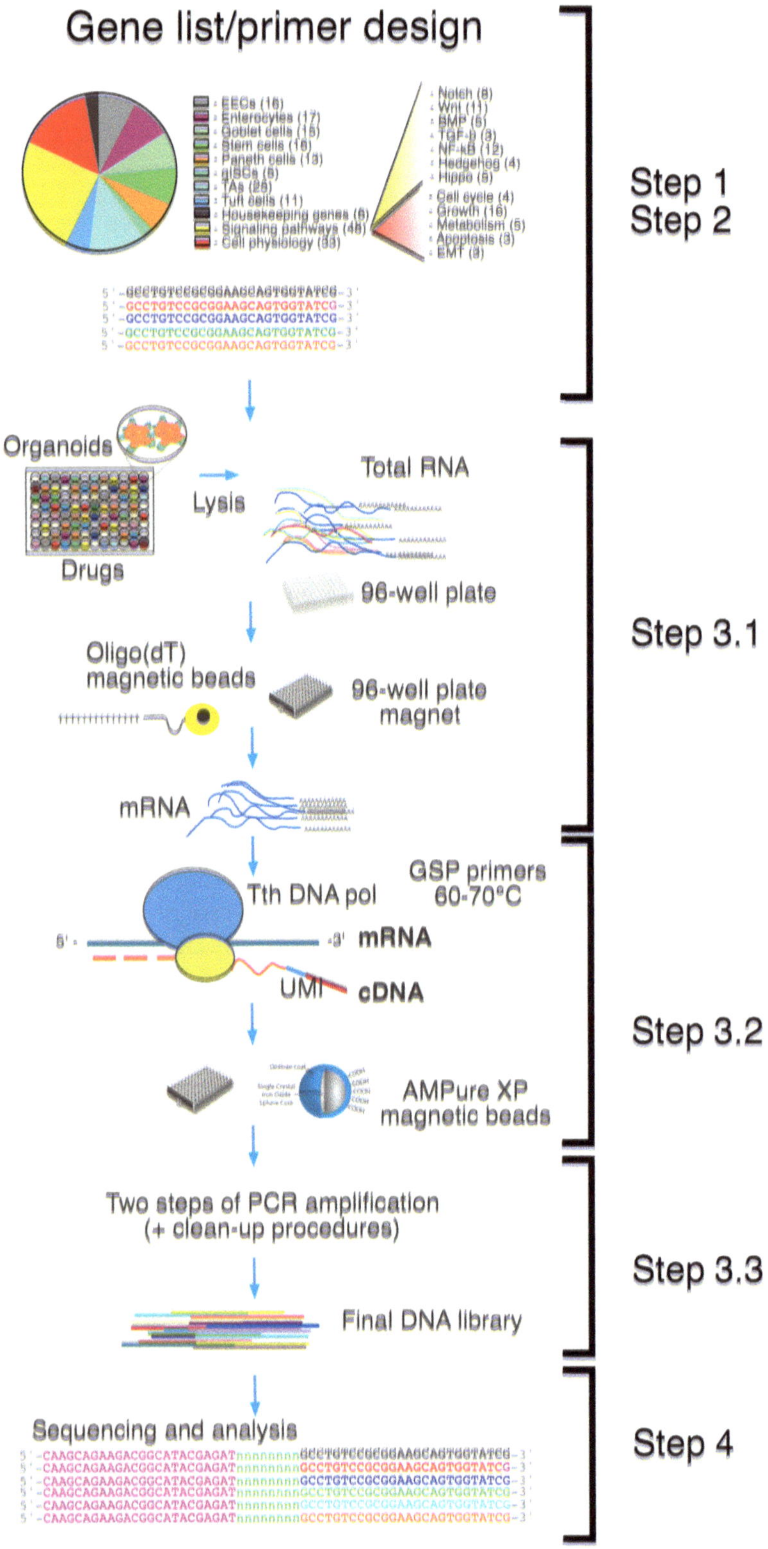

Fig. 2 Schematic view of high-throughput processing of drug-treated intestinal organoids. The first step (**steps 1, 2**) in the protocol is to design the panel of genes and corresponding primers. Further, in the screening, organoids are

1. The media from the 96-well plate is removed by inverting and flicking the plate. For lysing the organoids, 200 μl of lysis buffer (S1550S, NEB) is added to each well using a multichannel pipette; Matrigel is disrupted by mechanical force using 200 μl tips and a multichannel pipette.
2. Lysis buffer containing organoids is transferred to special 1.2 ml low-profile plates (AB-1127, Thermo Fisher Scientific), and an additional 100–200 μl of lysis buffer is added to each well. Organoids are disrupted by gently pipetting around 50 times. Here a **PAUSE** can be introduced: the lysates can be frozen at −80 °C and the protocol can be continued anytime later. Unfreezing of the lysates is performed on ice and takes time, approximately 1–2 h (*see* **Note 2**).
3. The 96-well plate is covered with aluminum sealing (6570, Corning), and put on a shaker at 300 rpm at RT for 10–15 min for efficient lysis.
4. mRNA binding step: 22 μl of oligo-d$(T)_{25}$ magnetic beads (S1550S, NEB) per sample is used. The mRNA isolation is performed according to the manufacturer's protocol using 200 μl of all wash buffers and 40–50 μl of elution buffer. A magnet (AM10027, Thermo Fisher Scientific, Magnetic Stand-96) is used for separating the magnetic beads during the washes, and a second magnet (12331D Thermo Fisher Scientific, DynaMag-96 Side) is used for separating the magnetic beads during the elution.
5. After successful washing steps, the mRNA-bead suspension is transferred to a non-skirted PCR plate (AB0600, Thermo Fisher Scientific) and incubated in a PCR machine at 55 °C for 140 s prior to elution (*see* **Note 3**). Eluent is transferred to a V-bottom 96-well plate (6,018,323, 96-well micro test plates, V-bottom, PP Ratiolab) and can be stored at −80 °C. Here a **PAUSE** can be taken, and the mRNA can be further proceeded at any time.

Fig. 2 (continued) seeded in a 96-well plate format in Matrigel, drug-treated, and then lysed (**step 3.1**). mRNA is isolated from lysates using oligo-dT magnetic beads in a 96-well plate format (**step 3.1**). Tth DNA polymerase is used for the RT reaction (**step 3.2**), and cDNA is purified with AMPure XP magnetic beads. After two rounds of PCR amplification (**step 3.3**), an additional cleanup procedure with AMPure XP magnetic beads is performed, and final libraries are pooled and purified twice on 2% agarose gel (not shown). The last step of the protocol is sequencing and analysis (**step 3.4**)

3.3.2 RT Reaction

RT reaction is carried out in 96-well plate format with Tth DNA polymerase (11,480,022,001, Roche) according to the manufacturer's protocol in a non-skirted PCR plate (AB0600, Thermo Fisher Scientific) with 2.5 units of enzyme per reaction and 1 μl of a 0.5 μM mixture of gene-specific reverse primers:

1. Prepare the mixture of 4 ul mRNA, 1 μl of a 0.5 μM mixture of gene-specific reverse primers, and the RT buffer in 17 ul total per reaction without the Tth enzyme. Preincubate a 96-well plate with 17 ul mRNA/primers/buffer mixture at 72 °C for 5 min without enzyme, and place on ice (preferably in 96-well plate cooling rack for an efficient cooldown).
2. Add the 2.5 units of Tth enzyme (0.5 ul) in 3 ul of 1× RT buffer and add to 17 ul preincubated reaction mix. Place the reaction at 72 °C for 1 min and 60 °C for 30 min (*see* **Note 4**).
3. Clean obtained cDNA transcripts with Agencourt AMPure magnetic beads (A63881, Beckman Coulter Inc.) according to the manufacturer's protocol at a 1:1.8 ratio of cDNA synthesis mixture/AMPure beads and eluted with 40 μl of elution buffer (*see* **Note 5**). Obtained cDNA can be stored at −20 °C. Here a **PAUSE** can be taken, and the cDNA can be further proceeded at any time.

3.3.3 Library Amplification (First- and Second-Stage PCR)

1. Use cleaned cDNA as a template for first-stage PCR and amplify with GoTaq G2 HS Polymerase (M7423, Promega). Use 2 μl of purified cDNA, 1 μl of 5 μM first-stage common primer, and 1 μl of a 0.5 μM mixture of gene-specific forward primers. PCR program can be found in a supplementary file at [20].
2. Purify obtained products with AMPure magnetic beads as described above in Subheading 3.3.2. Amplified first PCR products can be stored at −20 °C. Here a **PAUSE** can be taken, and the amplified first PCR products can be further proceeded at any time.
3. Second-stage PCR is carried out with GoTaq G2 HS Polymerase, 1 μl of each 5 μM Fw and Rev. second-stage indexing primers, and 2 μl of purified first-stage PCR product and amplified for 15–20 cycles with the following PCR program: 94 °C, 20 s; 68 °C, 15 s; and 72 °C, 10 s.
4. Pool separately amplified libraries together by plates (96 samples), and purify twice on 2% agarose gels using QIAquick Gel Extraction Kit (28,704, Qiagen) to extract 300 bp products according to the manufacturer's protocol (*see* **Note 6**). Amplified second PCR products can be stored at −20 °C. Here a **PAUSE** can be taken, and the amplified second PCR products can be further proceeded at any time.

3.4 RNA Sequencing and Analysis

1. Sequence pooled libraries at an average depth of 70,000 reads per sample on Illumina sequencer (e.g., NextSeq500, HiSeq 4000, or similar platforms) using a single-end sequencing mode with at least 40–50 or higher number of cycles: it's enough to cover 10 bp UMI sequence, 20 bp gene-specific primer sequence, and 10–20 bp of an inner part of the amplicons for specific alignment. **Importantly**, use PAGE-purified custom index primer, 5′- act ctg cgt tga tac cac tgc ttc cgc gga cag gc −3′, which has to be supplied to sequencer machine to be able to demultiplex the libraries using two indexes.
2. Proceed demultiplex reads with the custom script available at https://github.com/MaximNorkin91/Tornado-seq-protocol/. Briefly, the first 10 bases of each read containing UMI sequences are deduplicated with UMI-tools. Process deduplicated reads with the Bowtie (ver. 1.2.2.)/samtools (ver. 1.2.7) pipeline, and align them to a custom genome containing corresponding target sequences. Count aligned reads with Cufflinks (ver. 2.2.1) (*see* **Note** 7).

4 Notes

1. We found it to be nearly impossible to predict primer behavior in multiplex systems using only modeling software because the number of unspecific products depends on the expression level of the targeted as well as the nontargeted genes which can vary as a function of the organoid treatment. Therefore, optimal conditions should be achieved after a few iteration rounds of primer redesign and adjustments of primer concentrations as analyzed by RNA-seq.
2. While unfreezing the cell lysates, white salt precipitations can be formed at the bottom of the wells. Gently resuspend the lysate mixture with multichannel pipette to have a homogeneous solution. Avoid the long incubation time of the solution with separated phases, and resuspend the precipitate directly after the samples are thawed.
3. While 96-well plate is covered with aluminum sealing, it can be cut into two pieces, and each 48 samples can be proceeded subsequently to prevent annealing of the mRNA back to the beads as the sample cools down after 55 °C incubation.
4. Make sure to seal the aluminum foil tightly, to prevent evaporation during the RT reaction or PCR steps.
5. Lower ratios of cDNA synthesis mixture/AMPure beads can be used, e.g., 1:1, to enrich for longer amplicons. Using the lower ratios results in lower yields but higher purity. Ratios of 1:1.8 showed a good compromise in our hands.

6. A custom adjustments of library quantities can be performed after gel electrophoresis based on intensity of the corresponding bands (quantity of the DNA library).
7. Custom fasta and .GFF files have to be generated using the amplified sequences as a reference genome.

References

1. Barcellos-Hoff MH, Aggeler J, Ram TG, Bissell MJ (1989) Functional differentiation and alveolar morphogenesis of primary mammary cultures on reconstituted basement membrane. Development 105:223–235. https://doi.org/10.1242/dev.105.2.223
2. Petersen OW, Rønnov-Jessen L, Howlett AR, Bissell MJ (1992) Interaction with basement membrane serves to rapidly distinguish growth and differentiation pattern of normal and malignant human breast epithelial cells. Proc Natl Acad Sci U S A 89:9064–9068. https://doi.org/10.1073/pnas.89.19.9064
3. Bissell MJ, Kenny PA, Radisky DC (2005) Microenvironmental regulators of tissue structure and function also regulate tumor induction and progression: the role of extracellular matrix and its degrading enzymes. Cold Spring Harb Symp Quant Biol 70:343–356. https://doi.org/10.1101/sqb.2005.70.013
4. Simian M, Bissell MJ (2017) Organoids: a historical perspective of thinking in three dimensions. J Cell Biol 216:31–40. https://doi.org/10.1083/jcb.201610056
5. Sato T, Vries RG, Snippert HJ et al (2009) Single Lgr5 stem cells build crypt-villus structures in vitro without a mesenchymal niche. Nature 459:262–265. https://doi.org/10.1038/nature07935
6. Folkesson E, Niederdorfer B, Nakstad VT et al (2020) High-throughput screening reveals higher synergistic effect of MEK inhibitor combinations in colon cancer spheroids. Sci Rep 10:11574. https://doi.org/10.1038/s41598-020-68441-0
7. Du Y, Li X, Niu Q et al (2020) Development of a miniaturized 3D organoid culture platform for ultra-high-throughput screening. J Mol Cell Biol 12:630–643. https://doi.org/10.1093/jmcb/mjaa036
8. Michels BE, Mosa MH, Streibl BI et al (2020) Pooled in vitro and in vivo CRISPR-Cas9 screening identifies tumor suppressors in human colon organoids. Cell Stem Cell 26:782–792.e7. https://doi.org/10.1016/j.stem.2020.04.003
9. Ringel T, Frey N, Ringnalda F et al (2020) Genome-scale CRISPR screening in human intestinal organoids identifies drivers of TGF-β resistance. Cell Stem Cell 26:431–440.e8. https://doi.org/10.1016/j.stem.2020.02.007
10. Lukonin I, Serra D, Challet Meylan L et al (2020) Phenotypic landscape of intestinal organoid regeneration. Nature 586:275–280. https://doi.org/10.1038/s41586-020-2776-9
11. Brandenberg N, Hoehnel S, Kuttler F et al (2020) High-throughput automated organoid culture via stem-cell aggregation in microcavity arrays. Nat Biomed Eng 4:863–874. https://doi.org/10.1038/s41551-020-0565-2
12. Boehnke K, Iversen PW, Schumacher D et al (2016) Assay establishment and validation of a high-throughput screening platform for three-dimensional patient-derived colon cancer organoid cultures. SLAS Discovery 21:931–941. https://doi.org/10.1177/1087057116650965
13. van de Wetering M, Francies HE, Francis JM et al (2015) Prospective derivation of a living organoid biobank of colorectal cancer patients. Cell 161:933–945. https://doi.org/10.1016/j.cell.2015.03.053
14. Schütte M, Risch T, Abdavi-Azar N et al (2017) Molecular dissection of colorectal cancer in pre-clinical models identifies biomarkers predicting sensitivity to EGFR inhibitors. Nat Commun 8:14262. https://doi.org/10.1038/ncomms14262
15. Verissimo CS, Overmeer RM, Ponsioen B et al (2016) Targeting mutant RAS in patient-derived colorectal cancer organoids by combinatorial drug screening. elife 5:e18489. https://doi.org/10.7554/eLife.18489
16. Ye C, Ho DJ, Neri M et al (2018) DRUG-seq for miniaturized high-throughput transcriptome profiling in drug discovery. Nat Commun 9:4307. https://doi.org/10.1038/s41467-018-06500-x
17. Bush EC, Ray F, Alvarez MJ et al (2017) PLATE-Seq for genome-wide regulatory network analysis of high-throughput screens. Nat

Commun 8:105. https://doi.org/10.1038/s41467-017-00136-z

18. Simon JM, Paranjape SR, Wolter JM et al (2019) High-throughput screening and classification of chemicals and their effects on neuronal gene expression using RASL-seq. Sci Rep 9:4529. https://doi.org/10.1038/s41598-019-39016-5
19. Teder H, Koel M, Paluoja P et al (2018) TAC-seq: targeted DNA and RNA sequencing for precise biomarker molecule counting. NPJ Genom Med 3:34. https://doi.org/10.1038/s41525-018-0072-5
20. Norkin M, Ordóñez-Morán P, Huelsken J (2021) High-content, targeted RNA-seq screening in organoids for drug discovery in colorectal cancer. Cell Rep 35:109026. https://doi.org/10.1016/j.celrep.2021.109026
21. Brown SS, Chen Y-W, Wang M et al (2017) PrimerPooler: automated primer pooling to prepare library for targeted sequencing. Biol Methods Protoc 2:bpx006. https://doi.org/10.1093/biomethods/bpx006

Chapter 7

Epigenetic and Transcriptional Dynamics of Notch Program in Intestinal Differentiation

Shahadat Rahman, Xi Lan, Christopher Terranova, Rayan El-Kholdi, Omer H. Yilmaz, and Chia-Wei Cheng

Abstract

The equilibrium between stem cell self-renewal and differentiation followed by proper lineage specification of progenitor cells is considered imperative for maintaining intestinal homeostasis. In the hierarchical model, intestinal differentiation is defined by the stepwise acquisition of lineage-specific mature cell features, where Notch signaling and lateral inhibition instructively regulate the cell-fate decisions. Recent studies reveal a broadly permissive intestinal chromatin underlies the lineage plasticity and adaptation to diet mediated by Notch transcriptional program. Here, we review the conventional understanding of Notch programming in intestinal differentiation and describe how new data from epigenetic and transcriptional analyses may refine or revise the current view. We provide instructions on sample preparation and data analysis and explain how to use ChIP-seq and scRNA-seq in combination of lineage tracing assay to determine the dynamics of Notch program and intestinal differentiation in the context of dietary and metabolic regulation of cell-fate decisions.

Key words ChIP-seq, scRNA-seq, Notch, Cell-fate decisions

1 Introduction

In the conventional view, the stem-to-progenitor differentiation occurs as soon as the cell progeny leaves the stem cell zone in the middle of the crypt (i.e., transit-amplifying zone) [1], where the early differentiation is epigenetically modulated with the termination of stem cell-specific transcription and the initiation of transcriptional programs for lineage specification [2]. When moving out from the crypt zone, the transit-amplifying (TA) progenitor cells face a binary cell-fate decision to commit to either absorptive lineage or to secretory lineage [3]. During intestinal differentiation, the Notch signaling pathway has been viewed as a major cell-

Shahadat Rahman and Xi Lan contributed equallty to this work

Paloma Ordóñez-Morán (ed.), *Intestinal Differentiated Cells: Methods and Protocols*, Methods in Molecular Biology, vol. 2650,
https://doi.org/10.1007/978-1-0716-3076-1_7,

fate determinant, in part, by inhibiting the transcriptional program for secretory cell differentiation [3]. Recent studies of epigenetic and transcriptional analyses suggest a revised view that both the stem-to-TA differentiation and the TA lineage commitment processes are broadly permissive and highly dynamic in adult intestine, highlighting the need to refine our understanding of the mechanisms controlling the intestinal lineage commitment and plasticity [2, 4].

The dynamic nature of Notch signaling in intestinal differentiation has been extensively studied in the mammalian and *Drosophila* intestine. Although the exact mechanisms remain context-dependent, the epigenetic regulation at Notch target genes generally coordinated with the three molecular events: the repression, activation, and turnover of Notch signaling pathway (Fig 1). In the absence of a Notch signal, the transcription factor CSL (CBF-1, suppressor of hairless, Lag-2; aka RBP-j) recruits corepressor

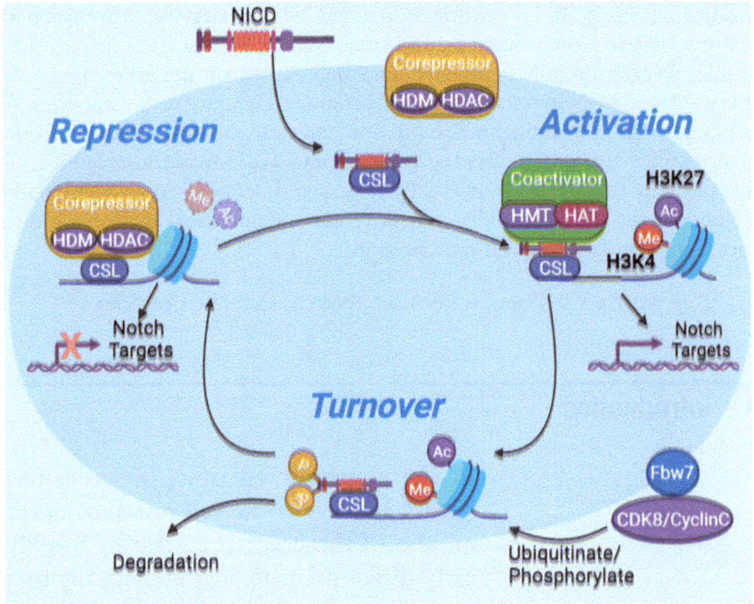

Fig. 1 Notch lateral inhibition model. Stochastic differences in Notch are reinforced between Notch-high and Notch-low cells. Upon binding of Notch ligand from Notch-low sending cells (left) to NICD receptor on Notch-high receiving cells (right), a cleavage event occurs. NICD receptor then translocates to the nucleus of the receiving cell to relieve repression of Notch targets, particularly Hes family proteins. Activation of Hes family proteins such as Hes1 inhibits transcription of Atoh1 and subsequently Delta ligand. The absence of Delta ligand in the receiving cell inhibits this pathway from being upregulated in the sending cell, concomitantly maintaining repression of Notch targets such as Hes1, transcription of Atoh1, and expression of Delta ligand

proteins as well as the histone demethylase (HDM) and histone deacetylase (HDAC) to form a complex that erases H3K4me3 and H3K27ac chromatin marks to prevent the transcription of Notch target genes (Fig 1, **repression**) [5–7]. Upon Notch activation, the cleaved Notch intracellular domain (NICD) can translocate to the nucleus and interact with CSL, which recruits histone acetyltransferases (HAT) and histone methyltransferase (HMT) to establish H3K27ac and H3K4me marks, resulting in an active chromatin state for Notch target gene transcription [7–10] (Fig 1, **activation**). The transcription activity ends rapidly with NICD degradation followed by the replacement of coactivators with corepressors. The turnover process is regulated by the cyclin C/Cdk8-mediated phosphorylation and E3 ubiquitin ligase Fbw7/Sel10-induced ubiquitinylation of NICD [11] (Fig 1, **turnover**). In this model, CSL statically occupies consensus binding sites while stochastically exchanging repressors for activators depending on the presence of NICD, leading to the dynamic switch of Notch program, which can be detected based on the change of histone marks.

Given the versatility of NICD-dependent activation, lateral inhibition, by which cells instruct adjacent cells to adopt a different fate than the instructing cell, is thought to be the regulatory mechanism that stabilizes the lineage commitment and cell-fate decision during intestinal differentiation [3, 12]. At the crypt base, the intestinal stem cells (ISCs) receive Notch signaling input through the direct contact with the Delta-like ligands from the niche Paneth cells. The activation of Notch in ISCs represses production of the Notch ligand (Delta) by inhibiting secretory-specific transcription factor Atoh1 and therefore reinforces the low Notch state of the Paneth cells (Fig 2, **sending cell**). Similarly, in the process of lineage commitment, the positive feedback loop amplifies any initial stochastic difference in Notch activity between adjacent bipotential progenitors and drives two neighboring cells into opposite Notch-level status and hence into distinct differentiation trajectories (Fig 2). Transcriptional analysis for Hes1 versus Atoh1 target gene expression, therefore, has been a useful approach for determining the mutually exclusive cell-fate decision modulated by Notch signaling and lateral inhibition.

Notably, chromatin immunoprecipitation and sequencing of the associated DNA fragments (ChIP-seq) analyses comparing the enhancer profiles of lineage-specific progenitors suggest that intestinal lineages in adult intestine remain plastic beyond the period of lateral inhibition and argue for a persistent role of Atoh1 in the lineage-committed cells [13]. These findings may explain why some differentiated crypt cells readily assume stem cell properties upon tissue damage and stress conditions such as inflammation and tumorigenesis [2, 4, 14]. In line with this possibility, we have recently reported that Lgr5+ ISCs rely on ketone body metabolite

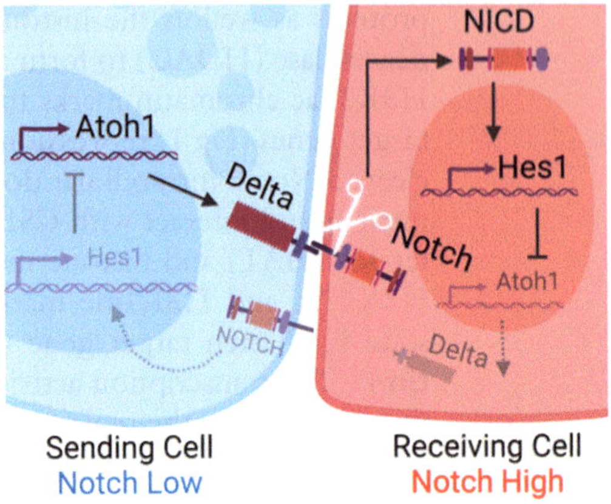

Fig. 2 Epigenetic switch of Notch program. In the repressed state, the transcription factor complex CSL binds a corepressor containing histone demethylase (HDM) and histone deacetylase (HDAC), erasing H3K4me3 and H3K27ac chromatin marks on Notch targets and inducing a heterochromatic state. In the active state, upon NICD activation and translocation to the nucleus, CSL detaches from the corepressor complex and binds to NICD. A coactivator containing histone acetyltransferases (HAT) and histone methyltransferases (HMT) is then recruited to this complex to establish H3K4me3 and H3K27ac chromatin marks on Notch targets, enabling their transcription. In the turnover phase, this process is then turned off by various E3 ubiquitin ligases including Fbw7 and Deltex, which ubiquitinate the NICD receptor and target it for proteasomal or lysosomal degradation, respectively. Kinases such as cyclin C/Cdk8 are also able to phosphorylate NICD receptor and target it for degradation to turn off this process. Turnover of the coactivator complex leads to induction of the repressed state with binding of CSL to the corepressor complex

beta-hydroxybutyrate (βOHB)-mediated HDAC inhibition to reinforce its Notch-high state and that loss of βOHB results in stem-to-Paneth differentiation [15]. Here, we will describe how we use (i) ChIP-seq analysis to determine βOHB-mediated activation of Notch (Fig 3 and 4) and (ii) lineage tracing in combination with single-cell RNA-seq to demonstrate the perturbation of lateral inhibition and stem-to-Paneth differentiation upon loss of βOHB.

2 Materials

2.1 ChIP-seq Analysis for H3K27ac Profiling

1. Hanks' Balanced Salt Solution (HBSS).
2. 1% formaldehyde.
3. Phosphate buffered saline (PBS).
4. Protease inhibitor cocktail.
5. I-ChIP buffer (12 mM Tris-HCI, pH 8.0, 6 mM EDTA, pH 8.0, 0.1× PBS, 0.5% SDS).

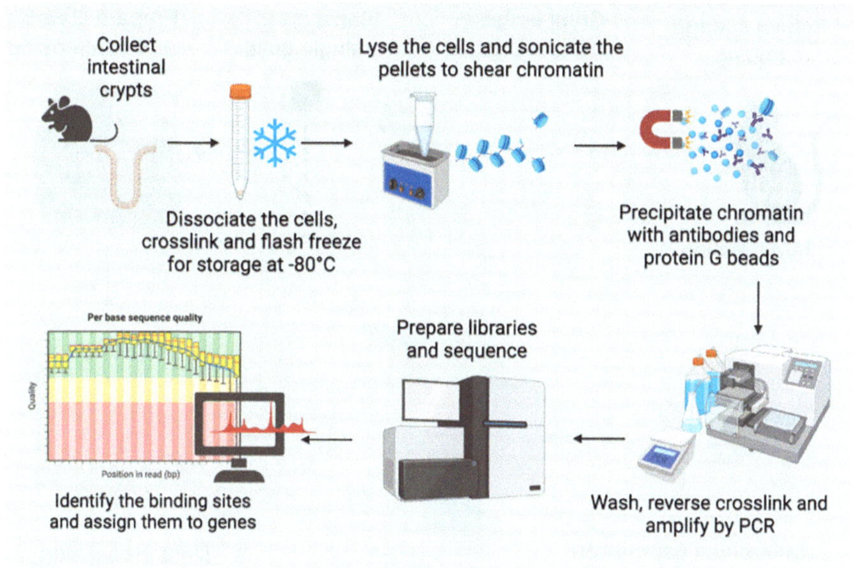

Fig. 3 ChIP-seq analysis for H3K27ac profiling. The intestinal crypts are collected from mice and then processed for cross-linking. The resulting pellet can be stored at −80 °C after flash freezing. The cells are then lysed and then sonicated for chromatin fragmentation and then incubated with protein G magnetic beads and the antibodies for immunoprecipitation. The next day, the washed and eluted chromatin is de-cross-linked and the DNA is purified for PCR amplification. Libraries are sequenced on Illumina's sequencing devices, and the data are processed to identify the binding sites and to assign them to genes

6. 1.5 mL polystyrene sonicator tubes.
7. Binding buffer (PBS + 0.1% TWEEN-20 + 0.2% BSA).
8. Protein G magnetic Dynabeads.
9. RIPA wash buffer (10 mM Tris-HCI, pH 8.0, 1 mM EDTA, pH 8.0, 140 mM NaCl, 1% Triton X-100, 0.1% SDS, 0.1% DOC).
10. High-salt RIPA buffer (RIPA buffer with 500 mM NaCl).
11. LiCl buffer (10 mM Tris-HCI, pH 8.0, 1 mM EDTA, pH 8.0, 250 mM LiCl, 0.5% NP-40, 0.5% DOC).
12. Direct elution buffer (10 mM Tris-HCI, pH 8.0, 5 mM EDTA, pH 8.0, 300 mM NaCl, and 0.5% SDS).
13. Proteinase K.
14. RNase A.

2.2 Single-Cell RNA-seq for Lateral Inhibition and Stem-to-Paneth Differentiation

1. Tamoxifen (100 mg/mL, in corn oil with 10% EtOH).
2. HBSS.
3. Gel bead and chip kits (10× Genomics, Pleasanton, CA).
4. 0.4% BSA-PBS.

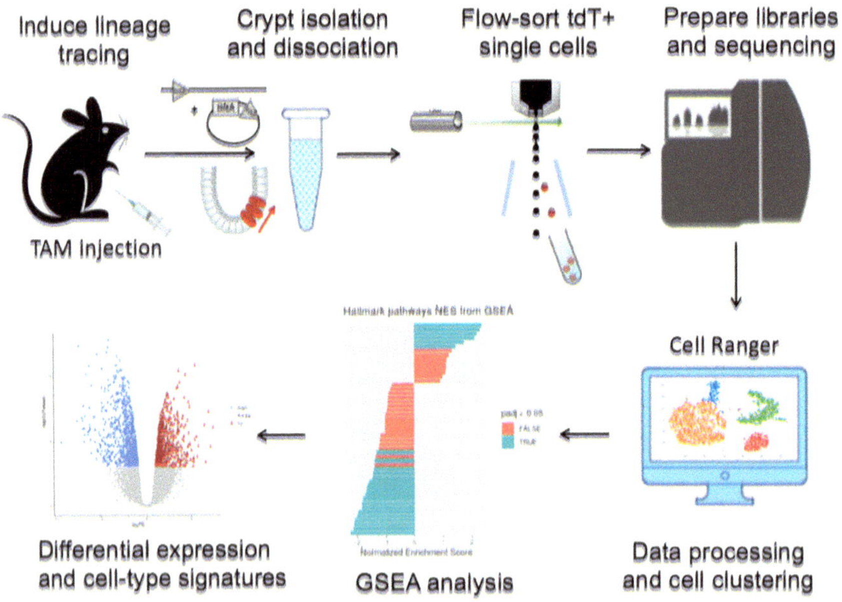

Fig. 4 Lineage tracing and single-cell RNA-seq for lateral inhibition and stem-to-Paneth differentiation. Mice are injected with tamoxifen (TAM) to induce gene deletion and tdTomato (tdT) expression for Lgr5 lineage tracing. Crypts are isolated and dissociated for sorting tdT+ single cells. The sorted cells are then processed by the 10× Genomics Chromium Controller to create single-cell libraries. The libraries are sequenced on Illumina's platform. Data is processed by alignment to mm10 mouse genome and unique molecular identifier (UMI). Gene set enrichment analysis (GSEA) is performed for up- or downregulation of signaling pathways in specific cell cluster due to loss of the targeted gene in the Lgr5-derived tdT+ cells. Finally, the differential expressed cell-type signature genes in the stem cell cluster are identified and visualized using the volcano plot

3 Methods

3.1 ChIP-seq Analysis for H3K27ac Profiling

3.1.1 Processing Intestinal Crypts for Cross-Linking

1. Collect intestinal crypts from mice fasted for 24 h (high βOHB) or fed ad libitum (low βOHB) into a 15 mL conical centrifuge tube.
2. Disassociate intestinal crypts manually in 2 mL HBSS by pipetting.
3. Add another 8 mL HBSS and continue to disassociate the intestine crypts in a homogenizer.
4. Cross-link the crypts using 1% formaldehyde supplemented with HBSS.
5. Rotate the tube for 10 min at 37 °C.
6. Add 200 μL 0.125 M glycine per 3 mL of the crypt mixture for quenching, and continue to shake for 5 min at 37 °C (*see* **Note 1**).
7. Centrifuge the tube at 4 °C for 5 min at 900 × *g*.

8. Remove the supernatant and wash with 5 mL cold PBS added with 1× protease inhibitor cocktail.
9. Centrifuge at 4 °C for 5 min at 900 × *g*, then remove the supernatant.
10. Freeze the pellets at −80 °C for the next processing step.

3.1.2 Lysis, Sonication, and Antibody Preparation

11. Thaw the pellets, and add 300 μL I-ChIP buffer with protease inhibitors, lysing for 30 min on ice.
12. During crypt lysing, pre-warm the water bath. Put the tubes in the water bath, sonicating the small intestinal crypts for 60 cycles with the condition of 30 s on/off to obtain the shear length of chromatin fragments at ~250–500 bp.
13. To prepare the protein G magnetic beads, use 1 mL binding buffer to wash 20 μL beads, which is for one antibody per crypt sample (*see* **Note 2**).
14. Couple the protein G magnetic Dynabeads to antibodies of interest for at least 1 h at 4 °C. For crypts in each tube, prepare 3 μg antibody with 20 μL beads into 100 μL binding buffer at 4 °C for 2 h, and incubate by revolving.
15. Determine the quality of samples as previously described [16].
16. Centrifuge the sonicated material at 4 °C for 15 min. Then transfer the supernatant to new tubes.
17. Take 10% from the supernatant solution as the input control for later usage, and store at 4 °C.

3.1.3 Chromatin Immunoprecipitation

18. Dilute the chromatin lysate five times with ChIP dilution buffer.
19. Put the tubes containing the incubated antibody-magnetic beads on a magnetic stand, and discard the supernatant.
20. Wash with 700 μL ChIP dilution buffer mixed with protease inhibitors and remove the supernatant (*see* **Note 2**).
21. Resuspend the beads with 240 μL ChIP dilution buffer.
22. Add 30 μL bead solution into each crypt sample tube.
23. Add the sonicated fragments into the tube with protein G beads coupled with antibody and turn over the tube with a rotator overnight at 4 °C.

3.1.4 Washing and Reverse Cross-Linking

24. Transfer the immune complexes to a 96-well plate and put it on a magnet. Wait for 30 s until the beads adhere, then remove the supernatant.
25. Wash beads five times by adding 150 μL cold RIPA buffer. Wait for 30 s until the beads adhere to the magnet after each wash.

26. Wash the beads twice with high-salt RIPA buffer.
27. Wash the beads twice with LiCl buffer.
28. Add the input control from **step 17** into the 96-well plate with the DNA fragment samples.
29. For reverse cross-linking, add 50 μL direct elution buffer with proteinase K and RNaseA per ChIP and input sample at 65 °C overnight.
30. Amplify DNA sequences by a PCR thermal cycler with the following condition: 37 °C for 4 h, 50 °C for 4 h, and 65 °C for 8–16 h.
31. Clean up the eluted DNA using solid-phase reversible immobilization (SPRI) beads (Beckman-CoμLter).

3.1.5 ChIP-Sequencing Library Preparation

32. Perform library preparation use enzymes from New England Biolabs for DNA end-repair, A-base addition, adaptor ligation, U excision, and PCR enrichment.
33. Barcode ChIP libraries use TruSeq DNA LT adapters.
34. Perform sequencing use HiSeq 2000 (Illumina).

3.1.6 ChIP-seq Data Processing

35. Process raw fastq reads with a Snakemake-based pipeline using FastQC (http://www.bioinformatics.babraham.ac.uk/projects/fastqc/).
36. Align uniquely mapped reads to the mm9 reference genome using Bowtie version 1.1.2.
37. Remove duplicate reads using SAMBLASTER before compression to bam files.
38. Downsample the aligned reads to 20 million per condition.
39. Sort and index the reads for each mark using samtools version 1.5.
40. Visualize ChIP-seq libraries on the IGV genome browser with deepTools version 2.4.0.
41. Scale the bam files to reads per kilobase per million (RPKM) to generate bigWig files.
42. Generate super ChIP-seq tracks by merging, sorting, and indexing replicate bam files using samtools and scaled to RPKM using deepTools.

3.1.7 Identification of ChIP-seq Binding Sites

43. Use model-based analysis of ChIP-seq (MACS) version 1.4.2 peak calling algorithm with a p-value threshold of 1e-5 to identify H3K27ac enrichment over "input" background.
44. Identify consensus replicate sites and unique "fed (low βOHB)" and "fasted (high βOHB)" sites for H3K27ac

using the concatenate, cluster, and subtract tools from the Galaxy/Cistrome web-based platform.

45. First, generate a consensus peak set by clustering intervals of replicate peaks that directly overlapped by a minimum of 1 bp. Next, generate a shared peak set by clustering intervals of fed consensus peaks that directly overlapped fasted consensus peaks by a minimum of 1 bp.
46. Identify unique peaks by subtracting the total number of peaks in each condition by the shared peak set.

3.1.8 Assigning Binding Sites to Genes

47. Obtain the list of Ensembl genes from the UCSC Table Browser (http://genome.ucsc.edu/). Define proximal promoters as ±5 kb from the transcription start site (TSS). Define the gene body as all genic regions outside of the +5 kb promoter region. Define intergenic regions as all regions outside of both the proximal promoter and gene body.
48. Assign H3K27ac peaks to genes if they overlapped the promoter by a minimum of 1 bp.
49. Identify H3K27ac enhancers as all sites outside of the proximal promoter.

3.2 Lineage Tracing and Single-Cell RNA-seq for Lateral Inhibition and Stem-to-Paneth Differentiation

3.2.1 Sample Preparation and Sequencing

1. Induce Lgr5 lineage tracing by injecting Lgr5-Cre^{ERT2}; tdTomato^{LSL} mice with tamoxifen (100 mg/kg body weight, i.p.). Alternatively, perform lineage tracing in combination of Hmgcs2 deletion using Hmgcs2$^{loxp/loxp}$; Lgr5-Cre^{ERT2}; tdTomato^{LSL} mice (*see* **Note 3**).
2. Perform crypt isolation and dissociation for cell sorting as previously described [17].
3. Flow-sort 2 × 10^5 cells tdTomato+ single cells into ice-cold 50 μl of 0.4% BSA-PBS (*see* **Note 4**).
4. Process the sorted cells through the Chromium single-cell platform using the Chromium Single Cell 3′ Library, Gel Bead and Chip Kits (10× Genomics, Pleasanton, CA) (*see* **Note 5**).
5. Add 7000 cells to each channel of a chip (estimated recovery rate L1, 500–2500 cells).
6. Sequence libraries on an Illumina NextSeq 500.

3.2.2 Data Processing and Cell Clustering

7. Perform alignment to mm10 mouse genome and unique molecular identifier (UMI) collapsing using the Cellranger toolkit (version 1.3.1, 10× Genomics).
8. Perform variable gene selection as previously described [18], resulting in a set of 672 variable genes.
9. Perform dimensionality reduction using the R package "Seurat" version 2.3.4.

10. Select 13 principal components by inspection of the "knee" in a scree plot.
11. To avoid missing rarest cells in some batches due to random sampling effects this problem, we performed an initial round of unsupervised clustering using *k*-nearest neighbor (*k*NN) graph-based clustering, implemented in Seurat using the "FindClusters" function, using a resolution parameter of 1.
12. Next, perform batch correction within each identified cluster (which contained only transcriptionally similar cells) ameliorating problems with differences in abundance.
13. Perform batch correction using ComBat as implemented in the R package sva, using the default parametric adjustment mode.
14. Re-ran PCA and *k*NN-based clustering to identify the final clusters (resolution parameter = 0.25).

3.2.3 GSEA Analysis for Lateral Inhibition

15. Perform gen set enrichment analysis (GSEA) for gene expression change due to loss of HMGCS2 in stem cell cluster (i.e., HMGCS2 WT stem vs. HMGCS2 KO stem).
16. In R, use "phyper," to assign "Notch inhibition responses (GSE51398: Transcripts enriched in Rbpj-/- and DBZ Sec--Pro)" and "Atohl deletion target (GSE51398: Transcripts enriched in Atohl-/- crypt cells)" as "Pathways" and test for enrichment, respectively. Set padj<0.05 as significant.
17. Plot the results using ggplot (), with labs (labs(x = "Pathway", y = "Normalized Enrichment Score").

3.2.4 Differential Expression of Cell-Type Signature Genes: Stem-to-Paneth Transition

18. Identify maximally specific genes for cell types by performing differential expression tests between each pair of clusters for all possible pairwise comparisons.
19. For a given cluster, filter putative signature genes using the maximum FDR Q-value and ranked by the minimum log2 fold-change of means (across the comparisons).
20. Obtain cell-type signature genes for the initial droplet-based scRNA-seq data were using a maximum FDR of 0.001 and a minimum log2 fold-change of 0.1.
21. Test differential expression using a two-part "hurdle" model to control for both technical quality and mouse-to-mouse variation, using the R package MAST, and p values for differential expression were computed using the likelihood-ratio test.
22. Perform multiple hypothesis testing correction by controlling the false discovery rate (FDR) using the R function "p.adjust."
23. Visualize results of assigned signature gene (i.e., Paneth cell) and cluster (i.e., stem) using "volcano" plots constructed using "ggplot2."

4 Notes

1. For method 3.1, glycine solution can be kept for one month once made, as the pH of it changes over time.
2. When using magnet beads, pay attention not to disturb the beads while washing.
3. For the intestinal tissue used in method 3.2, collect 3–5 cm intestinal tissues per mouse for histology, which can be used for validating the efficiency of gene deletion and lineage tracing.
4. For method 3.2, precoating the collection tubes with 5% BSA for 1 h prior to the cell sorting can improve recovery of the cells sorted for scRNA-seq.
5. Use RBC lysis buffer and trypan blue exclusion to optimize sorting live cells without red blood cell contamination in samples for scRNA-seq.

References

1. Beumer J, Clevers H (2021) Cell fate specification and differentiation in the adult mammalian intestine. Nat Rev Mol Cell Biol 22(1): 39–53. https://doi.org/10.1038/s41580-020-0278-0
2. Verzi MP, Shivdasani RA (2020) Epigenetic regulation of intestinal stem cell differentiation. Am J Physiol Gastrointest Liver Physiol 319(2):G189–G196. https://doi.org/10.1152/ajpgi.00084.2020
3. Sancho R, Cremona CA, Behrens A (2015) Stem cell and progenitor fate in the mammalian intestine: Notch and lateral inhibition in homeostasis and disease. EMBO Rep 16(5): 571–581. https://doi.org/10.15252/embr.201540188
4. Capdevila C, Trifas M, Miller J, Anderson T, Sims PA, Yan KS (2021) Cellular origins and lineage relationships of the intestinal epithelium. Am J Physiol Gastrointest Liver Physiol 321(4):G413–G425. https://doi.org/10.1152/ajpgi.00188.2021
5. Liefke R, Oswald F, Alvarado C, Ferres-Marco-D, Mittler G, Rodriguez P, Dominguez M, Borggrefe T (2010) Histone demethylase KDM5A is an integral part of the core Notch-RBP-J repressor complex. Genes Dev 24(6): 590–601. https://doi.org/10.1101/gad.563210
6. Oswald F, Winkler M, Cao Y, Astrahantseff K, Bourteele S, Knochel W, Borggrefe T (2005) RBP-Jkappa/SHARP recruits CtIP/CtBP corepressors to silence Notch target genes. Mol Cell Biol 25(23):10379–10390. https://doi.org/10.1128/MCB.25.23.10379-10390.2005
7. Borggrefe T, Oswald F (2016) Setting the stage for Notch: the drosophila Su(H)-hairless repressor complex. PLoS Biol 14(7): e1002524. https://doi.org/10.1371/journal.pbio.1002524
8. Oswald F, Tauber B, Dobner T, Bourteele S, Kostezka U, Adler G, Liptay S, Schmid RM (2001) p300 acts as a transcriptional coactivator for mammalian Notch-1. Mol Cell Biol 21(22):7761–7774. https://doi.org/10.1128/MCB.21.22.7761-7774.2001
9. Wilson JJ, Kovall RA (2006) Crystal structure of the CSL-Notch-mastermind ternary complex bound to DNA. Cell 124(5):985–996. https://doi.org/10.1016/j.cell.2006.01.035
10. Nam Y, Sliz P, Song L, Aster JC, Blacklow SC (2006) Structural basis for cooperativity in recruitment of MAML coactivators to Notch transcription complexes. Cell 124(5): 973–983. https://doi.org/10.1016/j.cell.2005.12.037
11. Borggrefe T, Oswald F (2009) The Notch signaling pathway: transcriptional regulation at Notch target genes. Cell Mol Life Sci 66(10): 1631–1646. https://doi.org/10.1007/s00018-009-8668-7
12. Bonis V, Rossell C, Gehart H (2021) The intestinal epithelium – fluid fate and rigid structure from crypt bottom to villus tip. Front Cell Dev Biol 9:661931. https://doi.org/10.3389/fcell.2021.661931

13. Kim TH, Li F, Ferreiro-Neira I, Ho LL, Luyten A, Nalapareddy K, Long H, Verzi M, Shivdasani RA (2014) Broadly permissive intestinal chromatin underlies lateral inhibition and cell plasticity. Nature 506(7489):511–515. https://doi.org/10.1038/nature12903
14. de Sousa EMF, de Sauvage FJ (2019) Cellular plasticity in intestinal homeostasis and disease. Cell Stem Cell 24(1):54–64. https://doi.org/10.1016/j.stem.2018.11.019
15. Cheng CW, Biton M, Haber AL, Gunduz N, Eng G, Gaynor LT, Tripathi S, Calibasi-Kocal-G, Rickelt S, Butty VL, Moreno-Serrano M, Iqbal AM, Bauer-Rowe KE, Imada S, Ulutas MS, Mylonas C, Whary MT, Levine SS, Basbinar Y, Hynes RO, Mino-Kenudson M, Deshpande V, Boyer LA, Fox JG, Terranova C, Rai K, Piwnica-Worms H, Mihaylova MM, Regev A, Yilmaz OH (2019) Ketone body signaling mediates intestinal stem cell homeostasis and adaptation to diet. Cell 178(5):1115–1131. https://doi.org/10.1016/j.cell.2019.07.048
16. Terranova C, Tang M, Orouji E, Maitituoheti M, Raman A, Amin S, Liu Z, Rai K (2018) An integrated platform for genome-wide mapping of chromatin states using high-throughput ChIP-sequencing in tumor tissues. J Vis Exp 134. https://doi.org/10.3791/56972
17. Gebert N, Rahman S, Lewis CA, Ori A, Cheng CW (2021) Identifying cell-type-specific metabolic signatures using transcriptome and proteome analyses. Curr Protoc 1(9):e245. https://doi.org/10.1002/cpz1.245
18. Kowalczyk MS, Tirosh I, Heckl D, Rao TN, Dixit A, Haas BJ, Schneider RK, Wagers AJ, Ebert BL, Regev A (2015) Single-cell RNA-seq reveals changes in cell cycle and differentiation programs upon aging of hematopoietic stem cells. Genome Res 25(12):1860–1872. https://doi.org/10.1101/gr.192237.115

Chapter 8

Defining Anorectal Transition Zone Heterogeneity Using Single-Cell RNA Sequencing

Loucinė Mitoyan, Charlyne Gard, Sébastien Nin, Béatrice Loriod, and Géraldine Guasch

Abstract

Special regions called transition zones (TZs) are found at numerous places in the body. TZs represent the junction between two different types of epithelia and are located between the esophagus and the stomach, in the cervix, in the eye, and between the anal canal and the rectum. TZ is a heterogeneous population, and the detailed characterization of its populations requires an analysis at the single-cell level. In this chapter, we provide a protocol to do single-cell RNA sequencing primary analysis of anal canal, TZ, and rectum epithelia.

Key words Anorectal transition zone, Epithelial cells, Cellular heterogeneity, Single-cell RNA sequencing, Intestine, Rectum

1 Introduction

Transition zones (TZs), such as anorectal and gastroesophageal TZs, are special regions located between two different types of epithelia and frequently associated with pathologies such as ulcers and cancer [1–4]. Previous studies have shown the presence of cells with stem cell properties in those regions [5–7]. Nevertheless, the role of TZ cells in maintaining surrounding epithelia during homeostasis and injury is poorly known. We hypothesized that TZ stem cells may contribute to maintaining surrounding epithelia during homeostasis and upon injury. Using a mouse model of lineage tracing, we recently showed that TZ stem cells maintain the squamous epithelium of the anal canal during homeostasis. Upon wounding, a small TZ population identified as hybrid cells, which present molecular characteristics of both squamous and

The authors "Loucinė Mitoyan" and "Charlyne Gard" contributed equally to this work.

Paloma Ordóñez-Morán (ed.), *Intestinal Differentiated Cells: Methods and Protocols*, Methods in Molecular Biology, vol. 2650, https://doi.org/10.1007/978-1-0716-3076-1_8,

glandular epithelium, showed an unexpected plasticity that allowed them to expand their fate and participate in the regeneration of glandular epithelium of the rectum [3]. Most of the time, different cell types can coexist within the same population, forming subpopulations with distinct roles [8–11]. Therefore, a given population of cells may have different transcriptional profiles. We then used single-cell RNA sequencing technology to decipher anorectal TZ heterogeneity and characterize cells displaying a multipotent phenotype [3]. In contrast to single-cell analysis, many pieces of information can be missed with bulk assays (such as RNAseq): one cell type can dominate the rest of the population; consequently, it is not possible to know if a modification in the expression profile is a consequence of a change in cell composition or transcriptome regulation [12]. Also, it does not consider the identification of the different subpopulations, so correlations between gene expression and subpopulation can be missed. Single-cell RNA sequencing can estimate the level of gene expression of a single cell in a given population. In this chapter, we will focus on how to extract RNA and sequence anorectal TZ cells at the single-cell level.

2 Materials

Cells processing for single-cell sequencing

2.1 Cell Counting

1. Logos Biosystems, LUNA-FL™ Dual Fluorescence Cell Counter, L20001.
2. Logos Biosystems, PhotonSlide™ 50 Slides, L12005.
3. Logos Biosystems, Propidium Iodide Stain, F23003.

2.2 Chromium Single-Cell RNA Sequencing (10× Genomics)

1. 10× Genomics, Chromium Controller and Next GEM Accessory Kit, 1000204.
2. 10× Genomics, Chromium Next GEM Single Cell 3′ GEM, Library and Gel Bead Kit v3.1, 4 rxns PN-1000128.
3. 10× Genomics, Chromium Next GEM Chip G Single Cell Kit, 48 rxns, PN-1000120.
4. 10× Genomics, Single Index Kit T Set A, 96 rxns PN-1000213.
5. Bio-Rad, C1000 Touch thermal cycler with 96-deep well reaction module, 1851197.
6. StarLab, PCR tubes 0.2 mL eight-tube strips, I1402-3700.
7. Eppendorf, DNA LoBind tubes 1.5 mL, 022431021.
8. Eppendorf, DNA LoBind tubes 2.0 mL, 022431048.
9. Dutscher, Falcon® clarified polypropylene 50 mL conical bottom tube, 352070.

10. Dutscher, Falcon® clarified polypropylene 15 mL conical bottom tube, 352095.
11. Nuclease-free water.
12. Thermo Fisher Scientific, low TE buffer (10 mM Tris–HCl pH 8.0, 0.1 mM EDTA), 12090-015.
13. Ethanol 80%.
14. Beckman Coulter, SPRIselect Reagent Kit, B23318.
15. Bio-Rad, 10% Tween-20, 1662404.
16. BioVision, glycerol (50%) sterile solution 100 mL, B1012.
17. Qiagen, Qiagen Buffer EB, 19086.
18. Fisher Scientific, Scientific Industries SI™ Vortex-Genie™ 2, SI-0256.
19. VWR, Microcentrifugeuse MiniStar Silverline, rotor for four strips 8 × 0.2 mL, 521-2848.
20. VWR, Microcentrifugeuse MiniStar Silverline, rotor for eight tubes of 1.5/2.0 mL, 521-2847.
21. Fisher Scientific, Argos Technologies™ Polystyrene Pipetting Reservoir Basins, 03-391-536.
22. Eppendorf, Eppendorf Thermomixer Compact 5350 Mixer.
23. Rainin, Pipet-Lite Multi Pipette L8-50XLS+, 17013804*.
24. Rainin, Pipet-Lite Multi Pipette L8-200XLS+, 17013805*.
25. Rainin, Pipet-Lite LTS Pipette L-2XLS+, 17014393*.
26. Rainin, Pipet-Lite LTS Pipette L-10XLS+, 17014388*.
27. Rainin, Pipet-Lite LTS Pipette L-20XLS+, 17014392*.
28. Rainin, Pipet-Lite LTS Pipette L-200XLS+, 17014391*.
29. Rainin, Pipet-Lite LTS Pipette L-1000XLS+, 17014382*.
30. Rainin, Tips LTS 200UL Filter RT-L200FLR, 30389240*.
31. Rainin, Tips LTS 1ML Filter RT-L1000FLR, 30389213*.
32. Rainin, Tips LTS 20UL Filter RT-L10FLR, 30389226*.
33. Wide-bore tips.

*According to manufacturer's recommendation.

2.3 Quantification and Quality Control

1. Advanced Analytical, Fragment Analyzer Automated CE System – 12 caps, FSv2-CE2F.
2. Agilent Technologies, Kit d'Analyse de Fragment NGS Sensibilité Standard, 1pb–6000 pb, 500 samples, DNF-473-0500.
3. Thermo Fisher Scientific, Qubit 2.0 fluorometer, Q32866.
4. Life Technologies, Qubit dsDNA HS assay kit, 100, Q32851.
5. Life Technologies, Qubit assay tubes kit of 500, Q32856.

2.4 Sequencing

1. Illumina, NextSeq 500 System, SY-415-1001.
2. Illumina, NextSeq 500/550 High-Output Kit v2.5 (75 cycles), 20024906.
3. Illumina, NextSeq PhiX Control Kit, FC-110-3002.
4. NaOH, 0.2 M.
5. Tris–HCl, 0.2 M.

2.5 Bioinformatic Analysis

1. Cell Ranger 6.0.2.
2. FastQC 0.11.9.
3. Fastq Screen 0.14.
4. MultiQC 1.9.

3 Methods

3.1 Single-Cell RNA Sequencing

At this step, the cells are transferred to the sequencing platform (TGML) as quickly as possible to prevent cell death and perform the single-cell RNA sequencing. After checking the cell concentration, cells are isolated using the Chromium Controller (10× Genomics). The messenger RNAs are reverse transcribed, the cDNAs are amplified, and the libraries are sequenced.

3.2 Cell Counting

The cell count is performed with *LUNA-FL™ Dual Fluorescence Cell Counter (Logos Biosystems)*:

1. Preserve enough cell suspension to perform the single-cell RNA sequencing. We recommend having 16.5 μL of cell suspension for a concentration of 1000 cells per microliter, to perform it, according to the manufacturer's instructions.
2. For cell counting, 20 μL of cell suspension is needed. Most of the time, the volume of cells is limited. Thus, if you cannot have 20 μL, resuspend the remaining volume of cell suspension in your specific medium to reach 20 μL.
3. Mix gently the 20 μL of cell suspension with 2 μL of propidium iodide (PI).
4. Take 20 μL of the mix with cells and PI and load one of the wells of the PhotonSlide™.
5. Put the PhotonSlide™ on the *LUNA-FL™ Dual Fluorescence Cell Counter*, on the side of the well.
6. Choose the program *Fluorescence cell counting* and count the cells.
7. If you diluted your cells to reach 20 μL (**step 1**), do not forget to compute the dilution factor to have the correct count.

Recommended viability to proceed with Chromium Controller is >95%, but >80% gives good results also (*see* **Note 1**).

3.3 Chromium Single-Cell RNA Sequencing

The Chromium Controller technology developed by 10× Genomics is currently the one that allows isolating the largest number of cells [13]. This technique is optimized for the transcriptional study of 500–10,000 single cells via a microfluidic platform, allowing the manipulation of liquid in droplet form in the nanoliter range. This is a droplet-based method: the flows are controlled in a chip containing microchannels and allow encapsulating single cells in a droplet with a barcoding system, which will permit knowing the cellular origin of the transcripts that will be studied [14].

For this part of the method, the manufacturer's protocol is strictly followed, so in this section each step will be summarized and explained briefly.

3.4 GEM Generation and Barcoding

A pool of approximately 3,500,000 10× barcodes is sampled separately to index the transcriptome of each cell. This is done by partitioning thousands of cells into nanoliter droplets called GEM (Gel beads in EMulsion), where all recovered DNA/RNA molecules will share a common barcode 10×. These will allow associating the reads to its original GEM. Each GEM will be composed, in theory, of a single cell, a gel bead, and retrotranscription reagents. GEMs are generated by combining in a Chromium Chip B 10× barcoded gel beads, a master mix, and partitioning oil (Fig. 1). Cells are delivered into the Chip at a limited concentration to minimize the number of GEMs containing cell duplicates (90–99% of the generated GEMs contain no cells). At this step, be careful to resuspend cells with wide-bore pipette tips before loading the chip, to avoid damaging cells.

At the end of the partitioning, we obtain an emulsion composed of three possible configurations of GEMs: singletons which are GEMs containing a single cell, doublets or multiplets containing two or more cells, and negatives which contain no cells. The emulsion must look like the one in Fig. 2a. (*See* **Note 2** if the emulsion looks like the one in Fig. 2b.)

Then, the emulsion is transferred to a thermal cycler where the cells break by increasing the temperature and the gel bead releases the sequences present on its surface. From this, the retrotranscription creates a complete cDNA strand library that will have been generated from the polyadenylated mRNAs, all of which will contain a cell barcode (Fig. 3).

3.5 Post-GEM-RT Cleanup and cDNA Amplification

After retrotranscription, the GEMs are broken up, and silane magnetic beads are used to purify the barcoded cellular products (mRNA). The cDNA sequences are then amplified by PCR for library construction using a template switch oligo (Fig. 3).

Then, *total cDNA yield* (ng) is computed based on the measurement on *Qubit fluorometer 2.0.* This number is important to know how many PCR cycles are needed for cDNA amplification. At last, the quality of cDNA will be analyzed with the *Fragment*

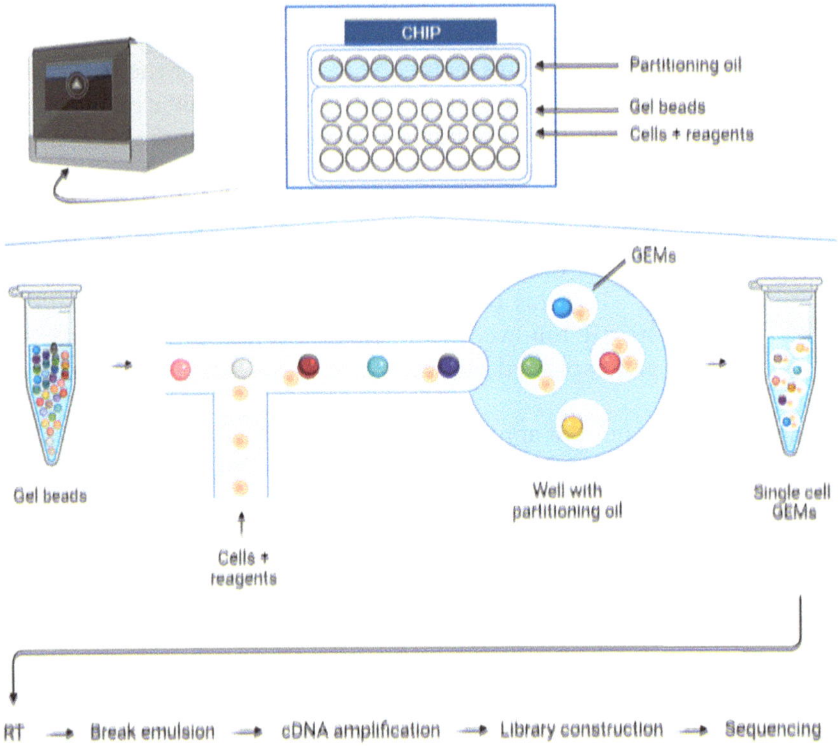

Fig. 1 Single-cell sequencing with 10× Genomics chromium. RT = retrotranscription. GEMs = Gel beads in EMulsion. The 10× Genomics single-cell sequencing method is a droplet-based sequencing microfluidic system that generates an emulsion of aqueous droplets in oil. Each droplet of oil generated will theoretically contain a cell, a gel bead, and reagents that will be used for reverse transcription of mRNA. These droplets are named GEMs. At the end of the emulsion partition, three types of GEMs configurations are possible. Negative, no cells; singlet, one cell; doublet or multiplet, two or more cells. From these GEMs a library will be prepared and sequenced at the desired depth, depending on the biological question. (This schema was created with Biorender.com)

Analyzer, with the sample diluted to an optimal concentration of 3 ng/μL. The traces of cDNA should look like the traces in Fig. 4.

Quality control with Qubit fluorometer 2.0 and Fragment Analyzer is performed according to the manufacturer's instructions.

3.6 3′ Gene Expression Library Construction

The complete cDNA sequences from the previous step undergo enzymatic fragmentation, an end repair and a-tailing, and a double-size selection, in order to optimize their size. The read 2 (Fig. 3) is added during the step of end repair. Next, the adaptor P5 (Fig. 3) is added during adaptor ligation. Then, the index and the adaptor P7 are added during the sample index PCR step (Fig. 3). Read 1 (Fig. 3) is added during the GEM incubation.

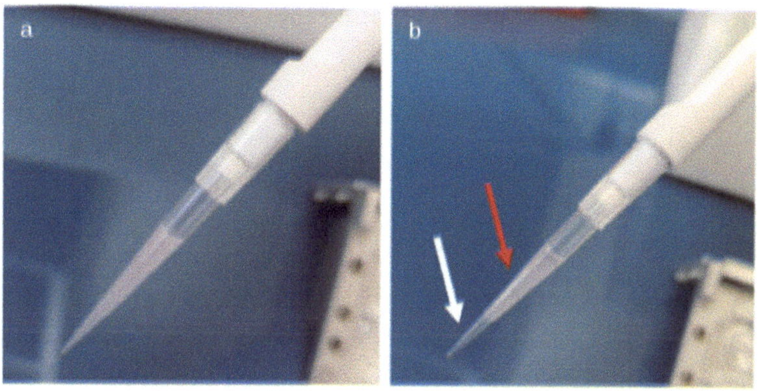

Fig. 2 Emulsion after partitioning. (**a**) Normal emulsion, with a homogeneous phase. (**b**) Abnormal emulsion after partitioning. Two phases are visible: the opaque one (red arrow) corresponds to the gel beads in emulsion and the transparent one (white arrow) which is specific to a clog

The estimation of library constructs concentration is done by measurement on the Qubit fluorometer 2.0. The fragment size of the library constructs will be analyzed with the Fragment Analyzer, with the sample diluted to an optimal concentration of 3 ng/μL. The traces of the library fragments should show enrichment at 400 bp (Fig. 5). (*See* **Note 3** if there is an additional peak.)

Quality control with Qubit fluorometer 2.0 and Fragment Analyzer is performed according to manufacturer's instructions.

3.7 Sequencing

Sequencing was performed on the Illumina NextSeq 500 using the NextSeq 500/550 High-Output v2.5 Kit according to the supplier's recommendations. The sequencing is done in paired end with single indexing. The cycles are distributed as follows: Read 1 = 28 cycles/Index i7 = 8 cycles/Read 2 = 55 cycles (*see* Fig. 3 and **Note 4**):

1. Estimate the molarity of the library with the data of the library quantification control made by the Fragment Analyzer.
2. Put distilled water in a dedicated tray.
3. Place the flow cell at least 30 min at room temperature (but 1 h max).
4. Take the reagent cartridge and the HT1 dilution buffer out of −20 °C and thaw them by floating in the water.
5. Dilute the library sample to the concentration of 4 nM if possible, otherwise 2 nM, 1 nM, or 0.5 nM. If you have several library samples, dilute them at the same concentration.
6. Pool the libraries if you have several samples (according to the number of reads expected for each sample).

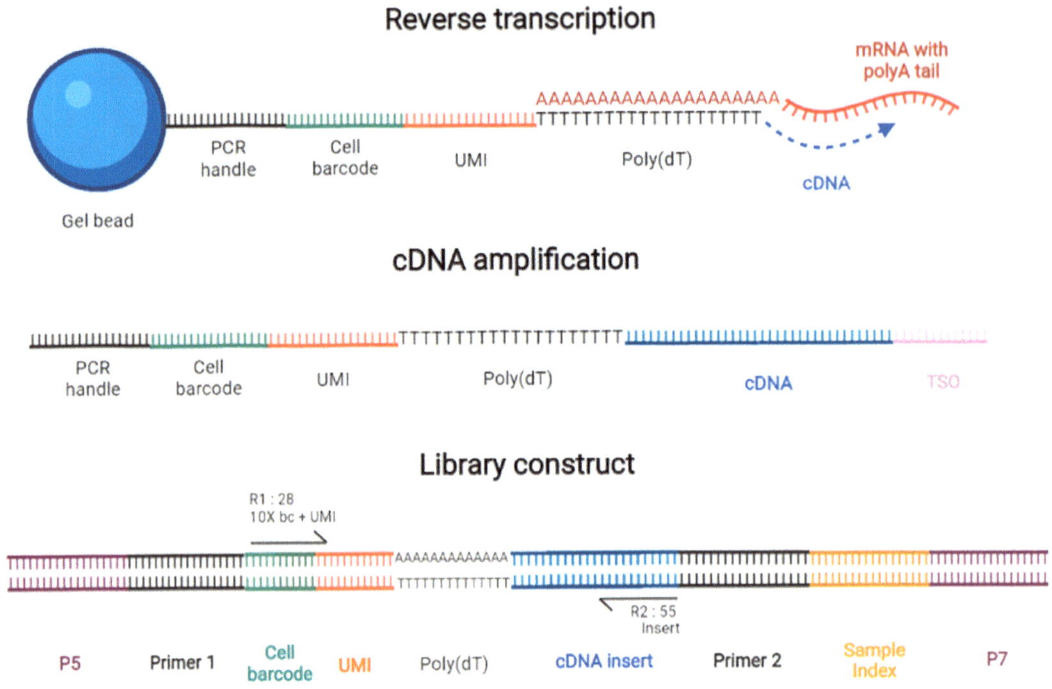

Fig. 3 Library preparation. The gel bead has numerous nucleotide sequences on its surface that are composed of the following: 10× barcodes of 16 nucleotides, allowing the identification of the original GEM, each bead has a unique one; UMI (unique molecular identifier) sequence of 12 nucleotides, which is a sequence that will be unique, it will allow the correction of amplification biases; and poly(dT) tail of 30 nucleotides, which will allow the capture of mRNA. *Reverse transcription*. Incubation of GEMs produces a complete barcoded cDNA from the polyadenylated mRNA. After reverse transcription, the GEMs are broken, and silane magnetic beads are used to purify the barcoded cell products. Then, the cDNA sequences are amplified by PCR for library construction. *cDNA amplification*. TSO, template switch oligo. During DNA amplification, a TSO will allow the amplification of all DNA strands. TSO is added during the reverse transcription. *Library constructs*. The library is composed of sequences containing the following: a cDNA from a transcript; two primers, primer 1 (added during the incubation step of the GEMs) and primer 2; an index for sample identification (composed of four unique oligonucleotides to balance the base composition of the sample index during sequencing); and adapters, p5 and p7 which will be used for the "bridge" amplification during sequencing, these adapters are added by ligation. *Sequencing*. Read 1 (R1): 28 cycles to identify the cell barcode and the UMI (16 nt + 12 nt). Read 2 (R2): 55 cycles, for transcripts. Sample index: 8 cycles for the index (8 nt). (This schema was created with Biorender.com)

7. Denature during 5 min in the appropriate volume of 0.2 M NaOH (Table 1).
8. Stop denaturation by adding the appropriate volume of 0.2 M Tris–HCl + HT1 buffer (Table 1).
9. Mix 1183 μL of HT1 and 117 μL of the denatured fluid on the flow cell (do not vortex; homogenize by inversion) to reach a concentration of 1.8 pM.
10. Remove 1 μL from the tube and add 1.2 μL of 20 pM denatured Ψx. Do not vortex, and homogenize by inversion.

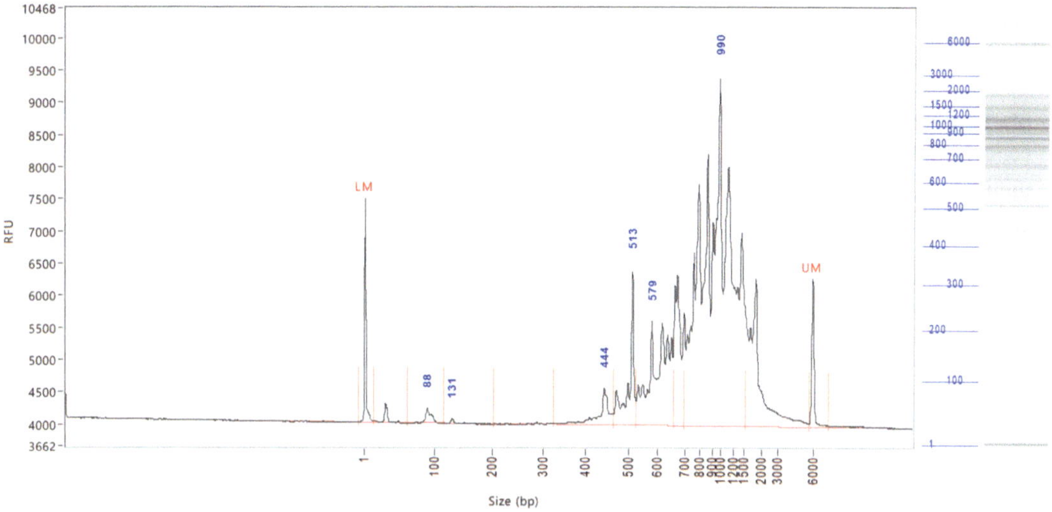

Fig. 4 Traces of cDNA after GEM generation and RT-PCR. LM = lower marker; UM = upper marker. (Traces are obtained with a Fragment Analyzer with the software Prosize 3.0)

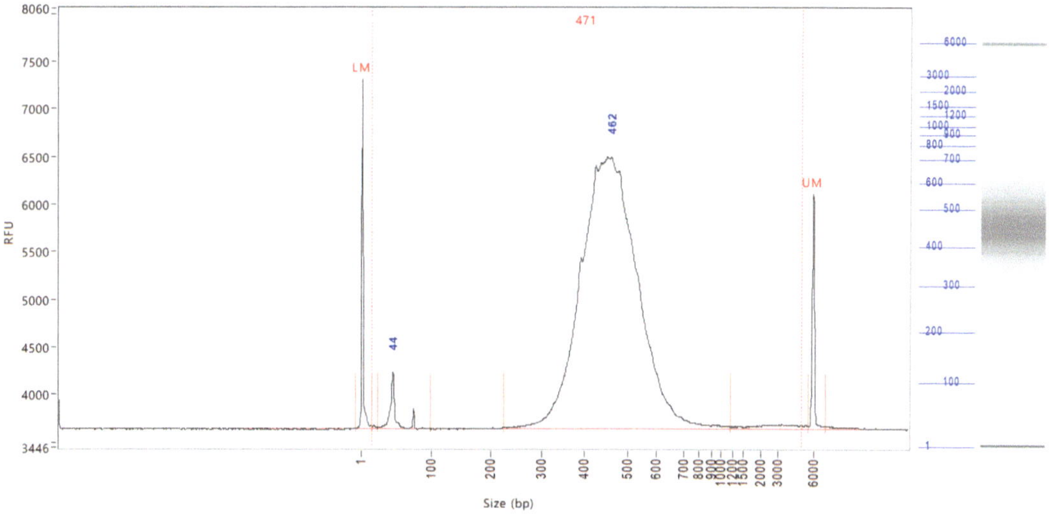

Fig. 5 Traces of the fragments of a ScRNAseq library. LM = lower marker; UM = upper marker. (Traces are obtained with a Fragment Analyzer with the software Prosize 3.0)

11. Retrieve the cassette (HORC) and check that the buffers are thawed, and homogenize the buffers by inverting (5 times minimum).
12. Drill with clean pipette tips the loading hole: "Load Libraries Here."
13. Load the entire denatured library/HT1/Ψx mix (1.3 mL) (*see* **Note 5**).
14. Load the Illumina NextSeq 500 with the reagents according to the manufacturer's recommendation.

Table 1
Concentration of reagents for denaturation according to the library molarity

Starting library concentration (nM)	Volume of the library for denaturation (μL)	Volume of 0.2 M NaOH (μL)	Volume of 0.2 M Tris–HCl, pH 7 (μL)	Volume of prechilled HT1 (μL)
4	5	5	5	985
2	10	10	10	980
1	20	20	20	970
0.5	40	40	40	880

Mitoyan et al. [3]

3.8 Bioinformatic Analysis

In this article, only the primary and the secondary analyses are summarized. A tertiary analysis is necessary to analyze the data but it is out of focus here.

The raw data obtained after sequencing with Illumina technology are BCL files for Base Call format. For these files to be processed bioinformatically, they must be converted into FASTQ files (Fig. 6). This step is performed with the *Cell Ranger mkfastq* tool from 10× Genomics whose output is directly compatible with the use of their *Cell Ranger count* tool. Cell Ranger is an analysis pipeline specifically designed to process single-cell data generated using 10× Genomics protocols. It can perform the following steps (Fig. 6):

- Primary analysis
  1. Demultiplexing of BCL files and conversion to FASTQ format (mkfastq)
- Secondary analysis
  2. Quality control of reads (FastQC, Fastq Screen)
  3. Alignment of reads to a reference genome (STAR)
  4. Quantification of UMIs

For quality control the MultiQC tool was used. It allows you to collect the quality control results in an HTML report per sequencing run. Cell Ranger count takes FASTQ files as input and generates as output (non-exhaustive): a gene/cell-matrix containing the UMI count for each gene in each cell, a *.cloupe* file that can be used with the Loupe Browser software, and an *HTML* report. On the latter, there are some relevant numbers to check to estimate the quality of the data (Fig. 7):

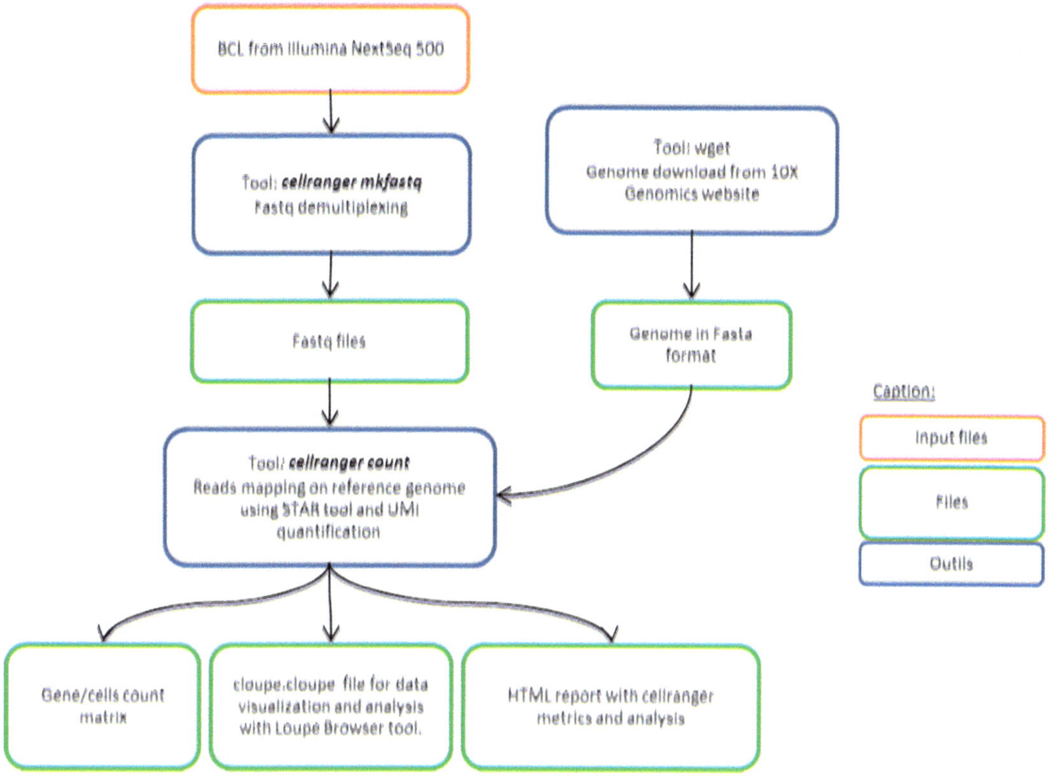

Fig. 6 TGML scRNA-seq data processing workflow. The workflow used to process the data converts BCL files into fastq files using cellranger mkfastq. Obtained reads are then mapped onto mouse genome mm10 using cellranger count. This tool also quantifies UMI of mapped reads and its output files can be used in Loupe Browser or in R for further analysis

- The estimated number of cells is usually between 40% and 60% of the loaded cell number. For instance, if 10,000 cells were loaded, a range between 4000 and 6000 cells is expected (*see* **Note 6**).
- Mean reads per cell are ideally above 50,000 mean reads per cell (*see* **Note 7**).
- Median genes per cell correspond to the median of genes detected per cell-associated barcode and depend on your cell type.
- The barcode rank plot (right side of Fig. 7) shows the quality of the sample. Ideally, the plot decreases steeply between cells (blue part) and the background noise (gray part).

Fig. 7 Screenshot of the summary page of the web summary. This file is an HTML report and summarizes the data of single-cell sequencing performed with 10× Technology. This document includes two tabs: the "summary" including data about the quality of the experiment and the "analysis" including a visual representation of the data like projection t-SNE and table of top expressed genes

4 Notes

1. If viability is lower than 80%, it is recommended to enrich the sample with viable cells: FACS, *Miltenyi Biotec* Dead Cell Removal Kit.

   It is extremely recommended to test the cellular viability in your medium before performing chromium single-cell RNA sequencing; indeed the risk is to have numerous cells fragment if the medium is not suitable for the cells.

2. A clog results in a nonhomogeneous emulsion after a chromium run. The manufacturer recommends rerunning the sample if a clog occurs during GEM generation. But, it is possible to continue if the sample is precious, the quality of the cells will not be affected, but the number of cells will be lower than expected.

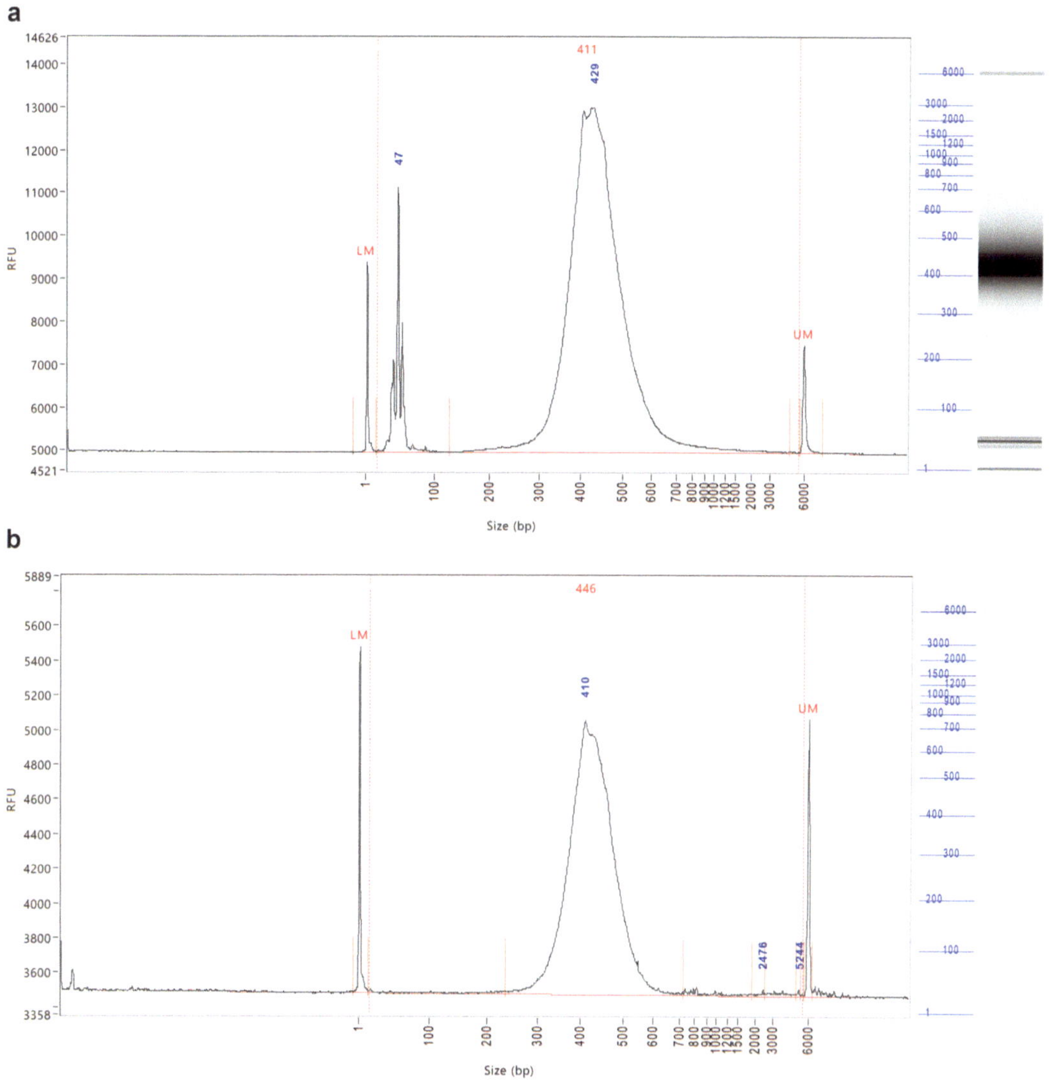

Fig. 8 Primer dimers. (**a**) Library traces showing primer dimers' peak at around 50pb before purification. (**b**) The same library after repurification. The primer dimers' peak was deleted after purification

3. An additional peak at 45 pb on the library traces (Fig. 8a) corresponds to dimer-primer (DP). These fragments must be eliminated because they could interfere with the sequencing. To eliminate this DP, the library must be purified by repeating **step 3.6** (Post Sample Index PCR Double Sided Size Selection—SPRIselect) of the 10× protocol. Note that a loss of DNA is expected after purification (about 40% according to the manufacturer) (Fig. 8b).
4. We choose to attribute 55 cycles to Reads 2 instead of 91 cycles (such as the manufacturer's recommendation). This

configuration is commonly used for this type of experiment because it makes it possible to reduce the price of the sequencing.

5. PhiX concentration can be elevated if your samples present low diversity or are rich in GC content. It can also be used as a control for troubleshooting cluster generation problems, in order to know if an error is due to library preparation.
6. Cell fragments can result in more cells being detected than loaded/expected. In this case, the medium may not be suitable, so it is necessary to find one that causes less fragmentation. Another solution is to reduce the volume of cells if you cannot remove all the fragments.
7. If this number is significantly below 50,000 mean reads per cell, this means that the sequencing depth is too low. To increase it, rerun the sequencing.

Acknowledgments

This work was supported by ANR grant #ANR20-CE13-0009-01 (G.G.) and partly supported by research funding from the Canceropôle Provence-Alpes-Côte d'Azur, Institut National du Cancer and Région Sud, grants from the Excellence Initiative of Aix-Marseille University A*Midex, "Investissement d'avenir" (CapoStromEx) and Inserm Plan Cancer AAP single cell (G.G.). L.M. is a recipient of the French ministerial research fellowship and the Ligue Nationale Contre le Cancer fellowship. High-throughput sequencing was performed at the TGML platform, supported by grants from Inserm, GIS IBiSA, Aix-Marseille Université, and ANR-10-INBS-0009-10. The authors would like to thank Lisa Bargier and Dr. Véronique Chevrier for reading and corrections.

References

1. Wang X et al (2011) Residual embryonic cells as precursors of a Barrett's-like metaplasia. Cell 145:1023–1035
2. Guasch G et al (2007) Loss of TGFβ signaling destabilizes homeostasis and promotes squamous cell carcinomas in stratified epithelia. Cancer Cell 12:313–327
3. Mitoyan L et al (2021) A stem cell population at the anorectal junction maintains homeostasis and participates in tissue regeneration. Nat Commun 12:2761
4. Herfs M et al (2012) A discrete population of squamocolumnar junction cells implicated in the pathogenesis of cervical cancer. Proc Natl Acad Sci 109:10516–10521
5. Amitai-Lange A et al (2015) Lineage tracing of stem and progenitor cells of the murine corneal epithelium: lineage tracing of limbal and corneal epithelial cells. Stem Cells 33:230–239
6. Mcnairn AJ, Guasch G (2011) Epithelial transition zones: merging microenvironments, niches, and cellular transformation. Eur J Dermatol 21:21–28
7. Runck LA, Kramer M, Ciraolo G, Lewis AG, Guasch G (2010) Identification of epithelial label-retaining cells at the transition between the anal canal and the rectum in mice. Cell Cycle 9:3111–3117

8. Tasic B (2018) Single cell transcriptomics in neuroscience: cell classification and beyond. Curr Opin Neurobiol 50:242–249
9. Navin NE (2014) Cancer genomics: one cell at a time. Genome Biol 15:452
10. Dalerba P et al (2011) Single-cell dissection of transcriptional heterogeneity in human colon tumors. Nat Biotechnol 29:1120–1127
11. Treutlein B et al (2014) Reconstructing lineage hierarchies of the distal lung epithelium using single-cell RNA-seq. Nature 509:371–375
12. Trapnell C (2015) Defining cell types and states with single-cell genomics. Genome Res 25:1491–1498
13. Svensson V, Vento-Tormo R, Teichmann SA (2018) Exponential scaling of single-cell RNA-seq in the past decade. Nat Protoc 13: 599–604
14. Macosko EZ et al (2015) Highly parallel genome-wide expression profiling of individual cells using nanoliter droplets. Cell 161:1202–1214

Part III

In Vitro Culture and Applications

Chapter 9

Directed Differentiation of Murine and Human Small Intestinal Organoids Toward All Mature Lineages

A. Martinez-Silgado, J. Beumer, and H. Clevers

Abstract

Intestinal organoids are three-dimensional structures derived from tissue-resident adult stem cells. These organoids recapitulate key aspects of epithelial biology and can be used to study homeostatic turnover of the corresponding tissue. Organoids can be enriched for the various mature lineages which allows studies of the respective differentiation processes and of the diverse cellular functions. Here we describe mechanisms of intestinal fate specification and how these can be exploited to drive mouse and human small intestinal organoids into each of the functionally mature lineages.

Key words Organoids, Small intestine, Differentiation, Adult stem cell-derived organoids, Intestinal lineages

1 Introduction

The epithelium of the small intestine is organized in crypt-villus structures composed of a single layer of cells. Differentiated cells populate the villi structures which protrude toward the lumen, thereby increasing the absorptive surface area. Multiple crypts surround the base of each villus. At the crypt bottom, protected from chemical and mechanical insults, reside the long-lived intestinal stem cells (ISCs), also known as crypt base columnar cells (CBCs) [1]. This multipotent stem cell population continuously divides, giving rise to a rapidly proliferating transit-amplifying (TA) compartment which occupies the remainder of the crypts. The progeny of the TA cells will differentiate into all mature cells of the epithelium. These can be either absorptive cells (enterocytes and M cells) or secretory cells (Paneth cells, goblet cells, enteroendocrine cells, and tuft cells). Differentiated cells typically migrate up the villi in a journey that can take 3–5 days for some cells and weeks

A. Martinez-Silgado and J. Beumer contributed equally.

Paloma Ordóñez-Morán (ed.), *Intestinal Differentiated Cells: Methods and Protocols*, Methods in Molecular Biology, vol. 2650,
https://doi.org/10.1007/978-1-0716-3076-1_9,

for others. Upon reaching the villus tip, mature cells are extruded to the lumen as they are replaced by newly produced cells.

Unlike the rest of the differentiated cells, Paneth cells migrate toward the crypt bottom, where they are found interspersed between the CBCs and are marked by LGR5 [2, 3]. These secretory cells produce antimicrobial peptides and niche factors crucial for stem cell homeostasis: Wnt3, essential to maintain stemness, and proliferation-promoting EGF. In addition to these secreted factors, Paneth cells express the Notch ligands DLL1 and DLL4 on their surface [4]. These membrane-bound ligands activate the transmembrane Notch receptors in adjacent stem cells, a signal in addition to WNTs required to maintain stem cell multipotency.

Differentiation of stem cell daughter cells initiates when cells inactivate WNT or Notch signaling. When stem cells stochastically lose access to Notch ligands, differentiation to one of the secretory lineages will be initiated. These secretory progenitor cells in turn activate expression of Notch ligands, in order to stimulate Notch signaling in adjacent progenitors. These Notch-activated progenitor cells remain proliferative and fated toward the absorptive lineage. This process termed lateral inhibition ensures a fixed ratio of secretory and the more numerous absorptive cells, as a single Notch ligand-expressing cell can support signaling to multiple adjacent cells [5, 6]. Enterocytes are thus specified by the loss of WNT signaling, while receiving active Notch signals. Within the secretory lineage, differential activities of MAPK and WNT signaling dictate further specification. Paneth cells migrate down toward the crypt bottom and mature under the influence of high WNT levels [6–8]. Goblet cells and enteroendocrine cells specify independently from WNT [9], where high MAPK activity biases differentiation toward goblet cells over enteroendocrine cells [10]. The production of tuft cells, involved in intestinal immunity, is regulated predominately through immune cell-derived cytokines [11]. M cells comprise a lineage unique to the epithelium overlaying Peyer's patches, large lymphoid follicles in the intestinal mucosa. As all other cell types mentioned here, M cells derive from LGR5$^+$ CBCs; their production is dictated by secreted RANKL, a member of the TNF superfamily [12].

Several stromal cell populations play an important role in regulating the homeostasis of the intestinal niche [13–16]. Underneath the ISCs, stromal cells act as a source of WNT ligands and of the WNT potentiator and Lgr5-ligand R-spondin [17]. Mesenchymal cells found in this location also produce the BMP antagonist Gremlin1, unlike stromal cells and epithelial cells located in the villus, which express BMP agonists [18]. The resulting BMP gradient plays important roles for cell maturation along the crypt-villus-axis: the high BMP levels in the villus inhibit proliferation and drive differentiation [19] (Fig. 1). On top of this, functional zonation occurs over the crypt-villus axis in all of these lineages: For instance,

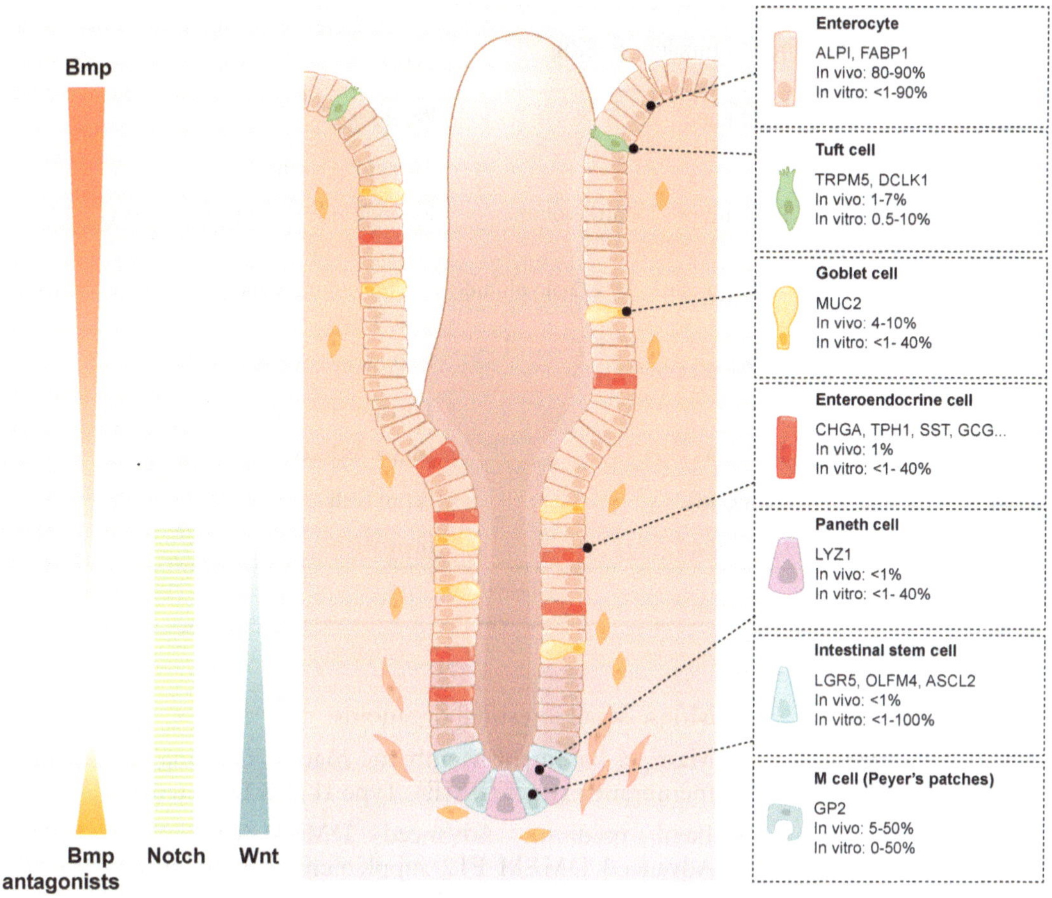

Fig. 1 Diagram depicting morphogen gradients and indicating cell-type abundance in small intestinal epithelium and organoid cultures

BMP induces a switch in expressed hormones in enteroendocrine cells when these cells move up along the crypt-villus axis [20].

Based on the knowledge of in vivo stem cell niche factors and the identification of LGR5 as a marker for CBCs [2], we were able to grow murine intestinal stem cells ex vivo using a simple cocktail of the WNT amplifier R-spondin1, epithelial growth factor (EGF), and the BMP inhibitory protein Noggin [21]. The resulting long-term expanding structures, termed organoids, self-organize into three dimensions and recapitulate the main features of the gut epithelium. Two years later, we reported a similar protocol for growing human intestinal organoids [22]. Here, the addition of WNT agonists appeared to be critical, as human epithelial niche cells do not produce sufficient WNTs to sustain stem cell renewal. Directed differentiation methods enriching for individual lineages have thereafter been published and are described in the current protocol. These differentiation cocktails have been established for all murine lineages. For human organoids, all lineages can be produced currently with the exception of tuft and Paneth cells.

Table 1
Mouse small intestinal growth medium (mENR)

Reagent	Final conc	Provider
AdDF +++	NA	Thermo Scientific
R-spondin3 or	R-spondin3: 1% final volume	R-spondin3: U-Protein Express (UPE)
R-spondin1	R-spondin1: 10% final volume	R-spondin1: conditioned medium made in-house
Noggin	0.5% final volume	U-Protein Express (UPE)
B27 supplement	1×	Thermo Scientific
N-Acetyl-L-cysteine	1.25 mM	Sigma-Aldrich
EGF	50 ng/mL	PeproTech
Primocin (facultative)	0.1 mg/mL	InvivoGen

2 Materials

2.1 Mouse Small Intestinal Organoid Maintenance and Differentiation

1. Mouse small intestinal organoids.
2. Matrigel Basement Membrane Matrix (Corning) or basement membrane extract (BME), Type II (R&D Systems).
3. Basal medium: Advanced DMEM/F12 +++ (AdDF). Advanced DMEM F12 supplemented with 10 mM HEPES, 2 mM GlutaMAX, 100 U/mL penicillin, and 100 μg/mL streptomycin.
4. Suspension cell culture plates (24-well plates and 12-well plates).
5. Mouse small intestinal growth medium (*see* Table 1).

See **Note 1** about in-house R-spondin1-conditioned medium.

2.2 Human Small Intestinal Organoid Maintenance and Differentiation

1. Human small intestinal organoids.
2. Matrigel Basement Membrane Matrix (Corning) or basement membrane extract (BME), Type II (R&D Systems.
3. Basal medium: Advanced DMEM/F12 +++ (AdDF). Advanced DMEM F12 supplemented with 10 mM HEPES, 2 mM GlutaMAX, 100 U/mL penicillin, and 100 μg/mL streptomycin.
4. Suspension cell culture plates (24-well plates and 12-well plates).
5. Human small intestinal stem cell medium (*see* Table 2).

Table 2
Human small intestinal stem cell medium or expansion medium

Reagent	Final conc	Provider
AdDF +++	NA	Thermo Scientific
R-spondin3 or	R-spondin3: 2% final volume	R-spondin3: U-Protein Express (UPE)
R-spondin1	R-spondin1: 20% final volume	R-spondin1: conditioned medium made in-house
Noggin	2% final volume	U-Protein Express (UPE)
B27 supplement without vitamin A	1×	Thermo Scientific
Nicotinamide	10 mM	Sigma-Aldrich
N-Acetyl-L-cysteine	1.25 mM	Sigma-Aldrich
hEGF	50 ng/mL	PeproTech
SB202190 (P38 inhibitor)	3 μM	Sigma-Aldrich
Prostaglandin E2	1 μM	Tocris
A83-01 (TGF-B inhibitor)	500 nM	Tocris
WNT surrogate or	WNT surrogate: 0.15 nM	WNT surrogate: U-Protein Express
WNT-conditioned medium (CM)	WNT CM: 50% final volume	WNT CM: in-house
Primocin (facultative)	0.1 mg/mL	InvivoGen

See **Note 2** about possible variations of the IF medium composition.

2.3 Mouse and Human Media Variations

1. For MEK signaling inhibition, use PD0325901 (Sigma-Aldrich): 1 μM for murine organoids, 100 nM for human organoids (*see* **Note 3**) [10].
2. For WNT signaling inhibitor, use IWP-2 (Stemgent): 5 μM.
3. For WNT signaling activation, use Chir: 5 μM.
4. For Notch signaling inhibition, use DAPT: 10 μM.
5. For BMP signaling activation, use human recombinant BMP2 (50 ng/mL, PeproTech) and BMP4 (50 ng/mL, PeproTech), and withdraw Noggin from the culture medium [20].
6. For M cell induction, use 100 ng/mL RankL (BioLegend) [12, 24].
7. For tuft cell induction, use 10 ng/mL IL-13 (BioLegend) (*see* **Note 4**) [25].

2.4 RNA Isolation and Quantitative Real-Time PCR Assay (qPCR)

1. RNA isolation kit such as RNeasy Mini Kit (Qiagen).
2. cDNA synthesis kit such as GoScript Reverse Transcription System (Promega).
3. qPCR reagents such as SYBR green (Thermo Fisher).
4. Oligos for qPCR (*see* Tables 5 and 6).
5. Appropriate qPCR instrument.

2.5 Immunofluorescence Staining

1. Cell Recovery Solution (Corning).
2. 4% formaldehyde.
3. Triton X-100 (Sigma).
4. Donkey serum (Jackson ImmunoResearch).
5. Primary and secondary antibodies (*see* Table 7).
6. Nuclear dye.
7. Mounting medium (preferably nonhardening), for example, Vectashield (Vector).
8. Glass microscopy slides and coverslips (18 × 18 mm).
9. Syringe filled with vaseline (optional).
10. Clear nail polish.
11. Fluorescent microscope.

3 Methods

Murine intestinal organoids can be expanded in "mENR" medium (*see* Table 1), mimicking a minimal stem cell niche that allows for simultaneous organoid expansion and differentiation. These cultures contain a near-normal ratio of all different intestinal lineages, with the exception of the Peyer's Patch-restricted M cells.

Human intestinal organoids do not display definitive differentiation in standard culture conditions (defined as expansion medium; *see* Tables 2 and 4), as WNT signals are an obligate part of the culture medium. Hence, these cultures purely consist of progenitor cells. In the next sections, we will describe how near-normal ratios of intestinal lineages can be achieved using these human progenitor cell cultures as a starting point. The composition of the expansion medium of progenitor cells is described in the materials section. Alternatively, we describe how the murine and human intestinal organoids can be used for directed differentiation into one of the intestinal lineages. These directed differentiation protocols have been published for all murine lineages and for human organoids with the exception of tuft and Paneth cells. For maintenance and expansion of mouse and human intestinal organoids, refer to previous literature [21, 22, 26].

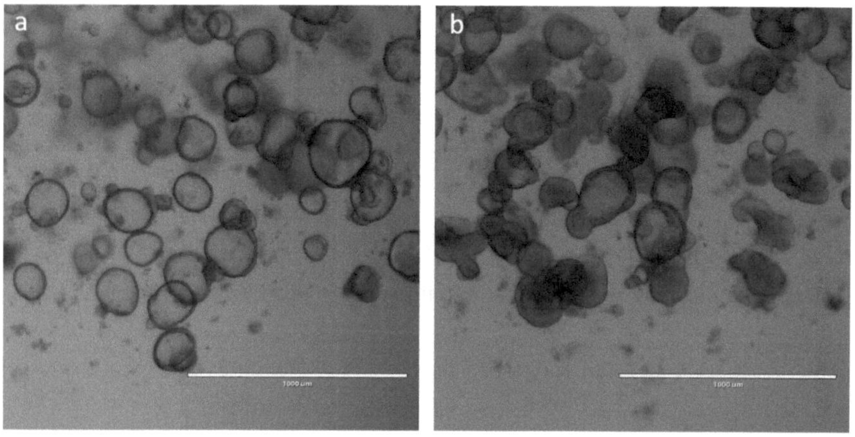

Fig. 2 Brightfield images of human small intestinal organoid cultures (ileum). Scale bar is 1 mm. (**a**) depicts organoids in expansion medium, showing growing and mostly cystic organoids. (**b**) depicts a day-5 enterocyte-differentiated culture, showing dense structures with halted proliferation

3.1 Near-Normal Ratios of Intestinal Lineages in Human Organoids

3.1.1 IF Medium Protocol

1. Organoids should be grown for 2–7 days after splitting in expansion medium (*see* Table 2), until these have reached a size between 100 and 200 μm. It is crucial that organoids remain cystic and do not turn dense (*see* **Note 6**). See Fig. 2a for an example of cystically growing organoids ready for differentiation.
2. Replace expansion medium with IF medium (*see* Table 3).
3. Replace the medium at least every 2 days. Add warm medium.
4. Differentiation is completed after 5 days. Organoid cultures can, however, be expanded long term in IF medium, maintaining both proliferating and differentiating cell types (Table 4).

3.2 Directed Differentiation of Murine and Human Intestinal Organoids

1. Mouse organoids should be grown for 2–7 days after splitting in mENR medium (*see* Table 1), until these have reached 100–200 μm and are successfully budding. The differentiation efficiency can be enhanced by first enriching mouse organoids for stem cells (*see* **Note 7**). Organoids can be taken from different parts of the gut (*see* **Note 8**).
2. Human organoids should be grown for 2–7 days after splitting in expansion medium (*see* Table 2), until these have reached a size between 100 and 200 μm. It is crucial that organoids remain cystic and do not turn dense (*see* **Note 6**). See Fig. 2a for an example of cystically growing organoids ready for differentiation.
3. Prepare the following mouse organoid differentiation media cocktails freshly (*see* subheading 2.3):

(a) Enterocytes: mENR/IWP2

(b) Goblet cells: mENR/DAPT+IWP2

Table 3
Human IF medium [23]

Reagent	Final conc	Provider
AdDF +++	NA	Thermo Scientific
WNT surrogate or	WNT surrogate: 0.15 nM	WNT surrogate: U-Protein Express
WNT-conditioned medium (CM)	WNT CM: 50% final volume	WNT CM: in-house
R-spondin3 or	R-spondin3: 2% final volume	R-spondin3: U-Protein Express (UPE)
R-spondin1	R-spondin1: 20% final volume	R-spondin1: conditioned medium made in-house
Noggin	2% final volume	U-Protein Express (UPE)
B27 supplement	1×	Thermo Scientific
N-Acetyl-L-cysteine	1.25 mM	Sigma-Aldrich
IGF-1	100 ng/mL	BioLegend
FGF2	50 ng/mL	PeproTech
Primocin (facultative)	0.1 mg/mL	InvivoGen

Table 4
Human ENR medium (hENR)

Reagent	Final conc	Provider
AdDF +++	NA	Thermo Scientific
R-spondin3 or	R-spondin3: 1% final volume	R-spondin3: U-Protein Express (UPE)
R-spondin1	R-spondin1: 10% final volume	R-spondin1: conditioned medium made in-house
Noggin	1% final volume	U-Protein Express (UPE)
B27 supplement	1×	Thermo Scientific
N-Acetyl-L-cysteine	1,25 mM	Sigma-Aldrich
EGF	50 ng/mL	PeproTech
Primocin (facultative)	0.1 mg/mL	InvivoGen

(c) EECs: mENR/DAPT+MEKi+IWP2

(d) Tuft cells: mENR + IL13

(e) Paneth cells: mENR/DAPT+CHIR

(f) M cells: mENR + RANKL

4. Prepare the following human organoid differentiation media cocktails freshly (*see* subheading 2.3):
   (a) Enterocytes: hENR/IWP2
   (b) Goblet cells: hENR/DAPT+IWP2
   (c) EECs: hENR/DAPT+MEKi+IWP2
   (d) M cells: hENR + RANKL
5. Prewarm media to 37 °C.
6. Add 0.5 mL of medium per well of a 24-well plate or 1 mL for wells in a 12-well plate.
7. If desirable, differentiation stage of EECs can be adjusted by BMP levels (*see* **Note 9**). For BMP activation use human recombinant BMP2 and BMP4, and remove Noggin from the culture medium. Wells are washed once with warm AdDF +++ for 30 min to prevent residual Noggin from inhibiting BMP signals.
8. Refresh media at least every 2 days.
9. Differentiation is completed after 5 days (in the case of tuft cells, 48-h differentiation suffices). The cells are ready for downstream functional assays. Figure 2b shows an example of enterocyte-differentiated human organoids.

3.3 qPCR-Mediated Validation of Differentiation

1. Remove media from the desired wells and resuspend gel-embedded organoids in the provided lysis buffer. 25 μL of gel-embedded organoids generates approximately 1 μg RNA.
2. Reverse transcribe RNA into cDNA using the GoScript Reverse Transcription System (or similar kits) following the manufacturer's protocol.
3. Run cDNA samples in duplicates or triplicates in a real-time PCR instrument (*see* Tables 5 and 6 for a list of qPCR oligos).

3.4 Immuno-\fluorescence-Mediated Validation of Differentiation

1. Remove culture media from the desired wells. To release organoids from BME or Matrigel (gel), collect organoids using 1 mL cold Cell Recovery Solution, and transfer to a 15 mL tube precoated with 1× PBS + 2% v/v donkey serum (*see* **Notes 10** and **11**).
2. Incubate organoids 30 min at 4 °C on a horizontal rotating platform until the gel is dissociated.
3. Spin down organoids for 5 min at 500× *g* at 4 °C. Wash organoids using 1 mL of 1 × PBS, spin down again, and remove supernatant. Repeat wash once more.

Table 5
Oligos for qPCR (mouse)

Cell type	Target name	Sequence fw	Sequence rev
Stem cells	Lgr5	ACCCGCCAGTCTCCTACATC	GCATCTAGGCGCAGGGATTG
Enterocytes	Alpi	GGCCATCTAGGACCGGAGA	TGTCCACGTTGTATGTCTTGG
Paneth cells	Lyz1	GGAATGGATGGCTACCGTGG	CATGCCACCCATGCTCGAAT
Goblet cells	Muc2	ATGCCCACCTCCTCAAAGAC	GTAGTTTCCGTTGGAACAGTGAA
Tuft cells	Dclk1	TCCACCGGAATTGAACTCGG	GGGAGCGAACAGTCTCAGA
M cells	Gp2	CCTGCGTTCTGACACTG	GCCGTGCAGGTTATCA
EECs (general)	Chga	CAGCTCGTCCACTCTTTCCG	CCTCTCGTCTCCTTGGAGGG
EECs (enterochromaffin cells)	Tph1	ACGTCGAAAGTATTTTGCGGA	ACGGTTCCCCAGGTCTTAATC
EECs (X cells)	Ghrl	CTGAGCTCCTGACAGCTTGA	ACCCAGAGGACAGAGGACAA
EECs (L cells)	Gcg	CTTCCCAGAAGAAGTCGCCA	GTGACTGGCACGAGATGTTG
EECs (L cells)	Nts	TGCTGACCATCTTCCAGCTC	GAATGTAGGGCCTTCTGGGT
EECs (K cells)	Gip	AACTGTTGGCTAGGGGACAC	TGATGAAAGTCCCCTCTGCG
EECs (D cells)	Sst	GACCTGCGACTAGACTGACC	CCAGTTCCTGTTTCCCGGTG

4. Add 1 mL 4% formaldehyde to fix organoids for 1 h at room temperature (RT).
5. Spin down organoids for 5 min at 500× *g* at RT. Wash organoids using 1 mL of 1× PBS, spin down again, and remove supernatant.
6. Add 1 mL blocking solution: 2% v/v donkey serum in 1× PBS and incubate for 15 min at RT. Afterward, spin organoids down as before and remove supernatant.

Table 6
Oligos for qPCR (human)

Cell type	Target name	Sequence fw	Sequence rev
Stem cells	LGR5	CTCCCAGGTCTGGTGTG TTG	GAGGTCTAGGTAGGAGG TGAAG
Enterocytes	ALPI	TGAGGGTGTGGCTTACCAG	GATGGACGTGTAGGC TTTGCT
Paneth cells	LYZ	CTTGTCCTCCTTTCTG TTACGG	CCCCTGTAGCCATCCA TTCC
Goblet cells	MUC2	GAGGGCAGAACCCGAAACC	GGCGAAGTTGTAG TCGCAGAG
Tuft cells	TRPM5	TTGCTGCCCTAG TGAACCAG	GCACGATGTCC TCCCAAGAG
M cells	GP2	AATCAAACCCATGCCATC TACAA	CACACTGACG TTCAGGGAACT
Enteroendocrine cells	CHGA	CAGCTCGTCCACTC TTTCCG	CCTCTCGTCTCC TTGGAGGG
EECs (enterochromaffin cells)	THP1	ACGTCGAAAGTA TTTTGCGGA	ACGGTTCCCCAGGTC TTAATC
EECs (MX cells)	GHRL	CAGGGGTTCAG TACCAGCAG	CCTCTTTGGCCTC TTCCCAG
EECs (MX cells)	MLN	ATGGTATCCCGTAAGGCTG TG	CTGGAGTTCGCCATAGG TGAA
EECs (L cells)	GCG	ACATTGCCAAACGTCACGA TG	TCTGCGGCCAAGTTC TTCAA
EECs (L cells)	NTS	TGCTTTAGATGGCTTTAGC TTGG	TTCCTGGATTAAC TCCCAGTGT
EECs (K cells)	GIP	GGATCTCATGCTAAGG TGAGC	GTCTTGTTGGTGAATC TTGTCCA
EECs (D cells)	SST	ACCCAACCAGACGGAGAA TGA	GCCGGGTTTGAG TTAGCAGA

7. Permeabilize organoids for 15 min - 1 h at RT using 0.5% Triton X-100 in 1× PBS. If staining nuclear proteins longer permeabilization is necessary. Remove supernatant after spinning down.
8. Add primary antibody at indicated concentrations (*see* Table 7) in 100 μL 2% donkey serum in 1× PBS. Incubate organoids ON on a horizontal rotating platform at 4 °C.

Table 7
Primary antibodies for immunofluorescence staining

Cell type	Target name	Species	Dilution	Supplier
Stem cells	OLFM4, on paraffin sections (*see* **Note 5**)	Rabbit	1:5000	Thermo Fisher Scientific
Enterocytes	APOA1	Rabbit	1:100	Thermo Fisher Scientific
Paneth cells	LYZ	Rabbit	1:500	Dako
Goblet cells	MUC2	Mouse	1:100	Abcam
Tuft cells	AVIL	Rabbit	1:100	Sigma
Enteroendocrine cells (EECs)	CHGA	Rabbit	1:100	Labnet
EECs (enterochromaffin cells)	5-HT	Rat	1:100	NovusBio
EECs (MX cells)	GHRL	Goat	1:100	Santa Cruz
EECs (MX cells)	MLN	Rabbit	1:100	Sigma
EECs (L cells)	GLP-1	Goat	1:100	Santa Cruz
EECs (L cells)	NTS	Rabbit	1:100	Santa Cruz
EECs (K cells)	GIP	Rabbit	1:500	Abcam
EECs (D cells)	SST	Goat	1:100	Santa Cruz

9. Wash organoids three times using 1 mL of 1× PBS.
10. Add secondary antibody, usually at a concentration of 1:1000 and DAPI (2 μg/mL) in 100 μL 2% donkey serum in 1× PBS for 1–2 h in the dark at RT in a rotating platform.
11. Spin down and wash three times. In the last step, carefully remove as much supernatant as possible.
12. To prevent the coverslip from disrupting organoids, apply four small amounts of vaseline where the coverslip will be placed. To apply vaseline, fill a syringe with the vaseline and attach a 200 mL pipette tip using Parafilm (*see* Fig. 3).
13. Cut the front end of a 200 μL pipette tip. Using this tip, resuspend organoids in 30 μL mounting solution and apply to the glass microscope slide.
14. Carefully place coverslip on top, and do not press.
15. Seal all coverslip edges using clear nail polish.
16. Image.

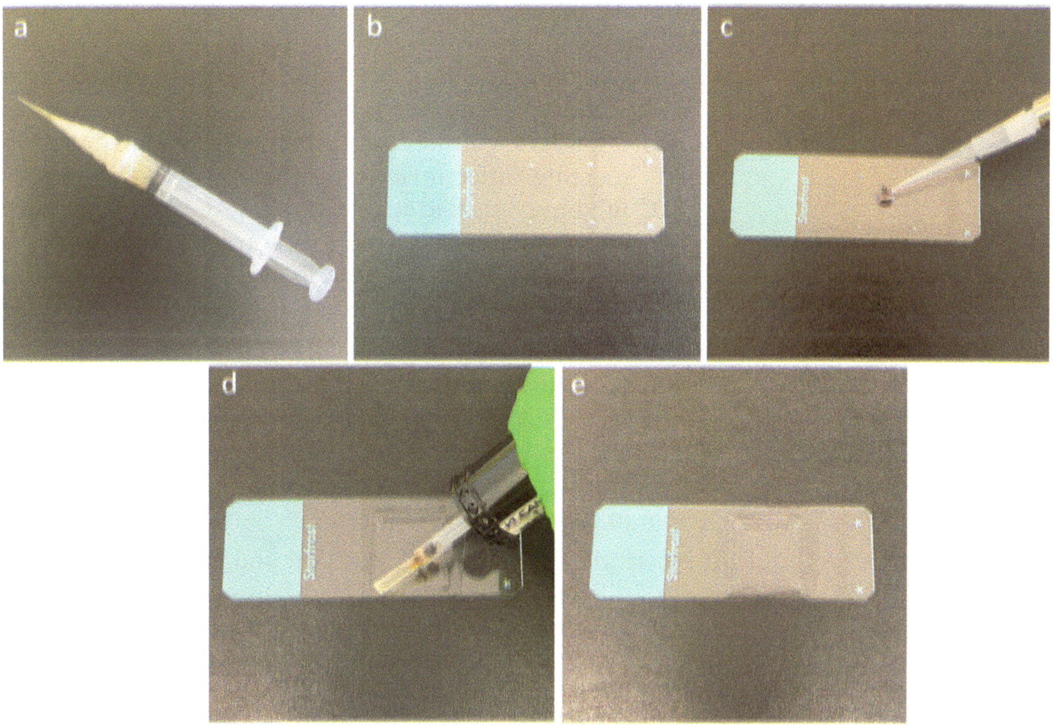

Fig. 3 Mounting procedure for immunofluorescence protocol

4 Notes

1. For a detailed protocol on WNT- and R-spondin-conditioned media preparation, we refer the reader to Pleguezuelos-Manzano et al. (2020) [26].
2. In the original paper from the Sato lab [23], recombinant gastrin is used in the culture medium. We did not observe substantial difference in the efficiency of differentiation in the absence of gastrin. Of note, this original protocol uses conditioned medium of WNT bound to a serum carrier protein called afamin.
3. MEK inhibition is toxic to cells at high concentrations. Some organoid lines are more sensitive than others. If toxicity is an issue, the inhibitor can be titrated to a lower concentration.
4. A protocol for tuft cell and Paneth cell enrichment in human organoids has not been published yet.
5. OLFM4 presents a broader expression than LGR5, including early progenitor cells. LGR5 remains challenging to stain due to low expression levels of the protein.

6. Murine organoid cultured in default medium (mENR) displays crypt-like compartments with multipotent stem cells and Paneth cells and differentiated compartments. As these organoids contain differentiated cells that cannot be specified to particular lineages, cultures can first be enriched for stem cells before initiating differentiation protocols. A combination of HDAC inhibitor valproic acid (VPA) and the WNT activator CHIR (GSK3 antagonist) allows the propagation of organoids containing nearly purely of stem cells [9].
7. It is important to consider from which part of the small intestine organoid cultures are established. The major biological differences in the epithelium along this axis are epigenetically imprinted and (largely) maintained in long-term organoid cultures. For example, enteroendocrine cells hormone expression varies greatly along this axis, and these differences remain present in corresponding cultures.
8. It is essential that organoids are cystic and growing well at the start of differentiation, as this is indicative of a pool of immature progenitor/stem cells that retain multipotency. If organoids do appear dense and display reduced growth, premature differentiation might have occurred. Lack of remaining undifferentiated cells will reduce the ability to further direct differentiation to specific lineages.
9. Enteroendocrine cells express different hormones along the crypt-villus axis, dependent on a BMP signaling morphogen gradient. Activation of BMP signaling in organoids induces the expression of villus-enriched hormones, including neurotensin and secretin, while repressing crypt-enriched hormones, including GLP-1 (encoded by *GCG)* and substance P (encoded by *TAC1)*.
10. During the immunofluorescence procedure, it could be of interest to preserve the extracellular matrix for some applications. If this is the case, PBS can be used instead of Cell Recovery Solution. Resuspend the harvested organoids in PBS, spin down once, and remove PBS almost completely, resuspending organoids with remaining PBS, and add formaldehyde.
11. Once organoids are removed from the gel, they can adhere to the sides of the tube and the pipette tips. To avoid loss of material, it is recommendable to precoat the tubes and tip with PBS + 2% donkey serum or PBS + 0.1% bovine serum antigen. Low binding tubes can also be used.

Acknowledgments

The development of the methods was supported by Netherlands Organ-on-Chip Initiative (024.003.001) from the Netherlands Organisation for Scientific Research (NWO) funded by the Ministry of Education, Culture and Science of the government of the Netherlands (A.M.S. and H.C.), the Oncode Institute (partly financed by the Dutch Cancer Society), the European Research Council under ERC Advanced Grant (Guthormones; nr 101020405) (J.B. and H.C.), and NETRF/Petersen Accelerator (J.B. and H.C.).

References

1. Cheng H, Leblond CP (1974) Origin, differentiation and renewal of the four main epithelial cell types in the mouse small intestine. V. Unitarian Theory of the origin of the four epithelial cell types. Am J Anat 141(4): 537–561. https://doi.org/10.1002/aja.1001410407
2. Barker N et al (2007) Identification of stem cells in small intestine and colon by marker gene Lgr5. Nature 449(7165):1003–1007. https://doi.org/10.1038/nature06196
3. Batlle E et al., β-catenin and TCF mediate cell positioning. p. 13
4. Sato T et al (2011) Paneth cells constitute the niche for Lgr5 stem cells in intestinal crypts. Nature 469(7330):415–418. https://doi.org/10.1038/nature09637
5. van Es JH et al (2005) Notch/gamma-secretase inhibition turns proliferative cells in intestinal crypts and adenomas into goblet cells. Nature 435(7044):959–963. https://doi.org/10.1038/nature03659
6. VanDussen KL et al (2012) Notch signaling modulates proliferation and differentiation of intestinal crypt base columnar stem cells. Development 139(3):488–497. https://doi.org/10.1242/dev.070763
7. Bastide P et al (2007) Sox9 regulates cell proliferation and is required for Paneth cell differentiation in the intestinal epithelium. J Cell Biol 178(4):635–648. https://doi.org/10.1083/jcb.200704152
8. Mori-Akiyama Y et al (2007) SOX9 is required for the differentiation of paneth cells in the intestinal epithelium. Gastroenterology 133(2):539–546. https://doi.org/10.1053/j.gastro.2007.05.020
9. Yin X, Farin HF, van Es JH, Clevers H, Langer R, Karp JM (2014) Niche-independent high-purity cultures of Lgr5+ intestinal stem cells and their progeny. Nat Methods 11(1): 106–112. https://doi.org/10.1038/nmeth.2737
10. Basak O, Beumer J, Wiebrands K, Seno H, van Oudenaarden A, Clevers H (2017) Induced quiescence of Lgr5+ stem cells in intestinal organoids enables differentiation of hormone-producing enteroendocrine cells. Cell Stem Cell 20(2):177–190.e4. https://doi.org/10.1016/j.stem.2016.11.001
11. Gerbe F et al (2016) Intestinal epithelial tuft cells initiate type 2 mucosal immunity to helminth parasites. Nature 529(7585):226–230. https://doi.org/10.1038/nature16527
12. de Lau W et al (2012) Peyer's patch M cells derived from Lgr5+ stem cells require SpiB and are induced by RankL in cultured "Miniguts". Mol Cell Biol 32(18):3639–3647. https://doi.org/10.1128/MCB.00434-12
13. Farin HF, Van Es JH, Clevers H (2012) Redundant sources of Wnt regulate intestinal stem cells and promote formation of Paneth cells. Gastroenterology 143(6):1518–1529.e7. https://doi.org/10.1053/j.gastro.2012.08.031
14. Valenta T et al (2016) Wnt ligands secreted by subepithelial mesenchymal cells are essential for the survival of intestinal stem cells and gut homeostasis. Cell Rep 15(5):911–918. https://doi.org/10.1016/j.celrep.2016.03.088
15. Stzepourginski I et al (2017) CD34+ mesenchymal cells are a major component of the intestinal stem cells niche at homeostasis and after injury. Proc Natl Acad Sci U S A 114(4): E506–E513. https://doi.org/10.1073/pnas.1620059114
16. Aoki R et al (2016) Foxl1-expressing mesenchymal cells constitute the intestinal stem cell niche. Cell Mol Gastroenterol Hepatol 2(2):

175–188. https://doi.org/10.1016/j.jcmgh.2015.12.004

17. de Lau W et al (2011) Lgr5 homologues associate with Wnt receptors and mediate R-spondin signalling. Nature 476(7360): 293–297. https://doi.org/10.1038/nature10337
18. McCarthy N et al (2020) Distinct mesenchymal cell populations generate the essential intestinal BMP signaling gradient. Cell Stem Cell 26(3):391–402.e5. https://doi.org/10.1016/j.stem.2020.01.008
19. Haramis A-PG et al (2004) De novo crypt formation and juvenile polyposis on BMP inhibition in mouse intestine. Science 303(5664): 1684–1686. https://doi.org/10.1126/science.1093587
20. Beumer J et al (2018) Enteroendocrine cells switch hormone expression along the crypt-to-villus BMP signalling gradient. Nat Cell Biol 20(8):909–916. https://doi.org/10.1038/s41556-018-0143-y
21. Sato T et al (2009) Single Lgr5 stem cells build crypt-villus structures in vitro without a mesenchymal niche. Nature 459(7244):262–265. https://doi.org/10.1038/nature07935
22. Sato T et al (2011) Long-term expansion of epithelial organoids from human colon, adenoma, adenocarcinoma, and Barrett's epithelium. Gastroenterology 141(5): 1762–1772. https://doi.org/10.1053/j.gastro.2011.07.050
23. Fujii M et al (2018) Human intestinal organoids maintain self-renewal capacity and cellular diversity in niche-inspired culture condition. Cell Stem Cell 23(6):787–793.e6. https://doi.org/10.1016/j.stem.2018.11.016
24. Rouch JD et al (2016) Development of functional microfold (M) cells from intestinal stem cells in primary human enteroids. PLoS One 11(1):e0148216. https://doi.org/10.1371/journal.pone.0148216
25. Howitt MR et al (2016) Tuft cells, taste-chemosensory cells, orchestrate parasite type 2 immunity in the gut. Science 351(6279): 1329–1333. https://doi.org/10.1126/science.aaf1648
26. Pleguezuelos-Manzano C, Puschhof J, van den Brink S, Geurts V, Beumer J, Clevers H (2020) Establishment and culture of human intestinal organoids derived from adult stem cells. Curr Protoc Immunol 130(1):e106. https://doi.org/10.1002/cpim.106

Chapter 10

Modeling Notch Activity and Lineage Decisions Using Intestinal Organoids

Yifan Qiu, Sabrina K. Phanor, Subin Pyo, and Chia-Wei Cheng

Abstract

Organoid cultures have been developed to model intestinal stem cell (ISC) function in self-renewal and differentiation. Upon differentiation, the first fate decision for ISC and early progenitors to make is between secretory (Paneth cell, goblet cell, enteroendocrine cell, or tuft cell) and absorptive (enterocyte and M cell) lineages. Using genetic and pharmacological approaches, in vivo studies in the past decade have revealed that Notch signaling functions as a binary switch for the secretory vs. absorptive lineage decision in adult intestine. Recent breakthroughs in organoid-based assays enable real-time observation of smaller-scale and higher-throughput experiments in vitro, which have begun contributing to new understandings of mechanistic principles underlying intestinal differentiation. In this chapter, we summarize the in vivo and in vitro tools for modulating Notch signaling and assess its impact on intestinal cell fate. We also provide example protocols of how to use intestinal organoids as functional assays to study Notch activity in intestinal lineage decisions.

Key words Notch, Lineage, Organoids

1 Introduction

In the stem cell zone model, the cellular hierarchy of the intestinal epithelium begins at the crypt-base intestinal stem cells (ISCs) [1]. ISCs are able to self-renew or differentiate into secretory (Paneth cell, goblet cell, enteroendocrine cell, or tuft cell) or absorptive (enterocyte and M cell) lineages. The lineage differentiation occurs through a well-organized migration along the crypt-villus axis, which is coordinated with stepwise changes of Notch activity (*see* Fig. 1). Notch activation in ISC begins with the binding of the ligands expressed by Paneth cells (e.g., Dll1 and Dll4) to the receptors on the ISC surface (e.g., Notch1 and Notch2), which triggers two proteolytic cleavage events at the Notch receptor, first

Yifan Qiu and Sabrina K. Phanor contributed equally to this work.

Paloma Ordóñez-Morán (ed.), *Intestinal Differentiated Cells: Methods and Protocols*, Methods in Molecular Biology, vol. 2650, https://doi.org/10.1007/978-1-0716-3076-1_10,

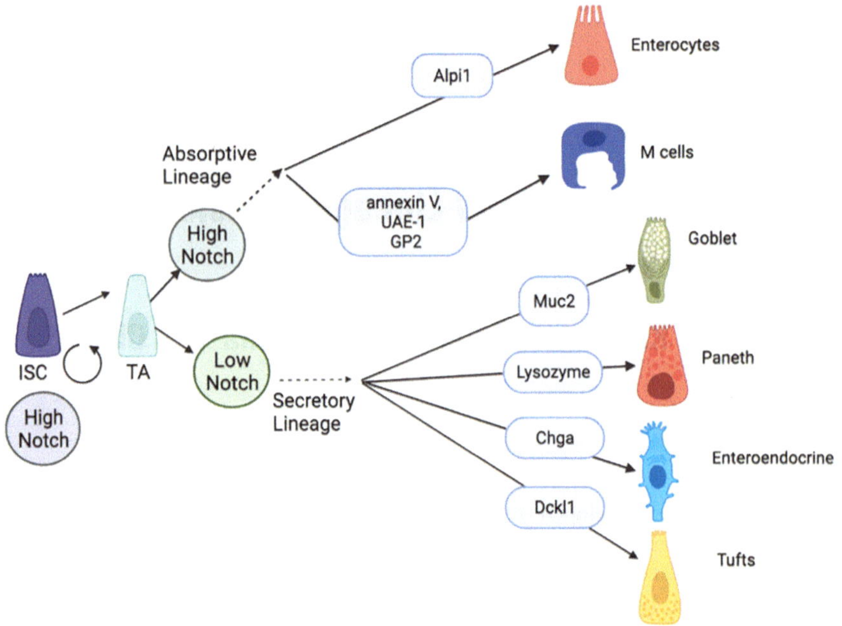

Fig. 1 Notch activity in intestinal stem cell self-renewal and differentiation. Stem cells rely on high Notch activity to maintain the self-renewal capacity, whereas transit-amplifying (TA) cells based on Notch activity make lineage decisions. TA cells with high Notch generate absorptive lineage: enterocytes and M cells. TA cells with low Notch differentiate into the secretory lineage consisting of goblet, Paneth, enteroendocrine, and tuft cells. Each cell population expresses unique genes and proteins indicated as the cell identity marker in the figure in both lineages

by ADAM metalloproteases and then by γ-secretase. The cleaved intracellular domain of the Notch receptor (NICD) then translocates to the nucleus and engages RBPJ (recombination signal binding protein for immunoglobulin kappa J region) to activate transcription of Notch target genes, including the HES1 transcription factor that promotes self-renewal of ISCs and determines lineage decisions of transit-amplifying (TA) progenitor cells (*see* Fig. 2).

The current understanding regarding Notch-signaling-mediated intestinal cell fate decisions mostly comes from perturbation studies targeting Notch ligands, receptors, and downstream effectors, respectively. Inhibition of Notch signaling by blockade of Notch ligands (e.g., Dll1 and Dll4) [2], deletion of Notch receptors (e.g., Notch1 and Notch2) [3], prevention of Notch receptor cleavage (e.g., ADAM10, γ-secretase) [4, 5], or ablation of transcriptional effector RBPJ [6] leads to a consistent reduction of ISCs and absorptive lineage cells and expansion of secretory cell populations such as lysozyme+ or MMP7+ Paneth and MUC2+ goblet cells and ChgA+ endocrine cells. Notably, the increase of mucin-secreting goblet cells based on Alcian blue staining is highlighted across all the studies. On the other hand, forced activation of Notch using the NICD-overexpression mouse model results in a

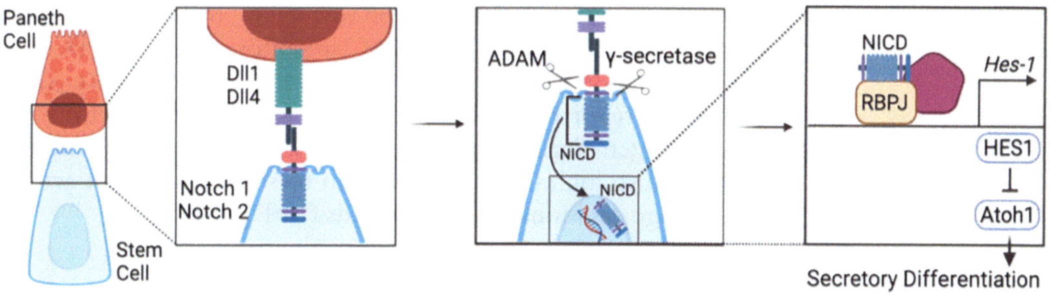

Fig. 2 Key steps of Notch activation. Ligands (Dll1, Dll4) from Paneth cells bind to receptors (Notch1, Notch2) on intestinal stem cells. ADAM-10 and γ-secretase cleave Notch receptors, separating the Notch intracellular domain (NICD) from the extracellular domain. NICD then travels to the nucleus of the stem cell and binds to the RBPJ complex. The binding of the NICD onto RBPJ then recruits other coactivators to initiate the transcription of the Notch target gene, such as Hes1. HES1 subsequently inhibits the transcription of Atoh1 that otherwise promotes secretory differentiation

reduction of Alcian blue+ goblet cell and an upregulation of cell proliferation [3]. In line with these findings, Notch1 lineage tracing and Hes1-GFP reporter mouse models demonstrate exclusive Notch activity in crypt stem cells and absorptive progenitors [7]. Many of the genetic and pharmacological tools used for manipulating or monitoring Notch signaling have not yet been examined in the setting of organoid-based assays.

Given that organoids are developed for modeling ISC function, a mature organoid should display crypt-villus structures and contain all major differentiated cell types that can be generated by the stem cells. Indeed, recent studies have reported that Paneth cells, goblet cells, and enterocytes can be generated in primary intestinal organoids with standard ENR media culture without signaling induction or genetic modification. Specifically, the presence of Paneth cells in organoids has been demonstrated based on lysozyme expression using qRT-PCR and immunofluorescence (IF) [8, 9]. The expression of goblet cells in the intestinal organoids was demonstrated by the detection of Muc2+ cells using IF [10]. Enterocytes were also found in regular intestinal organoid cultures based on mRNA and protein levels of Alpi [11]. In contrast, the generation of enteroendocrine (EEC), tuft cells, and M cells in organoids are more evident under the induced conditions. For example, inhibitions of Wnt, Notch, and Mek are necessary to promote the EEC marker gene ChgA+ expression [8]. Additional cytokine treatments are required to induce the generation of Dckl1+ tuft cells, independent of the activation of Atoh1 [12]. Induction of nuclear factor kappa-B ligand (RANKL) is essential to induce the generation of M cells in organoids [13].

We recently reported that Lgr5+ ISC specifically expresses HMGCS2 to produce ketone body metabolite, which reinforces Notch signaling by inhibiting class I HDACs [14]. We showed that loss of HMGCS2 compromises Notch signaling leading to secretory expansion, which can be rescued by treatments of either ketone body metabolite or HDAC inhibitors or by NICD overexpression in organoids [14]. Here, we will describe how we (i) use Hes1-GFP reporter in combination with γ-secretase inhibitor treatment, (ii) genetically engineer the NICD overexpression, and (iii) perform histological Alcian blue stain to study the Notch activity and lineage decisions using organoids (*see* Fig. 3).

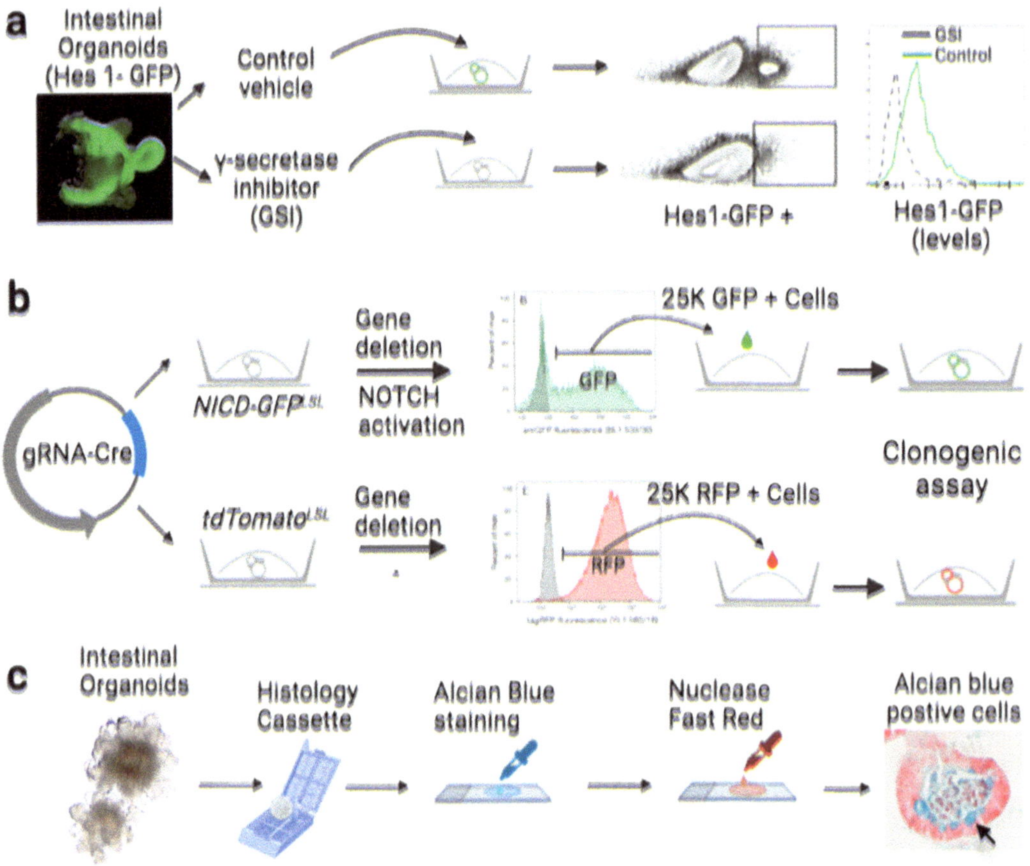

Fig. 3 Applications of intestinal organoids in studying Notch activity and lineage decisions. (**a**) Related to Method 3.1. Monitoring Notch activity in organoids. Treat intestinal organoids cultured from intestinal crypts of Hes1-GFP reporter mice with γ-secretase inhibitor or vehicle (control) to calibrate the relative Notch activity of experiment samples. Quantify Hes1-GFP at single-cell level based on positivity and intensity using flow cytometry analysis. (**b**) Related to Method 3.2. Schematic of rescuing experiment using genetically engineered organoid cells. Induce target gene (e.g., HMGCS2) deletion and Cre-mediated gene expression simultaneously using gRNA-Cre vector and LoxP-Stop-LoxP (LSL) organoids. Organoids cultured from intestinal crypts of NICD-GFPLSL and tdTomatoLSL mice are both transfected. Equal numbers of transduced cells are sorted and used to perform the functional clonogenicity assay. Compare the outcome of NICD-GFP+ versus tdTomato+ cells to determine the rescuing effects of NICD on target gene deletion in organoids. (**c**) Related to Method 3.3. Applications of organoid histology. Intestinal organoids are collected and processed for histology through fixation in PFA (paraformaldehyde) and HistoGel embedding. Processing, sectioning, and deparaffinization as regular histology protocols. Perform Alcian blue staining and counterstains with fast red to visualize mucin-secreting (Alcian blue+) goblet cells in organoids

2 Materials

2.1 Assess Notch Signaling Using Hes1-GFP Reporter and γ-Secretase Inhibitor Treatment

1. WRN (Wnt3a, R-spondin-3, and Noggin)-conditioned medium.
2. EN medium (specific growth factor cocktail consisting of EGF and BMP antagonist, Noggin).
3. γ-Secretase inhibitor MK-0752 (Cayman Chemical Company, #471905-41-6).
4. LoBind microcentrifuge tubes (Eppendorf, #022431081).
5. 7-Aminoactinomycin D (7-AAD) (Thermo Fisher, #A1310).
6. S-MEM (Life Technologies, #11380-037).
7. TrypLE Express (Gibco, #12604013).
8. 40 μm cell strainer.
9. FACS tube (BD#352235).

2.2 Genetic Engineering for Gene Deletion and NICD Overexpression

1. WRN (Wnt3a, R-spondin-3, and Noggin)-conditioned medium.
2. Cre carrying vector construct (VB180615–1103gue).
3. Lipofectamine™ Transfection Reagent (#18324012).
4. ENRY medium (ENR media plus Y-27632, a ROCK inhibitor).
5. ENR medium (specific growth factor cocktail consisted of EGF, the BMP antagonist Noggin, and R-spondin).
6. Matrigel (Corning, #356231).

2.3 Histological Alcian Blue Staining

1. HistoGel (HG-4000-012).
2. PBS.
3. Alcian blue staining solution (Alfa Aesar, Cat #50596508).
4. Nuclear fast red staining solution (Millipore Sigma, Cat # 6409-77-4).
5. Xylene.
6. 6.100% ethanol.
7. 80% ethanol.
8. Distilled water.
9. Tris-buffered saline, 0.1% Tween® 20 detergent (TBST).
10. Hydrophobic Barrier PAP Pen.
11. 4% PFA.
12. Cryomolds.
13. Deionized water.
14. Surgical blade.

15. 10% formalin.
16. Tissue cassettes.
17. 0.5% eosin.
18. 70% ethanol.

3 Methods

3.1 Assess Notch Activity Using Hes1-GFP Reporter and γ-Secretase Inhibitor Treatment

1. Perform primary organoid culture using intestinal crypts from Hes1-GFP reporter mice as previously described [15, 16].
2. Expand primary organoids in WRN-conditioned medium [17].
3. Hes1-GFP organoids were passaged 2–3 times before cryopreserving the organoids [18].
4. Revive the organoids with WRN-conditioned media for 3 days.
5. To reduce Notch-independent Wnt/beta-catenin-mediated Hes1 expression, passage revived organoids in EN medium (*see* **Note 1**) [19].
6. Treat organoids with γ-secretase inhibitor (10 μM), compounds of interest (e.g., beta-hydroxybutyrate, HDAC inhibitors), and vehicle (e.g., DMSO, water), respectively. Use 1000 organoids for each treatment/condition.
7. Use fluorescent microscope to check the Hes1-GFP expression after treatment.
8. Upon detection of Hes1-GFP reduction, collect organoids for flow cytometry analysis (*see* **Note 2**).
9. To break up the Matrigel dome and organoids, pipette up and down approximately 20 times per well.
10. Collect organoids with the EN media and Matrigel into Eppendorf tubes.
11. Centrifuge the collected organoids with culture medium and Matrigel (300*g* × 5 min).
12. Remove the supernatant.
13. Treat organoid pellet with dissociated TrypLE Express (incubate 32 °C, 5 min) (*see* **Note 3**).
14. Stop the dissociation by adding 5× volume of cold S-MEM and immediately place on ice.
15. Spin down the dissociated organoids (300*g* × 5 min at 4 °C).
16. Remove the supernatant and resuspend the cell pellet with 0.5 mL S-MEM containing 7AAD (1:5000).
17. Filter the cells using a 40 μm cell strainer.
18. Transfer the filtered cells and S-MEM plus 7AAD into FACS tube.

19. Use forward scatter (FSC) and side scatter (SSC) to gate for singlets.
20. Use 7AAD negativity to find viable cells.
21. Use Hes1-GFP intensity of vehicle control and GSI-treated organoids as calibrator to determine the relative Notch activity of organoids.

3.2 Genetic Engineering for Gene Deletion and NICD Overexpression

1. Perform primary organoid culture using intestinal crypts from LSL-tdTomato (JAX# 007909) and NICD-LSL-eGFP (JAX# 008159).
2. Expand primary organoids in WRN-conditioned medium [17].
3. Design and generate vector construct carrying Cre (VB180615–1103gue), and guide RNA targeting gene of interest (e.g., HMGCS2) (*see* **Note 4**).
4. Dissociate and transfect the LSL-tdTomato and NICD-LSL-eGFP organoids using lipophilic reagents [20].
5. Use fluorescent microscope to check the RFP and GFP expression after treatment.
6. Upon detection of RFP and GFP, collect organoids for flow cytometry and cell sorting.
7. Prepare sample for sorting as described in Subheading 3.1 (**steps 9–20**).
8. Sort equal numbers of viable RFP+ and GFP+ into ENRY medium (i.e., ENR media plus Y-27632, a ROCK inhibitor).
9. Spin down the sorted cells and remove the supernatant.
10. Resuspend 25K cells in 10 μL ENRY medium.
11. Plate the 25K cells onto unsolidified Matrigel dome and incubate 15 min at 37 °C.
12. Carefully add ENR medium into the well. Avoid touching the Matrigel dome.
13. Monitor organoid formation and quantify number and size of RFP+ and GFP+ organoids.
14. Determine if NICD overexpressing (GFP) rescue the effects of gene deletion (RFP) based on number and size of the organoids.

3.3 Histological Alcian Blue Staining (See Note 5)

Prepare organoids for histology sections:

1. Perform primary organoid culture using intestinal crypts from mice as previously described [15, 16], or perform treatments or gene editing as described in Subheadings 3.1 and 3.2.
2. Using bright-field microscope to observe organoid growth and budding.

3. Upon detecting the maturation of organoids (e.g., clear luminal space or secondary structure), collect organoids to process for histology (*see* **Note 6**).
4. Warm HistoGel in 65 °C in water bath for 1–2 h.
5. Remove the media from the wells without touching the Matrigel.
6. Wash the wells twice with 400 μL of PBS.
7. Add 400 μL of 4% PFA per well and incubate for 2 h at room temperature.
8. Place cryomolds on ice and add 150 μL warm HistoGel into one cryomold.
9. Place cryomold back on ice for 5–10 min.
10. Make sure to bring HistoGel back into the water bath.
11. Remove 4% PFA from all wells.
12. Wash the wells 3 times with 400 μL of deionized water.
13. Scrape off the Matrigel with 0.22 surgical blade and transfer into cryomold that has the HistoGel base.
14. Add 150 μL of warm HistoGel on top of the Matrigel in the cryomold to form a sandwich.
15. Cool on ice for 10–15 min.
16. Use a 0.11 surgical blade to release the gel from the mold.
17. Flip the HistoGel block into the tissue cassette.
18. Place the tissue cassette in 10% formalin for 16–20 h.
19. Transfer the cassette to container with 0.5% eosin in 70% ethanol for 1–2 days.
20. After 2 days, store the cassette in 70% ethanol until it is sent to histopathology.
21. To deparaffinize, prepare two xylene chambers.
22. Place organoids in one chamber of xylene for 5 min.
23. Switch to second xylene chamber for another 5 min.
24. Tap out excess xylene.
25. Move slides to 100% ethanol for 6 min and remove excess ethanol.
26. Move to 80% ethanol for 2 min and tap out excess.
27. Place in deionized water for 2 min.
28. Submerge slides in TBST.

Alcian blue staining:

29. Hydrate slides in distilled water.
30. Stain in the Alcian blue solution 30 min.
31. Wash in distilled water twice 5 min each.
32. Counterstain with nuclear fast red stain 1 min.

4 Notes

1. Wnt signaling pathway controls the cell fate decision cooperatively with Notch signaling. Low Notch along with high Wnt contributes to an increase in Paneth secretory cell proliferation. In contrast, low Notch signaling followed by low Wnt contributes to an increase of other secretory cells such as goblet, tuft, and enteroendocrine cells. High Notch and low Wnt allow for more absorptive cell proliferation, such as enterocytes and M cells. Studies using tumor spheroids or high-Wnt culturing systems should take the Wnt-Notch cross talk into account.
2. When preparing organoid samples for FACS analysis, be aware of the potential loss of cells (~30%) during washing and transferring. Calculate the minimal number of organoids required for each assay accordingly.
3. Organoid pellets may be hard to resuspend after TrypLE treatments. Use filter tips to prevent the clogging of pipetman by cell clumps.
4. Mammalian CRISPR gene editing vector carrying Cre (VB180615–1103gue) was designed with a guide RNA targeting HMGCS2 which can be replaced by another gRNA for the gene of interest.
5. Organoids sections prepared for Alcian blue staining can also be used for immunofluorescence (IF) and immunohistochemistry (IHC) analyses to demonstrate gene expression of differentiated cell populations.
6. When preparing organoids for histology, use the blunt-end pipette tips for transferring organoids to better preserve the morphology, especially the budding structure, of organoids.

References

1. Barker N, van Oudenaarden A, Clevers H (2012) Identifying the stem cell of the intestinal crypt: strategies and pitfalls. Cell Stem Cell 11(4):452–460. https://doi.org/10.1016/j.stem.2012.09.009
2. Pellegrinet L, Rodilla V, Liu Z, Chen S, Koch U, Espinosa L, Kaestner KH, Kopan R, Lewis J, Radtke F (2011) Dll1- and dll4-mediated notch signaling are required for homeostasis of intestinal stem cells. Gastroenterology 140(4):1230–1240. e1231–1237. https://doi.org/10.1053/j.gastro.2011.01.005
3. Demitrack ES, Gifford GB, Keeley TM, Carulli AJ, VanDussen KL, Thomas D, Giordano TJ, Liu Z, Kopan R, Samuelson LC (2015) Notch signaling regulates gastric antral LGR5 stem cell function. EMBO J 34(20):2522–2536. https://doi.org/10.15252/embj.201490583
4. Tsai YH, VanDussen KL, Sawey ET, Wade AW, Kasper C, Rakshit S, Bhatt RG, Stoeck A, Maillard I, Crawford HC, Samuelson LC, Dempsey PJ (2014) ADAM10 regulates Notch function in intestinal stem cells of mice. Gastroenterology 147(4):822–834 e813. https://doi.org/10.1053/j.gastro.2014.07.003
5. van Es JH, van Gijn ME, Riccio O, van den Born M, Vooijs M, Begthel H, Cozijnsen M, Robine S, Winton DJ, Radtke F, Clevers H (2005) Notch/gamma-secretase inhibition turns proliferative cells in intestinal crypts and

adenomas into goblet cells. Nature 435(7044): 959–963. https://doi.org/10.1038/nature03659

6. Riccio O, van Gijn ME, Bezdek AC, Pellegrinet L, van Es JH, Zimber-Strobl U, Strobl LJ, Honjo T, Clevers H, Radtke F (2008) Loss of intestinal crypt progenitor cells owing to inactivation of both Notch1 and Notch2 is accompanied by derepression of CDK inhibitors p27Kip1 and p57Kip2. EMBO Rep 9(4):377–383. https://doi.org/10.1038/embor.2008.7
7. Fre S, Hannezo E, Sale S, Huyghe M, Lafkas D, Kissel H, Louvi A, Greve J, Louvard D, Artavanis-Tsakonas S (2011) Notch lineages and activity in intestinal stem cells determined by a new set of knock-in mice. PLoS One 6(10):e25785. https://doi.org/10.1371/journal.pone.0025785
8. Basak O, Beumer J, Wiebrands K, Seno H, van Oudenaarden A, Clevers H (2017) Induced quiescence of Lgr5+ stem cells in intestinal organoids enables differentiation of hormone-producing enteroendocrine cells. Cell Stem Cell 20(2):177–190 e174. https://doi.org/10.1016/j.stem.2016.11.001
9. Heuberger J, Kosel F, Qi J, Grossmann KS, Rajewsky K, Birchmeier W (2014) Shp2/MAPK signaling controls goblet/paneth cell fate decisions in the intestine. Proc Natl Acad Sci U S A 111(9):3472–3477. https://doi.org/10.1073/pnas.1309342111
10. Kober OI, Ahl D, Pin C, Holm L, Carding SR, Juge N (2014) Gammadelta T-cell-deficient mice show alterations in mucin expression, glycosylation, and goblet cells but maintain an intact mucus layer. Am J Physiol Gastrointest Liver Physiol 306(7):G582–G593. https://doi.org/10.1152/ajpgi.00218.2013
11. Yin X, Farin HF, van Es JH, Clevers H, Langer R, Karp JM (2014) Niche-independent high-purity cultures of Lgr5+ intestinal stem cells and their progeny. Nat Methods 11(1): 106–112. https://doi.org/10.1038/nmeth.2737
12. Gracz AD, Samsa LA, Fordham MJ, Trotier DC, Zwarycz B, Lo YH, Bao K, Starmer J, Raab JR, Shroyer NF, Reinhardt RL, Magness ST (2018) Sox4 promotes Atoh1-independent intestinal secretory differentiation toward tuft and enteroendocrine fates. Gastroenterology 155(5):1508–1523. e1510. https://doi.org/10.1053/j.gastro.2018.07.023
13. de Lau W, Kujala P, Schneeberger K, Middendorp S, Li VS, Barker N, Martens A, Hofhuis F, DeKoter RP, Peters PJ, Nieuwenhuis E, Clevers H (2012) Peyer's patch M cells derived from Lgr5(+) stem cells require SpiB and are induced by RankL in cultured "miniguts". Mol Cell Biol 32(18): 3639–3647. https://doi.org/10.1128/MCB.00434-12
14. Cheng CW, Biton M, Haber AL, Gunduz N, Eng G, Gaynor LT, Tripathi S, Calibasi-Kocal-G, Rickelt S, Butty VL, Moreno-Serrano M, Iqbal AM, Bauer-Rowe KE, Imada S, Ulutas MS, Mylonas C, Whary MT, Levine SS, Basbinar Y, Hynes RO, Mino-Kenudson M, Deshpande V, Boyer LA, Fox JG, Terranova C, Rai K, Piwnica-Worms H, Mihaylova MM, Regev A, Yilmaz OH (2019) Ketone body signaling mediates intestinal stem cell homeostasis and adaptation to diet. Cell 178(5):1115–1131 e1115. https://doi.org/10.1016/j.cell.2019.07.048
15. Lim JS, Ibaseta A, Fischer MM, Cancilla B, O'Young G, Cristea S, Luca VC, Yang D, Jahchan NS, Hamard C, Antoine M, Wislez M, Kong C, Cain J, Liu YW, Kapoun AM, Garcia KC, Hoey T, Murriel CL, Sage J (2017) Intratumoural heterogeneity generated by Notch signalling promotes small-cell lung cancer. Nature 545(7654):360–364. https://doi.org/10.1038/nature22323
16. Cheng CW, Yilmaz OH, Mihaylova MM (2020) Strategies for measuring induction of fatty acid oxidation in intestinal stem and progenitor cells. Methods Mol Biol 2171:53–64. https://doi.org/10.1007/978-1-0716-0747-3_4
17. Powell RH, Behnke MS (2017) WRN conditioned media is sufficient for in vitro propagation of intestinal organoids from large farm and small companion animals. Biol Open 6(5):698–705. https://doi.org/10.1242/bio.021717
18. Clinton J, McWilliams-Koeppen P (2019) Initiation, expansion, and cryopreservation of human primary tissue-derived normal and diseased organoids in embedded three-dimensional culture. Curr Protoc Cell Biol 82(1):e66. https://doi.org/10.1002/cpcb.66
19. Boonekamp KE, Dayton TL, Clevers H (2020) Intestinal organoids as tools for enriching and studying specific and rare cell types: advances and future directions. J Mol Cell Biol 12(8): 562–568. https://doi.org/10.1093/jmcb/mjaa034
20. Schwank G, Andersson-Rolf A, Koo BK, Sasaki N, Clevers H (2013) Generation of BAC transgenic epithelial organoids. PLoS One 8(10):e76871. https://doi.org/10.1371/journal.pone.0076871

Chapter 11

Generation of Fetal Intestinal Organoids and Their Maturation into Adult Intestinal Cells

Masamichi Imajo, Akira Hirota, and Shinya Tanaka

Abstract

During embryonic development, the gut tube undergoes massive morphological changes from the simple tube structure composed of the pseudostratified epithelium into the mature intestinal tract composed of the columnar epithelium and characterized by the unique crypt-villus structures. In mice, maturation of fetal gut precursor cells into adult intestinal cells starts around embryonic day (E) 16.5, during which adult intestinal stem cells and their differentiated progenies are generated. In contrast to adult intestinal cells that form budding organoids containing both the crypt-like and villus-like regions, fetal intestinal cells can be cultured as simple spheroid-shaped organoids that show a uniform proliferation pattern. The fetal intestinal spheroids can undergo spontaneous maturation into adult budding organoids that contain intestinal stem cells and differentiated cells, including enterocytes, goblet, enteroendocrine, and Paneth cells, recapitulating intestinal cell maturation in vitro. Here, we provide detailed methods for establishment of fetal intestinal organoids and their differentiation into adult intestinal cells. These methods enable in vitro recapitulation of intestinal development and would be useful to reveal mechanisms that regulate the transition from fetal to adult intestinal cells.

Key words Intestinal epithelial cells, Organoid culture, Fetal intestine, Intestinal development, Intestinal stem cells

1 Introduction

During embryonic development, the intestine develops from the mid- and hindgut regions of the primitive gut tube [1, 2]. The gut tube forms a uniformly proliferative pseudostratified layer to rapidly expand the intestine in both length and diameter (from E9.5 to E14 mice) [1, 3]. Around E14, the pseudostratified layer is reorganized into the simple columnar epithelium, followed by formation of the protruding villus and intervillus regions [1, 3]. Cell proliferation becomes restricted to the intervillus region, which gradually forms the invagination called crypt. Crypt base columnar (CBC) cells expressing a stem cell marker gene, *Lgr5*, are generated in the crypt region from E16.5 to P5 and begin to produce differentiated

Paloma Ordóñez-Morán (ed.), *Intestinal Differentiated Cells: Methods and Protocols*, Methods in Molecular Biology, vol. 2650, https://doi.org/10.1007/978-1-0716-3076-1_11,

progenies including enterocytes, goblet, enteroendocrine, and Paneth cells [3, 4]. The transition from fetal to adult intestinal cells has been shown to be regulated by alternating roles of Wnt and YAP signaling that also play a critical role in homeostatic renewal of adult intestinal epithelium [4–7].

Similar to adult intestinal epithelial cells, fetal intestinal cells can be cultured in the form of three-dimensional organoid culture [4, 8]. Although adult small intestinal cells form budding organoids composed of both the crypt-like and villus-like domains, fetal intestinal cells form uniformly proliferative spheroids without any types of differentiated cells [4]. In vitro culture of cells from fetal intestine generates mixed populations of spheroids and budding organoids [4]. As fetal intestinal cells are maturated into adult intestinal cells during development, the population of spheroid-forming precursor cells decreases, whereas that of budding organoid-forming maturated cells increases [4]. Of note, fetal intestinal cells that form spheroids can be maturated into adult cells that form budding organoids under the in vitro culture condition [9]. This clearly indicates that fetal cells have ability to generate mature intestinal cells spontaneously and that the in vitro culture of fetal intestinal spheroids represents a remarkable model to study the mechanisms of intestinal development. In this chapter, we describe a detailed protocol to establish and culture fetal intestinal organoids from mouse embryos and to observe the transition from fetal to adult intestinal cells.

2 Materials

2.1 Dissection of Mouse Embryos and Isolation of Fetal Intestinal Cells

1. Surgical instruments (dissection scissors and forceps).
2. Plasticware: sterile 5 mL and 10 mL serological pipettes; 15 mL and 50 mL conical tubes; 1.5 mL microcentrifuge tubes; P20, P200, and P1000 pipette tips; 35 mm and 100 mm cell culture dishes.
3. Stereomicroscope.
4. Inverted microscope (for cell culture).
5. Phosphate-buffered saline (PBS):137 mM NaCl, 2.7 mM KCl, 10 mM Na_2HPO_4, 1.8 mM KH_2PO_4, pH 7.4.
6. 0.1% bovine serum albumin (BSA) in PBS.
7. 1 mM ethylenediaminetetraacetic acid (EDTA) in PBS.
8. 100 μm cell strainer.

2.2 Establishment and Passaging of Fetal Intestinal Spheroids

1. CO_2 incubator (for cell culture).
2. Refrigerated microcentrifuge (for microcentrifuge tubes).
3. Inverted microscope (for cell culture).

4. Plasticware: sterile 50 mL conical tubes; 1.5 mL microcentrifuge tubes; cut-down P200 pipette tips; uncut P20, P200, and P1000 pipette tips; 35 mm dishes; 24-well cell culture plate.
5. Matrigel, basement membrane matrix, growth factor reduced, phenol red-free (thawed at 4 °C overnight).
6. Advanced DMEM/F-12.
7. GlutaMAX (100× concentrate).
8. N2 supplement (100× concentrate).
9. B-27 supplement (50× concentrate).
10. 10 mM ROCK inhibitor Y-27632 stock solution in dimethyl sulfoxide (DMSO) (1000× concentrate).
11. 50 mg/mL recombinant mouse EGF protein dissolved in 0.1% (w/v) BSA in PBS (1000× concentrate).
12. 100 mg/mL recombinant mouse noggin protein dissolved in 0.1% (w/v) BSA in PBS (1000× concentrate).
13. 100 mg/mL recombinant mouse R-spondinl protein dissolved in 0.1% (w/v) BSA in PBS (1000× concentrate).
14. 10,000 U/mL penicillin-streptomycin (100× concentrate).
15. Culture medium: Advanced DMEM/F-12 containing N2 supplement (1×), B-27 supplement (1×), GlutaMAX (1×), penicillin-streptomycin (1×), recombinant mouse EGF (50 ng/mL), recombinant mouse R-spondin-1 (100 ng/mL), recombinant mouse noggin (100 ng/mL), and Y-27632 (10 μM).

2.3 Analysis of the Maturation Status of Cultured Fetal Intestinal Cells

1. Inverted microscope (for cell culture).
2. Thermal cycler (for cDNA synthesis).
3. Real-time PCR system.

3 Methods

The following protocol is for establishment of fetal intestinal spheroids from E15 mouse embryos and can be applied for other gestational days from E14 to P0 with slight modifications (3.1~3.2). Then, we provide detailed procedures to passage the established fetal intestinal spheroids and to observe their spontaneous differentiation into mature adult intestinal cells (3.3~3.4). These methods would help readers to recapitulate intestinal development by using organoid methodologies.

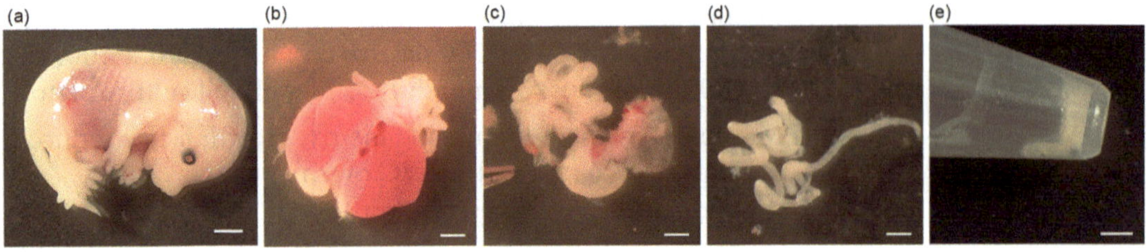

Fig. 1 Isolation of the intestinal tract from an E15 mouse embryo. Pictures showing a whole embryo (**a**) and the digestive tract dissected out from the embryo before (**b**) and after isolation of the liver (**c**), stomach, and mesentery (**d**). (**e**) Chopped fetal small intestine pieces in 1.5 mL microcentrifuge. Scale bars, 2 mm (a and e) and 1 mm (b–d)

3.1 Dissection of Mouse Embryos

1. Sacrifice a pregnant mouse according to the approved institutional guidelines, and immediately dissect out the mouse uterus by dissection scissors. We routinely use the C57BL/6 strain at gestational day 15, but the following procedures can be applied to other strains at various gestational days ranging from E14 to P0 with slight modifications. Figure 1 shows pictures of the dissected mouse embryo and fetal intestine at each step.
2. To isolate embryos, make an incision along the longitudinal axis of the uterus by dissection scissors, and further separate Reichert's membrane and visceral yolk sac from embryos. Gently transfer the exposed embryos into a 10 cm dish containing PBS by forceps (Fig. 1a).
3. Under a stereomicroscope, make an abdominal incision perpendicular to the body axis of the embryo, and open the peritoneal cavity using two pairs of forceps. Dissect out the whole gastrointestinal organs, including the stomach, small intestine, cecum, colon, liver, and pancreas, from mouse embryos.
4. By pulling and tearing tissues with two pairs of forceps, remove the organs other than the small intestine. First, remove large organs, such as the stomach, liver, and colon, and then remove small organs, such as the pancreas and mesentery (Fig. 1b–d).
5. Transfer the dissected small intestine into a 1.5 mL microcentrifuge tube (*see* **Note 1**).
6. Place the tube on ice until isolation and culture of intestinal epithelial cells. Use one tube for each embryo.

3.2 Isolation and Culture of Fetal Intestinal Epithelial Cells

1. By using dissection scissors, chop the small intestine in the microcentrifuge tube into pieces until it almost turns into paste (Fig. 1e).
2. Add 0.5 mL of 1 mM EDTA in PBS to the microcentrifuge tube, mix by pipetting several times with P1000 pipette tips, and incubate on ice for 30 min.

3. Vortex the tube vigorously until epithelial cells are separated from the intestinal specimens.
4. Using P20 pipette tips, collect 10–20 μL of the solution in 35 mm dish, and, under an inverted microscope, check whether epithelial cells are detached from the intestine pieces. Place the tube on ice until the intestine pieces settle down to the bottom of the tube.
5. Collect the supernatant with P1000 pipette tips, and filter it through a 100 μm cell strainer put on a 50 mL tube. Transfer the flowthrough into a new 1.5 mL microcentrifuge tube.
6. Centrifuge the above cell suspension at 200*g* for 5 min at 4 °C. Discard the supernatant.
7. Resuspend the pellet in 60 μL of cold Matrigel solution (*see* **Note 2**).
8. Plate 30 μL of the solution into the center of each well of a 24-well cell culture plate.
9. Incubate the culture plate for 10 min in a 37 °C, 5% CO_2 incubator.
10. Gently add 500 μL of the culture medium.
11. Culture cells in a 37 °C, 5% CO_2 incubator for 5–7 days. The culture medium should be replaced every other day. Y-27632 is added only for the first 2 days to prevent cellular anoikis.

3.3 Passaging of Fetal Intestinal Spheroids

1. Remove the culture medium in the wells to be passaged.
2. Using a 1 mL pipettor, add 1 mL of ice-cold 1 mM EDTA in PBS. Pipette the total solution up and down vigorously to solubilize Matrigel. Transfer the solution into a 1.5 mL microcentrifuge tube and place on ice.
3. Using P200 pipette tips, pipette the spheroid suspension up and down vigorously for 20–30 times to mechanically disrupt spheroids into smaller fragments.
4. Centrifuge the spheroid fragment suspension at 200*g* for 5 min at 4 °C. Discard the supernatant.
5. Resuspend the pellet in 90–150 μL of the cold Matrigel solution depending on the split ratio during passaging (*see* **Note 3**).
6. Plate 30 μL of the solution into the center of each well of a 24-well cell culture plate. Incubate the culture plate for 10 min in a 37 °C, 5% CO_2 incubator.
7. Gently add 500 μL of culture medium.
8. Culture cells in a 37 °C, 5% CO_2 incubator for 5–7 days as described in Subheading 3.2.

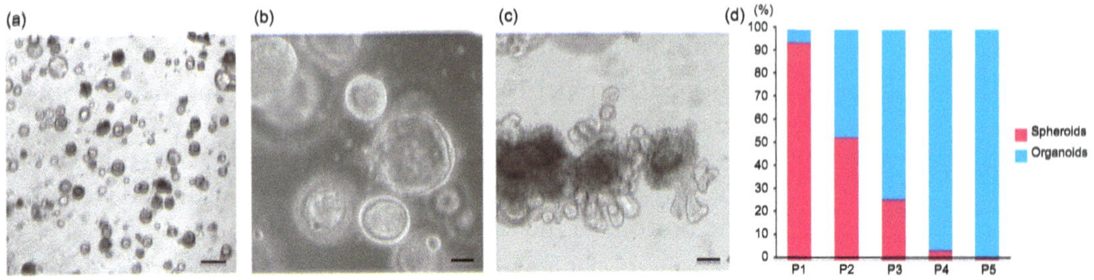

Fig. 2 Morphology of mouse fetal intestinal spheroids. Low- (**a**) and high-magnification (**b**) images of the mouse fetal intestinal spheroids. (**c**) Budding organoids spontaneously appeared during passaging of the fetal intestinal spheroids. (**d**) Time-dependent changes in the ratio of fetal small intestinal spheroids over budding organoids during passaging (P1~P5). Scale bars, 200 μm (a), 50 μm, and 100 μm (c)

3.4 Analysis of In Vitro Maturation of Fetal Intestinal Cells into Adult Intestinal Cells

1. The above procedures (Subheadings 3.1–3.3) enable establishment and culture of fetal intestinal spheroids from mouse embryos (Fig. 2). Although fetal cells initially form spheroids and show fetal intestine-specific gene expression profiles, they gradually lose fetal cell characteristics and maturate into adult intestinal cells, recapitulating intestinal development in vitro. The transition from fetal to adult intestinal cells can be observed morphologically and analyzed by gene expression profiling. The following are typical examples of such experiments.
2. To analyze morphological changes during in vitro culture, count the number of spheroids and budding organoids under an inverted microscope before each passaging (Fig. 2). As shown in Fig. 2d, the ratio of spheroids over budding organoids typically decreases during several passages. The increase in the number of budding organoids indicates maturation of fetal precursor cells into adult intestinal cells.
3. For gene expression analysis, collect a fraction of the cell suspension at each passaging. Centrifuge the suspension at 200*g* for 5 min at 4 °C and discard the supernatant.
4. The cell pellet should be frozen immediately and stored at − 80 °C until the analysis.
5. Total RNAs can be extracted, reverse-transcribed into cDNAs, and analyzed by real-time PCR by using commercially available kits. We routinely use primers for fetal intestine-specific genes, such as *Trop2*, *Cnx43*, *and Ly6a* (*see* **Note 4**), and also those for adult intestine-specific genes, such as *sucrase-isomaltase* (*Si*) and *Muc2*. Figure 3 shows the typical results of gene expression profiling of cultured fetal intestinal cells. Expression of fetal intestine markers rapidly decreases, whereas that of adult intestine markers oppositely increases during passaging. This clearly indicates that fetal intestinal cells can be maturated into adult intestinal cells under the in vitro organoid culture condition.

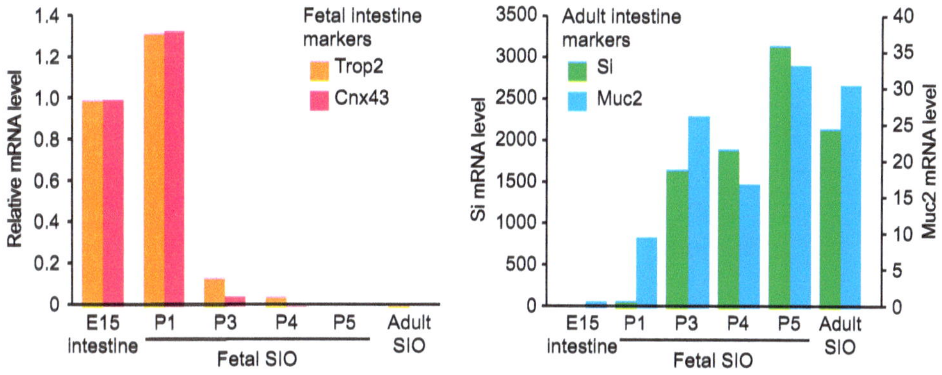

Fig. 3 Gene expression profiles of cultured fetal intestinal cells. Relative mRNA levels of fetal intestine markers (Trop2, Cnx43) and adult intestine markers (Si, Muc2) at each passage of fetal intestinal cells are shown. As reference, gene expression profiles of the fetal intestinal tract and adult small intestinal organoid (SIO) are shown

4 Notes

1. Before starting the cell culture procedures, plasticwares that come in contact with intestinal cells should be pre-wetted by 0.1% BSA in PBS, as adhesion of intestinal cells to their surface can reduce cell yield significantly. Pre-wetting should be performed immediately before the experiment.
2. We routinely use 60 μL of the Matrigel solution to culture intestinal cells taken from an E15 embryo. As the number of cells that can be isolated from one embryo can vary depending on the embryonic stage, the volume of the Matrigel solution used here should be adjusted when embryos at other embryonic days are used.
3. Split ratio should be determined depending on the number and size of the spheroids. As reference, we routinely passage at one-third to one-fourth dilutions. The exact dilution range should be determined by each experimenter, as it can vary depending on the duration of culture and subtle differences in experimental procedures (e.g., isolation efficiency of intestinal epithelial cells, cell viability, culture environment, etc.).
4. The sequences of primers for detection of fetal intestine-specific genes are as follows: Trop2-Fw, 5′-aaccactctgacctagac tccgag-3′; Trop2-Rv, 5′-tctgaatggtgggctcctcatagt-3′; Cnx43-Fw, 5′-tcctgctgatccagtggtacatct-3′; and Cnx43-Rv, 5′-ggacacc accagcatgaagatgat-3′.

Acknowledgments

This work was supported by the Takeda Science Foundation, Kowa Life Science Foundation, and Grant-in-Aid for Scientific Research (KAKENHI) on Innovative Areas, "Integrated analysis and regulation of cellular diversity" (18H05100), and for Scientific Research (C) (18K06929 and 22K06874) from MEXT (Ministry of Education, Culture, Sports, Science and Technology, Japan).

References

1. Noah TK, Donahue B, Shroyer NF (2011) Intestinal development and differentiation. Exp Cell Res 317:2702–2710
2. Zorn AM, Wells JM (2009) Vertebrate endoderm development and organ formation. Annu Rev Cell Dev Biol 25:221–251
3. Spence JR, Lauf R, Shroyer NF (2011) Vertebrate intestinal endoderm development. Dev Dyn 240:501–520
4. Mustata RC, Vasile G, Fernandez-Vallone V, Strollo S, Lefort A, Livert F et al (2013) Identification of Lgr5-independent spheroid-generating progenitors of the mouse fetal intestinal epithelium. Cell Rep 5:421–432
5. Yui S, Azzolin L, Maimets M, Pedersen MT, Fordham RP, Hansen SL et al (2018) YAP/TAZ-dependent reprogramming of colonic epithelium links ECM remodeling to tissue regeneration. Cell Stem Cell 22:35–49
6. Sprangers J, Zaalberg IC, Maurice MM (2021) Organoid-based modeling of intestinal development, regeneration, and repair. Cell Death Differ 28:95–107
7. Imajo M, Ebisuya M, Nishida E (2015) Dual role of YAP and TAZ in renewal of the intestinal epithelium. Nature Cell Biol 17:7–19
8. Fordham RP, Yui S, Hannan NRF, Soendergaard C, Madgwick A, Schweiger PJ et al (2017) Transplantation of expanded fetal intestinal progenitors contributes to colon regeneration after injury. Cell Stem Cell 13: 734–744
9. Navis M, Garcia TM, Renes IB, Vermeulen JLM, Meisner S, Wildenberg ME et al (2019) Mouse fetal intestinal organoid: new model to study epithelial maturation from suckling to weaning. EMBO Rep 20:e46221

Chapter 12

Visualization of Differentiated Cells in 3D and 2D Intestinal Organoid Cultures

Hikaru Hanyu, Shinya Sugimoto, and Toshiro Sato

Abstract

The intestinal epithelium maintains self-renewal and differentiation capacities via coordination of key signaling pathways, including the Wnt, bone morphogenetic protein (BMP), epidermal growth factor (EGF), and Notch signaling pathways. Based on this understanding, a combination of stem cell niche factors, EGF, Noggin, and the Wnt agonist R-spondin was shown to enable the growth of mouse intestinal stem cells and the formation of organoids with indefinite self-renewal and full differentiation capacity. Two small-molecule inhibitors, including a p38 inhibitor and a TGF-beta inhibitor, were added to propagate cultured human intestinal epithelium but at the cost of differentiation capacity. There have been improvements in culture conditions to overcome these issues. Substitution of the EGF and a p38 inhibitor with insulin-like growth factor-1 (IGF-1) and fibroblast growth factor-2 (FGF-2) enabled multilineage differentiation. Monolayer culture with mechanical flow to the apical epithelium promoted the formation of villus-like structures with mature enterocyte gene expression. Here, we summarize our recent technological improvements in human intestinal organoid culture that will deepen the understanding of intestinal homeostasis and diseases.

Key words Intestinal stem cells, LGR5, Organoids, Differentiation, Small intestine, Colon, Human, Immunostaining, Whole-mount staining

1 Introduction

The proliferation and differentiation of the intestinal epithelium are controlled by several niche (microenvironment) signals. In the human body, intestinal stem cells maintain cellular diversity by switching their state between self-renewal and differentiation in a manner driven by these signals. Well-known signaling pathways that regulate intestinal homeostasis are the Wnt, bone morphogenetic protein (BMP), epidermal growth factor (EGF), and Notch signaling pathways [1]. Wnt signals maintain stem cell fate and drive the proliferation of stem and transit-amplifying (TA) cells [2] and paradoxically drive the terminal differentiation of Paneth cells [3]. BMP signals are active in the villus compartment and

Paloma Ordóñez-Morán (ed.), *Intestinal Differentiated Cells: Methods and Protocols*, Methods in Molecular Biology, vol. 2650,
https://doi.org/10.1007/978-1-0716-3076-1_12,

negatively regulate stemness. Transient inhibition of BMP signaling by Noggin induces the formation of crypt-like structures along the flanks of the villi [4]. EGF signals have strong mitogenic effects on stem and TA cells and are essential for stemness. Notch signals also have a critical role in maintaining the undifferentiated state. Proliferative stem and TA cells differentiate into secretory lineage cells via blockade of Notch signaling [5].

Based on these findings regarding the stem cell niche, Sato et al. first revealed that a combination of niche factors, EGF, Noggin, and the Wnt agonist R-spondin is essential for culturing mouse intestinal stem cells in vitro [6]. The discovery of this system was the breakthrough in the culture of intestinal stem cells, named organoids. Subsequently, the culture conditions were adapted to long-term culture of human small intestine and colon cells by the addition of Wnt3a, an inhibitor of activin-like kinase 4/5/7 (A83-01), an inhibitor of the p38 mitogen-activated protein kinase (SB202190), and gastrin [7]. However, there was a limitation in that cultured organoids have low cellular diversity and low culture efficiency. Recently, we revealed that removal of the p38 inhibitor and addition of insulin-like growth factor-1 (IGF-1) and fibroblast growth factor-2 (FGF-2) (the IF condition) enabled us to maintain organoids with self-renewal and multidifferentiation capacities [8]. Under this IF condition, we observed most of the cell types identified in vivo, including stem cells, TA cells, early enterocytes, goblet cells, enteroendocrine cells, M cells, Paneth cells, and tuft cells, in the organoids.

Recent progress in organoid culture has been made not only in medium modifications but also in cell culture techniques. 2D intestinal organoid culture has been widely applied in several studies following the establishment of a primary mouse intestinal monolayer culture system [9]. Briefly, organoids have been seeded onto cell culture inserts coated with Matrigel™ and supplemented with optimized medium. For further differentiation, some studies have utilized air-liquid interface culture [10] or external flow [11, 12] systems. Further optimization of 2D culture systems for intestinal organoids with preserved cellular diversity will be a useful tool for understanding the intestinal epithelium.

In this protocol, we introduce recent methods for culturing 3D/2D human intestinal organoids and approaches to visualize differentiated cells by immunostaining.

2 Materials

2.1 Intestinal Organoid Culture (3D)

1. Matrigel™, basement membrane matrix, growth factor reduced (GFR), phenol red-free.
2. Basal culture medium: 10 mM HEPES, 100 U/mL penicillin, and 100 μg/mL streptomycin in advanced DMEM/F12 medium.
3. B27 supplement (50×).
4. N-Acetyl-L-cysteine [500× stock; 81.5 mg/mL in distilled water (500 mM)].
5. [Leu^{15}]-gastrin I (10,000× stock; 100 μM in 0.1% BSA/PBS).
6. GlutaMAX.
7. Afamin-Wnt3a serum-free conditioned medium prepared from a cell line [13] (*see* **Note 1**).
8. Recombinant human R-spondin (100× stock; 100 μg/mL in 0.1% BSA/PBS) (*see* **Note 2**).
9. Recombinant mouse Noggin (1000× stock; 100 μg/mL in 0.1% BSA/PBS).
10. Recombinant mouse EGF (10,000× stock; 500 μg/mL in 0.1% BSA/PBS).
11. Recombinant human IGF-1 (1000× stock; 100 μg/mL in 0.1% BSA/PBS).
12. Recombinant human FGF-2 (FGF-basic) (1000× stock; 50 μg/mL in 0.1% BSA/PBS).
13. A83-01 (1000× stock; 500 μM in DMSO).
14. Y-27632 (1000× stock; 10 mM in PBS).
15. Human WENRAIF medium: basal culture medium supplemented with 20% (vol/vol) afamin-Wnt3a serum-free conditioned medium, 1 μg/mL recombinant human R-spondin, 100 ng/mL recombinant mouse Noggin, 50 ng/mL EGF, 100 ng/mL recombinant human IGF-1, 50 ng/mL recombinant human FGF-basic (FGF-2), 500 nM A83-01, 1× B27 supplement, 1 mM N-acetyl-L-cysteine, and 10 nM [Leu^{15}]-gastrin I.
16. TrypLE Select Enzyme (10×), no phenol red (*see* **Note 3**).
17. Dulbecco's phosphate-buffered saline without Ca^{2+} and Mg^{2+} (DPBS).
18. 15 mL centrifuge tube.
19. Flat-bottom 48-well cell culture plate.
20. Cryopreserved organoids.

2.2 Intestinal Organoid Culture (2D)

1. Matrigel™.
2. Basal culture medium.
3. TrypLE Select Enzyme (10×), no phenol red.
4. DPBS.
5. Cell strainer (20 μm).
6. Human WENRAIF medium.
7. Cell culture inserts [24-well, 0.4 μm pores ($2 \times 10^6/cm^2$, transparent)].
8. Flat-bottom 24-well cell culture plate.

2.3 Whole-Mount Staining (3D)

1. Cell recovery solution.
2. DPBS.
3. Paraformaldehyde (PFA; 4%).
4. Triton™ X-100.
5. Blocking buffer (Power Block™).
6. Antibodies.
7. Hoechst 33342 (1000×; 10mg/mL).
8. Antifade mountant.
9. Low-protein-binding centrifuge tube (1.5 mL).
10. Rocking shaker.
11. Glass bottom dish.

2.4 Whole-Mount Staining (2D)

1. Basal culture medium.
2. PFA (4%).
3. DPBS.
4. Triton™ X-100.
5. Blocking buffer (Power Block™).
6. Antibodies.
7. Hoechst 33342 (1000×; 10mg/mL).
8. Antifade mountant.
9. Fine forceps.
10. Cell scraper.
11. Glass bottom dish.

Table 1
Culture medium components of human intestinal organoids

Reagents	Final concentration	WENRAIF medium *for maintenance*	WNRAIF (-E) medium *for differentiation*
Afamin-Wnt3A	20% (v/v)	+	+
EGF	50 ng/mL	+	–
Noggin	100 ng/mL	+	+
R-spondin1	1 μg/mL	+	+
A83-01	500 nM	+	+
IGF-1	100 ng/mL	+	+
FGF-2	50 ng/mL	+	+
Gastrin	10 nM	+	+
B27 supplement	1×	+	+
N-Acetyl-L-cysteine	1 mM	+	+

3 Methods

3.1 Intestinal Organoid Culture (3D)

1. Thaw Matrigel™ on ice (*see* **Note 4**). Prewarm 48-well plates at 37 °C.
2. Prepare cryopreserved organoids (*see* **Note 5**).
3. Quickly thaw cryopreserved organoids at 37 °C in a water bath and transfer into a 15 mL centrifuge tube.
4. Quickly add 10 mL of basal culture medium.
5. Centrifuge at 400×*g* for 3 min. Remove the supernatant.
6. Add Matrigel™ (20 μL/wells), and carefully resuspend the organoids, avoiding making bubbles.
7. Drop 20 μL of the crypt suspension in Matrigel™ (containing 1–2 × 10^3 cells) into the center of each well of a prewarmed 48-well plate (*see* **Note 6**).
8. Place the plate in a CO_2 incubator for 10 min at 37 °C under 5% CO_2.
9. After the Matrigel™ completely polymerizes, add 300 μL of optimized medium to each well (*see* Table 1) (*see* **Note 7**). To prevent anoikis, add 10 μM Y-27632 to the medium for the first 2–3 days.
10. Culture in a CO_2 incubator at 37 °C under 5% CO_2. Replace the medium every 2–4 days.

3.2 Passage of Intestinal Organoids (3D)

1. Thaw Matrigel™ on ice. Prewarm 48-well plates at 37 °C.
2. Remove the culture medium from the wells of the cell culture plates.
3. Collect the 3D intestinal organoids in a 15 mL centrifuge tube by pipetting with a 1:3 dilution of TrypLE Select (10×) (500 μL/well).
4. Incubate the tubes at 37 °C in a water bath for 10–20 min until the organoids are dissociated into single cells. During dissociation, gently pipette the solution every 5 min.
5. Add 10 mL of basal medium to the tube.
6. Centrifuge the crypts at 400×*g* for 3 min. Carefully remove the supernatant.
7. Count the cells and transfer the required volume of suspension (for 1–5 × 10^3 cells/well) into a new tube.
8. Repeat Subheading 3.1, **steps 5–10**.

3.3 Intestinal Organoid Culture (2D)

1. Before coating the cell culture inserts with Matrigel™, cool the inserts at −80 °C for 5 min or 4 °C for 30 min.
2. Prepare a 20% Matrigel™ solution in basal culture medium.
3. Set the cooled cell culture inserts into the 24-well plates.
4. Add 100 μL of the 20% Matrigel™ solution into the apical side of each cell culture insert, allowing it to coat the whole surface of the membrane, and then remove the solution (*see* **Note 8**).
5. Place the plate in an incubator for 5 min at 37 °C.
6. Wash the Matrigel™-coated cell culture inserts with basal culture medium.
7. Add 200 μL of basal culture medium into the apical side of each cell culture insert, and place the plates in a 37 °C incubator until the cells are seeded.
8. Collect the 3D intestinal organoids in a 15 mL centrifuge tube by pipetting with a 1:3 dilution of TrypLE Select (10×) solution.
9. Incubate the tubes at 37 °C in a water bath for 10–20 min until the organoids are completely dissociated into single cells. During dissociation, pipette the solution every 5 min.
10. Add basal medium and centrifuge at 400×*g* for 3 min. Remove the supernatant.
11. Resuspend the pellets in basal medium and filter them through a cell strainer. Count the cells by using a counting chamber (*see* **Note 9**).

12. Transfer the suspension (1.5–2 × 10^5 cells/well) in a new centrifuge tube and centrifuge at 400×*g* for 3 min. Remove the supernatant.
13. Resuspend the pellets in WENRAIF medium supplemented with 10 μM Y-27632 (200 μL/well).
14. Remove the basal culture medium from the cell culture inserts. Add 200 μL of the cell suspension into the apical side of each cell culture insert.
15. Leave the plates at room temperature for 10 min (*see* **Note 10**).
16. Add 700 μL of WENRAIF medium supplemented with 10 μM Y-27632 into the basal side of each cell culture insert (*see* **Note 11**).
17. Incubate in a CO_2 incubator (5% CO_2, 37 °C).
18. (optional) For further differentiation, incubate the plate on a rotary shaker at 150 rpm for 4 days to generate luminal flow at confluence (3–4 days post seeding).

3.4 Whole-Mount Staining (3D)

Differentiated cells can be visualized by whole-mount staining. Before staining, culture 3D intestinal organoids in 48-well cell culture plates at a density of 3 × 10^3 cells/well for 5–10 days:

1. Remove the culture medium from the wells of the cell culture plates.
2. Collect the 3D intestinal organoids in a 1.5 mL low-protein-binding centrifuge tube by pipetting with cell recovery solution (100 μL/well).
3. Place the tubes on ice for 30 min and pipette the solution every 10 min to depolymerize the gelled Matrigel™.
4. Allow the organoids to settle to the bottom of the tubes and carefully remove the supernatant (*see* **Note 12**).
5. Add DPBS into the tubes.
6. Allow the organoids to settle at the bottom of the tubes and carefully remove the supernatant (Fig. 1a, b).
7. Add 500 μL of 4% PFA solution into the tubes and incubate for 20 min at room temperature.
8. Wash with DPBS 3 times (repeat **steps 3.4 5–6**).
9. Add 1% Triton™ X-100 solution in DPBS and incubate for 10 min at room temperature.
10. Remove 1% Triton™ X-100 solution. Add 500 μL of blocking buffer and incubate for 10 min at room temperature.
11. Remove the blocking buffer. Add 200 μL of primary antibody solution in 0.2% Triton™ X-100. Incubate overnight on a rocking shaker at 4 °C (Fig. 1c) (*see* **Note 13**).

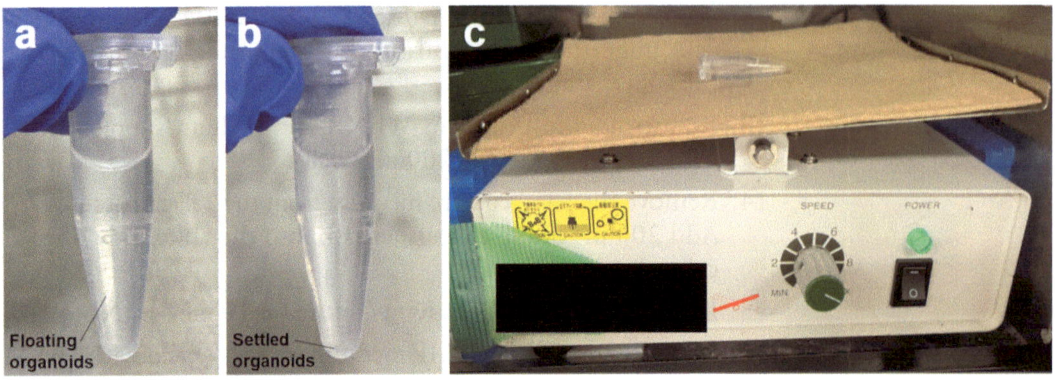

Fig. 1 Overviews of whole-mount staining (3D). (**a**) Organoids floating in the tube after the solution were pipetted. (**b**) Organoids settled to the bottom of the tube after allowing them to sink down for ~5 min. (**c**) Shaking the tube on the rocking shaker at 4 °C

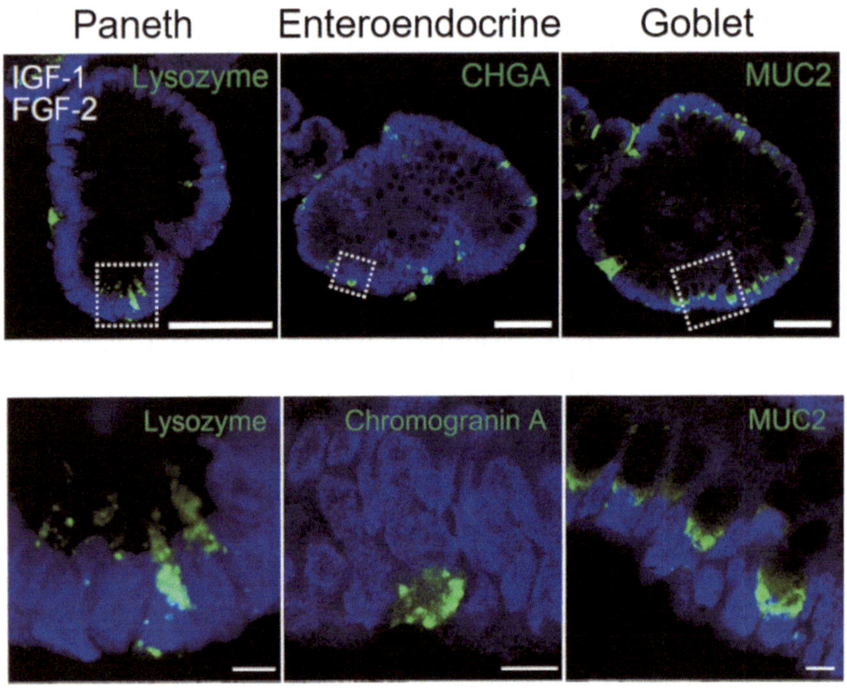

Fig. 2 Representative images of immunostained 3D organoids. Paneth, enteroendocrine, and goblet cells in 3D human small intestinal organoids, as detected by immunofluorescence. The inset shows a higher-magnification image. Scale bars, 5 μm (bottom) and 50 μm (top). (Data from Fujii et al. Cell Stem Cell. 2018 [8] are used with permission)

12. Remove the antibody solution. Wash with DPBS 3 times (repeat **steps 5–6** in Subheading 3.4).
13. Add 500 μL of secondary antibody solution in 0.2% Triton™ X-100 with Hoechst 33342. Incubate for 30 min on a rocking shaker at room temperature in the dark.

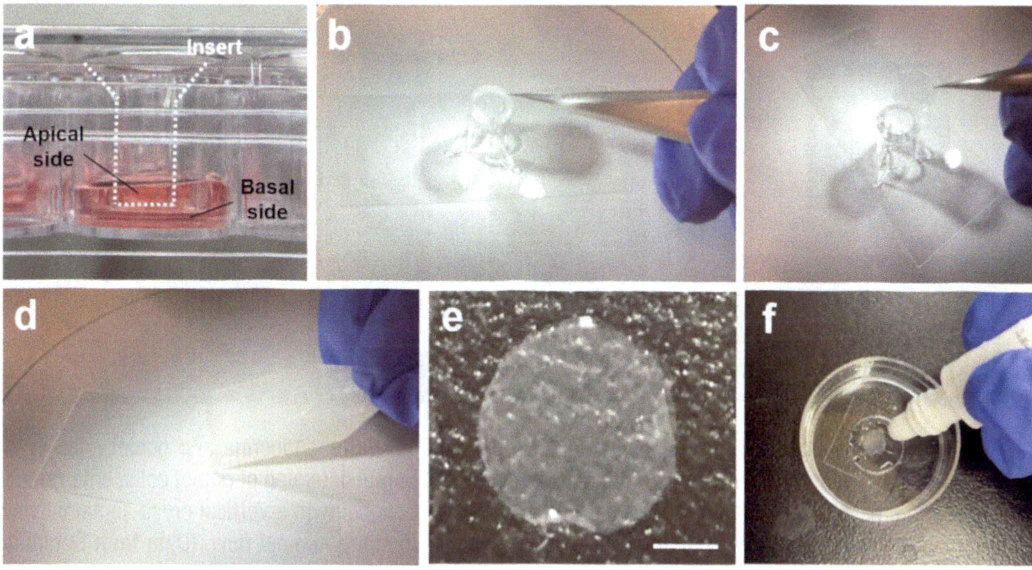

Fig. 3 Overviews of whole-mount staining (2D). (**a**) 2D culture of intestinal organoids using cell culture inserts. (**b**, **c**) Removal of the membrane from the insert frame with forceps. (**d**) Scraping of 2D organoids from the membrane with a cell scraper. (**e**) The layer of 2D organoids removed from the membrane. Scale bar, 200 μm. (**f**) Mounting 2D organoids on a glass bottom dish

14. Remove the antibody solution and wash with DPBS 3 times (repeat **steps 5–6** in Subheading 3.4).
15. Add 30 μL of antifade mountant and gently resuspend the organoids.
16. Add the suspension dropwise to the glass bottom dish and dry overnight at room temperature in the dark.
17. Acquire images via confocal or multiphoton microscopy (Fig. 2).

3.5 Whole-Mount Staining (2D)

Before staining, culture 2D intestinal organoids in 24-well cell culture inserts for ≥4 days (Fig. 3a):

1. Remove the culture medium from the wells of the cell culture inserts and plates.
2. Wash the apical side of each cell culture insert with basal culture medium.
3. Add 4% PFA into both sides of the cell culture inserts (e.g., 200 μL/well into the apical side and 500 μL/well into the basal side), and incubate for 20 min at room temperature.
4. Remove the 4% PFA and wash both sides of each cell culture insert with DPBS 3 times.
5. Add 1% Triton™ X-100 solution into both sides of each cell culture insert. Incubate for 10 min at room temperature.

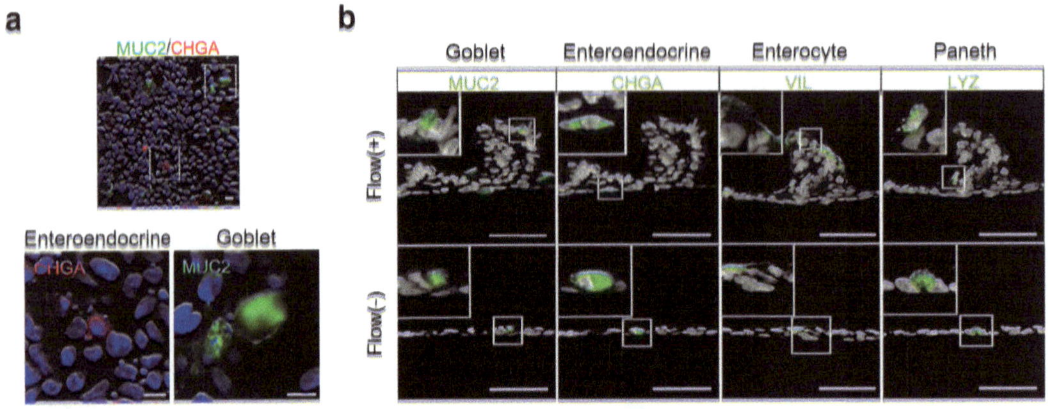

Fig. 4 Representative images of immunostained 2D organoids. (**a**) Enteroendocrine and goblet cells in 2D human colon organoids, as detected by immunofluorescence. (**b**) Immunostaining of goblet cells, enteroendocrine cells, enterocytes, and Paneth cells in paraffin-embedded sections, with a vertical cross-sectional view of 2D human duodenal organoids cultured with (top) or without (bottom) luminal flow. (Data from Sugimoto et al. Nature. 2021 [12] (b)). The inset shows a higher-magnification image. Scale bars, 10 μm (a) and 50 μm (b)

6. Remove 1% Triton™ X-100 solution. Add blocking buffer into both sides of each cell culture insert and incubate for 10 min at room temperature.
7. Remove the blocking buffer. Add primary antibody solution in 0.2% Triton™ X-100 into both sides of each cell culture insert. Incubate overnight at 4 °C.
8. Wash both sides of each cell culture insert with DPBS 3 times.
9. Add 500 μL of secondary antibody solution in 0.2% Triton™ X-100 with Hoechst 33342 into both sides of each cell culture insert. Incubate for 30 min at room temperature in the dark.
10. Wash both sides of each cell culture insert with DPBS 3 times.
11. Remove the membrane with the organoids from the culture inserts by cutting off the edge of the membrane (Fig. 3b, c).
12. Scrape the layer of 2D organoids from the membrane under a microscope (Fig. 3d, e) (*see* **Note 14**).
13. Place the layer of 2D organoids on a glass bottom dish and add antifade mountant (Fig. 3f).
14. Dry them overnight at room temperature in the dark.
15. Acquire images via confocal or multiphoton microscopy (Fig. 4a).

4 Notes

1. We used in-house-prepared afamin-Wnt3a serum-free conditioned medium at a final concentration of 20% as human WENRAIF medium. We confirmed that commercial afamin/Wnt3a CM (J2-001) at a final concentration of 10% can alternatively be used.
2. Recombinant R-spondin can be replaced with 10% (vol/vol) in-house-prepared conditioned medium produced by HEK293T cells [14].
3. Dilute TrypLE Select (10×) 3 times with DPBS (1:2). We used TrypLE Select instead of TrypLE Express since the organoids were easily dissociated into single cells.
4. Matrigel™ starts polymerizing and becomes a gel at room temperature. Keep it chilled on ice during handling.
5. Organoids can be cryopreserved 2–3 days after passage. For generation of organoids, please refer to the protocol we previously published [15].
6. Prewarming the 48-well plate allowed us to easily drop the Matrigel™ in a dome shape.
7. Removal of EGF from WENRAIF medium (WNRAIF) for 2 days allows a moderate level of cell differentiation. For further cell differentiation, it is known that specific cell differentiation can be induced by specific culture conditions. For example, removal of Noggin and addition of BMP-4 induce enterocyte differentiation; removal of EGF and addition of an EGFR inhibitor induce secretory linage cell, including enteroendocrine cell differentiation [16]; and addition of RANKL induces M cell differentiation [17]. Furthermore, a study showed that addition of a gamma-secretase inhibitor (DAPT) induced the differentiation of enterocytes, enteroendocrine cells, and goblet cells in human jejunal organoids [18]. For Paneth cell differentiation, several mechanical passages are required instead of a single passage conducted with TrypLE Select. However, the structure of villi cannot be reconstructed in 3D organoids even in culture with differentiation medium.
8. Optionally, the same 20% Matrigel™ solution can be reused to coat several cell culture inserts. The 20% Matrigel™ solution can be stored at −20 °C or 4 °C and reused for a few weeks.
9. Trypan blue is used to stain dead cells and accurately count live cells.
10. This step helps the cells evenly attach to the surface of the membrane by preventing the cells from clumping.

11. EGF is required for the initial few days to allow spreading of 2D cells. After the 2D cells become confluent, optimization of the medium allows cell differentiation similar to that in 3D culture. It has been reported that air-liquid interface culture promotes cell differentiation in 2D culture systems [10]. A recent study indicated that a 2D culture system can be applied to coculture with gut microbes [19].
12. Centrifugation at a high speed may disrupt the structure of organoids. If the organoids are still floating in the tubes, pipette the suspension again with fresh DPBS and wait to allow them to settle. It is acceptable to centrifuge the tubes at a low speed for a short time.
13. Representative markers of differentiated cells are described as follows: enteroendocrine cells, CHGA; Paneth cells, LYZ1; goblet cells, MUC2; tuft cells, DCLK1; and M cells, GP2. These markers can also be recognized by qPCR.
14. For whole-mount staining, the membrane should be removed from the 2D organoids before imaging to prevent diffuse reflection of light. For preparation of paraffin-embedded sections, keep the membrane to preserve the structure of the 2D organoids (Fig. 4b).

Acknowledgments

This work was in part supported by AMED (grant numbers JP21ek0109523 and JP21bm0704069), AMED-CREST (grant number JP18gm1210001), and JSPS KAKENHI (grant numbers JP21J21096, JP21K19540, JP20H03746, and JP17H06176). H.H. was supported by the Japan Society for the Promotion of Science Research Fellowships for Young Scientists.

References

1. Sato T, Clevers H (2013) Growing self-organizing mini-guts from a single intestinal stem cell: mechanism and applications. Science 340(6137):1190–1194
2. Korinek V, Barker N, Moerer P, van Donselaar E, Huls G, Peters PJ et al (1998) Depletion of epithelial stem-cell compartments in the small intestine of mice lacking Tcf-4. Nat Genet 19(4):379–383
3. Farin HF, Van Es JH, Clevers H (2012) Redundant sources of Wnt regulate intestinal stem cells and promote formation of Paneth cells. Gastroenterology 143(6):1518–1529. e7
4. Haramis A-PG, Begthel H, Van Den Born M, Van Es J, Jonkheer S, Offerhaus GJA et al (2004) De novo crypt formation and juvenile polyposis on BMP inhibition in mouse intestine. Science 303(5664):1684–1686
5. van Es JH, Van Gijn ME, Riccio O, Van Den Born M, Vooijs M, Begthel H et al (2005) Notch/γ-secretase inhibition turns proliferative cells in intestinal crypts and adenomas into goblet cells. Nature 435(7044):959–963
6. Sato T, Vries RG, Snippert HJ, Van De Wetering M, Barker N, Stange DE et al (2009) Single Lgr5 stem cells build crypt-villus structures in vitro without a mesenchymal niche. Nature 459(7244):262–265
7. Sato T, Stange DE, Ferrante M, Vries RG, Van Es JH, Van Den Brink S et al (2011) Long-

term expansion of epithelial organoids from human colon, adenoma, adenocarcinoma, and Barrett's epithelium. Gastroenterology 141(5): 1762–1772

8. Fujii M, Matano M, Toshimitsu K, Takano A, Mikami Y, Nishikori S et al (2018) Human intestinal organoids maintain self-renewal capacity and cellular diversity in niche-inspired culture condition. Cell Stem Cell 23(6): 787–793. e6
9. Moon C, VanDussen KL, Miyoshi H, Stappenbeck TS (2014) Development of a primary mouse intestinal epithelial cell monolayer culture system to evaluate factors that modulate IgA transcytosis. Mucosal Immunol 7(4): 818–828
10. Wang Y, Chiang I-L, Ohara TE, Fujii S, Cheng J, Muegge BD et al (2019) Long-term culture captures injury-repair cycles of colonic stem cells. Cell 179(5):1144–59. e15
11. Nikolaev M, Mitrofanova O, Broguiere N, Geraldo S, Dutta D, Tabata Y et al (2020) Homeostatic mini-intestines through scaffold-guided organoid morphogenesis. Nature 585(7826):574–578
12. Sugimoto S, Kobayashi E, Fujii M, Ohta Y, Arai K, Matano M et al (2021) An organoid-based organ-repurposing approach to treat short bowel syndrome. Nature 592(7852): 99–104
13. Mihara E, Hirai H, Yamamoto H, Tamura-Kawakami K, Matano M, Kikuchi A et al (2016) Active and water-soluble form of lipidated Wnt protein is maintained by a serum glycoprotein afamin/α-albumin. eLife 5:e11621
14. Ootani A, Li X, Sangiorgi E, Ho QT, Ueno H, Toda S et al (2009) Sustained in vitro intestinal epithelial culture within a Wnt-dependent stem cell niche. Nat Med 15(6):701–706
15. Sugimoto S, Fujii M, Sato T (2020) Organoid derivation and orthotopic xenotransplantation for studying human intestinal stem cell dynamics. Methods Mol Biol 2171:303–320
16. Basak O, Beumer J, Wiebrands K, Seno H, van Oudenaarden A, Clevers H (2017) Induced quiescence of Lgr5+ stem cells in intestinal organoids enables differentiation of hormone-producing Enteroendocrine cells. Cell Stem Cell 20(2):177–190. e4
17. de Lau W, Kujala P, Schneeberger K, Middendorp S, Li VS, Barker N et al (2012) Peyer's patch M cells derived from Lgr5+ stem cells require SpiB and are induced by RankL in cultured "miniguts". Mol Cell Biol 32(18): 3639–3647
18. Meran L, Massie I, Campinoti S, Weston AE, Gaifulina R, Tullie L et al (2020) Engineering transplantable jejunal mucosal grafts using patient-derived organoids from children with intestinal failure. Nat Med 26(10):1593–1601
19. Sasaki N, Miyamoto K, Maslowski KM, Ohno H, Kanai T, Sato T (2020) Development of a scalable coculture system for gut anaerobes and human colon epithelium. Gastroenterology 159(1):388–90.e5

Chapter 13

In Vitro Culture and Histological Evaluation of 3D Organotypic Cultures

William Dalleywater, Francesca Wheat, Declan Sculthorpe, Georgina Hyland, and Mohammad Ilyas

Abstract

Organotypic cultures allow cells to grow in a system which mimics in vivo tissue organization. Here we describe a method for establishing 3D organotypic cultures (using intestine as an example system), followed by methods for demonstrating cell morphology and tissue architecture using histological techniques and molecular expression analysis using immunohistochemistry, though the system is also amenable to molecular expression analysis, such as by PCR, RNA sequencing, or FISH.

Key words In vitro, 3D organotypic cultures, Intestine, Immunohistochemistry

1 Introduction

In vitro cell culture experiments use various assays to assess the behavior of cells in response to particular stimuli, microenvironments, or changes in cell populations. These usually relate to expression of particular molecules of interest, morphological analyses, and functional analyses. Three-dimensional tissue culture [1–6] has become a popular technique as it allows co-culture of multiple cellular populations, and thus hypothesis testing can take place in the type of 3D context naturally found in tissues. In order to fully interrogate these models, the molecular assays should be complemented with morphological analyses to assess changes in tissue organization and structure [7–9].

Here we describe an organotypic 3D culture technique which has been adapted to reflect tissue organization of cells in luminal organs but which can easily be adapted to other biological contexts (Fig. 1). The main benefit of this technique is the ability to study cells preserved in their morphological context (using simple histological techniques, immunohistochemistry, and in situ hybridization). We have used the technique to study the responses of

Paloma Ordóñez-Morán (ed.), *Intestinal Differentiated Cells: Methods and Protocols*, Methods in Molecular Biology, vol. 2650,
https://doi.org/10.1007/978-1-0716-3076-1_13,

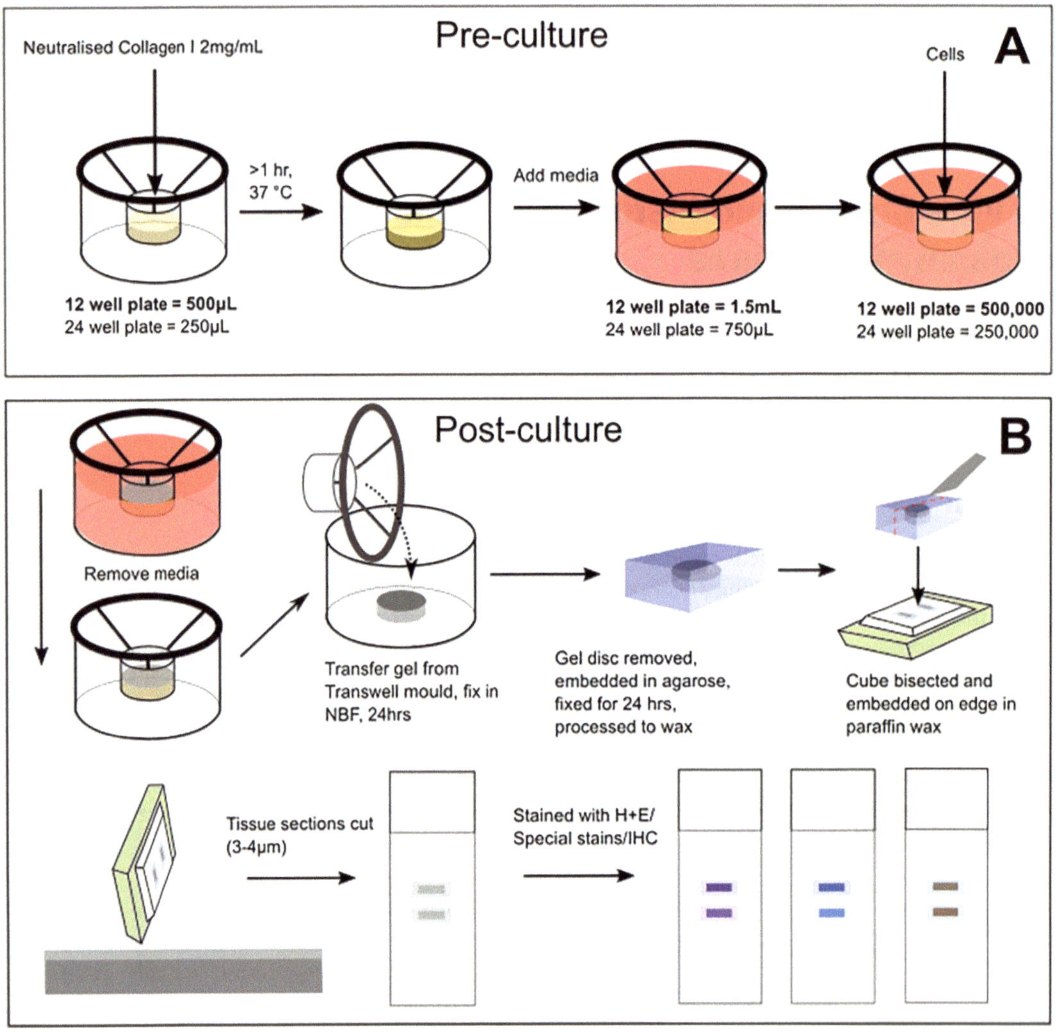

Fig. 1 An overview of the process of 3D cultures pre-culture (panel A) and post-culture histological techniques (panel B). Please see Subheading 3.1 for more details of the steps in panel A and 3.1 onward for details of the steps outlined in panel B

intestinal mucosa to exogenous growth factors and small molecules and to study tissue remodeling. The generated tissues can be processed very rapidly and can be stored indefinitely for further analysis at a later point. As well as being studied in their 3D context, the cells can also be easily retrieved for both bulk and single-cell RNA analyses. As the cells are grown as monolayers, they can be readily dissociated for separation for cell sorting to allow study of specific subpopulations.

2 Materials

2.1 3D Organotypic Culture (Collagen Gels)

1. Collagen I from rat's tail (3 mg/mL).
2. Sodium hydroxide (1 N).
3. DMEM/F12 1:1 medium with HEPES buffer.
4. Dulbecco's PBS (1×).
5. PBS (10×).
6. 12-well plate Transwell®.
7. Neutral-buffered formalin.
8. Tissue cassettes and lids.

2.2 Processing and Embedding Tissues

1. Automated tissue processor and racks.
2. Graded alcohols—ethanol or industrial methylated spirits [IMS] (50%, 70%, 90%, 100%).
3. Xylene.
4. Paraffin wax.
5. Embedding station, switched on before use to prepare cold and hot plates to correct temperature.
6. Embedding molds.
7. Heated forceps.
8. Paratrimmer or similar.

2.3 Microtomy

1. Microtome.
2. Feather-edge blades or similar microtome blades.
3. Water bath.
4. Slides (standard frosted slides and super-adherent slides such as polysine-coated slides or TOMO slides).
5. Tissue paper or slide drying rack.

2.4 Preparing Slides for Histological Techniques

1. Hot plate or slide oven.
2. Six large slide baths for dewaxing, three containing xylene and three containing 100% alcohol (ethanol or IMS).
3. Large container with distilled water.

2.5 H + E Staining

1. Coplin jars.
2. Regressive hematoxylin (e.g., Harris or Gill's hematoxylin).
3. Alcoholic eosin.
4. Tap water (for slide bluing).
5. Distilled water.

6. Acid-alcohol solution (1% concentrated hydrochloric acid in 95% ethanol—ensure the acid is added slowly to the alcohol solution).

2.6 Picrosirius Red Staining

1. Coplin jars.
2. Picrosirius red solution.
3. Stopwatch or timer.
4. Blotting paper.

2.7 Immunohistochemistry

1. Sequenza coverplates and racks.
2. Novolink immunohistochemistry kit (Novocastra).
3. Primary antibody (see recommendations in Table 1 for intestine-related antibodies).
4. Tris-buffered saline (TBS).
5. Tween-20.
6. Plastic container.
7. Plastic slide rack.
8. Microwave or pressure cooker.
9. Citrate buffer or Tris-EDTA buffer (*see* Subheading 3.6 for more information).
10. Distilled water.
11. Cold room or refrigerator.

Table 1
Some recommended antibodies for evaluating in vitro cell cultures of intestinal cell populations

Antibody	Manufacturer	Recommended dilution
Cdx-2 (EPR2764Y) (monoclonal)	Thermo fisher	1/100
SMA-α (ab5694) (polyclonal)	Abcam	1/400
E-Cadherin (24E10) (monoclonal)	Cell signalling	1/200
Villin (SP145) (monoclonal)	Abcam	1/200
Chromogranin A (ab45179) (polyclonal)	Abcam	1/500
Vimentin (D21H3) (monoclonal)	Cell signalling	1/200
MUC2 (EPR6145) (monoclonal)	Abcam	1/400
Nanog (D73G4) (monoclonal)	Cell signalling	1/100
Oct4 (2750) (monoclonal)	Cell signalling	1/100
Ki67 (MA5–14520) (monoclonal)	Thermo fisher	1/1000

2.8 Slide Dehydration and Mounting

1. Six large slide baths for dehydration, three containing xylene and three containing 100% alcohol (ethanol or IMS).
2. DPX.
3. Coverslips.
4. Wooden sticks or metal probe.
5. Fume hood.

3 Methods

3.1 3D Organotypic Culture

1. Collagen must be neutralized in order to form a gel. Type 1 rat-tail collagen (Life Technologies) is used at a stock concentration of 3 mg/mL. For each 1 mL of collagen gel required, 667 μL of stock are added to a tube on ice, for a final concentration of 2 mg/mL.
2. To this add 100 μL of 10× PBS.
3. In order to neutralize the solution, 0.025 μL of sterile 1 M sodium hydroxide are added for each microliter of collagen; for 1 mL of collagen gel, therefore, 16.7 μL of sodium hydroxide needs to be added.
4. Top up the solution with 1 mL of medium (DMEM/F12), which if neutralized correctly displays a slight pink color (*see* **Note 1**).
5. Use either 12-well or 24-well Transwell® plates (Corning) with 0.4 μm pore size membranes. The plates are prepared by adding 250 μL (24-well plate) or 500 μL (12-well plate) of ice-cold neutralized collagen into each Transwell® insert. Incubate the plates for 1 h in standard culture conditions, to allow the collagen to form a semisolid gel (*see* **Note 2**).
6. Once the collagen gel has solidified, cells can be added to the surface. Optimize the seeding density according to your cell line. A recommended starting point for optimization of the 12-well plate collagen gels is 500,000 cells per well. It can help to add ROCK inhibitor to the culture medium at passage to improve cell survival and attachment in this system. It is recommended to wait at least 24 h before media change and to change the media every 3–4 days. In our experience the cultures can be maintained for 3–4 weeks, but this will need to be optimized according to experimental endpoints and cell viability (*see* **Note 3**).
7. Following culture, media is aspirated from the wells and replaced with 10% neutral-buffered formalin (NBF) to fix the tissues. The gel can be carefully coaxed out of the Transwell mold into the well to allow it to be bathed in fixative (*see* **Note 4**).

8. Fix overnight at room temperature, for at least 18 h.
9. Remove the gel from the fixative and transfer to a small plastic or metal mold.
10. Warm a solution of 2% agarose to melt (approximately 60–70 °C).
11. Lay the molten agarose over the gel within its mold. Place on ice to solidify the agarose.
12. Bisect the gel/agarose roughly halfway through the gel.
13. After solidification, trim excess agarose (leave 5 mM margin around the gel).
14. Place the gel in a mold (*see* **Note 5**) with the bisected surface down. Pour molten agarose around the tissue and allow to solidify on ice.
15. Place the solidified gels into a tissue cassette, label, and place in NBF for a further 24 h before tissue processing.
16. The gels can now be processed to paraffin, sectioned, and stained using histological techniques described in subsequent sections.
17. Alternatively, the unfixed gels (at **step 15**) can be placed in RNA lysis buffer for downstream RNA analyses (*see* **Note 6**).

3.2 Tissue Processing and Embedding

1. It is recommended to use an automated tissue processor for tissue dehydration and wax penetration. Any standard program of 12–14-h length is sufficient for these tissues.
2. Load the cassettes into the tissue processor rack ensuring they are secure.
3. Start the tissue processor using a standard program.
4. Once the tissue processor has completed, transfer the cassettes into a molten wax bath in the tissue embedding station.
5. As the orientation has already been completed in Subheading 3.1, the tissues can simply be transferred from the cassette into the tissue mold with the previously bisected/cut surface facing down in the mold.
6. Move the mold to the hot plate and fill the mold with wax.
7. Apply gentle pressure to the tissue with heated forceps and transfer to the cold plate.
8. Allow the wax to solidify slightly (so that the tissue remains at the bottom of the mold), and then place the labeled cassette on top of the mold so that the bottom of the cassette is placed on top of the mold. Top up the mold with wax so that the base of the cassette is covered.

9. Place on the cold plate of the embedding station and leave to cool and solidify completely.
10. Once fully cooled, the cassette can be removed from the mold with gentle pressure.
11. Blocks should be free of extraneous wax round the edges of the cassette. Use the block iron [Paratrimmer] (carefully—very hot surface) pressing each of the four edges of the block for a few seconds until the wax melts off. Be careful to avoid pressing the main surfaces of the wax (i.e., the tissue surface and the opposite side) to the heat block.
12. The block can either be stored or sectioned with a microtome.

3.3 Microtomy

1. The following is a general protocol for tissue microtomy—please ensure all safety features are in operation and you are familiar with the microtome before use (*see* **Note 7**).
2. Turn on the water bath at least 15 min before you wish to commence section cutting. Ensure the water is clean/free of debris. Top up or replace the water in the bath with fresh distilled water. The water bath should be set to 48 °C.
3. Blocks should be kept on ice with the tissue surface face down for at least 15 min before section cutting (*see* **Note 8**).
4. *Safety*: Check whether there is a blade loaded by carefully revealing the blade by sliding the guard out. If not or it needs replacing, unlock the blade holder and remove the blade using forceps. Discard the blade in a sharps bin. Remove a fresh blade from the blade dispenser and load using forceps to slide the blade into place. Relock the blade and conceal the blade with the guard.
5. *Safety*: Lock the microtome wheel and place the guard over the blade.
6. First, align and trim the block. Set the microtome to 5 microns and place the block into position on the stage. Ensure the block holder arm is not overly extended. Unlock the microtome wheel and reveal the blade by sliding the guard away. Move the block closer to the blade by repositioning the blade holder assembly and sliding it so that it is slightly away from the block surface. Observe how the surface of the block aligns with blade, and roughly align the surface so that by eye the horizontal and vertical edges are roughly in equal alignment relative to the blade edge. Rotate the microtome wheel to move the block surface closer to the blade edge. When the block surface is just being cut by the blade, realign the block by looking at the block surface (*see* **Note 9**).

7. Once satisfied with the alignment, proceed to trim in the block until the desired tissue surface is reached. The tissue surface is reached when the wax surface is shiny, and the outline of the tissue can be seen when the block surface is held to the light. If the tissue has been embedded unevenly, it may be necessary to trim in further to full face in order to see the desired tissue structures.
8. Once trimmed in, return the tissue block to the ice box with the surface face down on the ice. Remember to use appropriate safety measures when removing the block from the microtome—locking the microtome wheel and placing the guard over the blade.
9. Wait 30 s–1 min for the block to cool. In the meantime, set the microtome to the required cutting depth (3–4 microns suggested). Check that the appropriate safety measures are in place, and replace the block in the microtome, ensuring it is in the same orientation as when trimming in.
10. Prepare the microtome for cutting by sliding the guard away and unlocking the wheel.
11. Rotate the microtome wheel to create a ribbon of sections. After cooling, the first few turns may produce no sections or incomplete sections as the tissue surface has retracted slightly. A ribbon will be formed of a few sections of wax/tissue the same shape as the block outline joined together. Forming a ribbon makes the sections easier to transfer to the water bath and is more efficient at generating sections, but may take some practice to master.
12. The ribbon can be transferred to the water bath by gripping one end with forceps and carefully lifting off the knife stage with a brush. Once lifted, move over to the water bath, and quickly lay the ribbon down, starting with the section furthest from the one held by the forceps.
13. Allow up to a minute for the sections to float on the water bath. Prepare slides by labeling in the meantime (*see* **Note 10**). To transfer the section to the slide, place the slide vertically in the water bath with the slide surface facing the section you want to lift. Bring the slide closer to the section and beginning at one edge of the section, allow this edge to stick to the slide. Lift the slide upward, gently angling the top of the slide toward the surface of the section. Stop lifting once the bottom edge of the section is attached to the slide. To separate the section from the rest of the ribbon, gently wiggle the slide until the ribbon detaches.
14. Stand the slide on some tissue paper or in a slide rack to allow the water to drain off. Repeat the lifting process for the remainder of the sections in the ribbon (*see* **Note 11**).

3.4 Preparing Slides for Histology

Turn on the slide oven or hot plate, and allow to warm for at least 15 minutes. Set the temperature to approximately 60 °C:

1. Once at temperature, place the slides in the oven. Allow the tissue/wax to melt completely and to remain in the oven or on the hot plate for at least 30 min.
2. Remove the slides from the oven and place in a slide rack.
3. Proceed to deparaffinize/rehydrate the tissue by moving through sequential xylene/alcohols:
   (a) Xylene 1, 5 min.
   (b) Xylene 2, 5 min.
   (c) Xylene 3, 5 min.
   (d) Ethanol/IMS 1, 2 min.
   (e) Ethanol/IMS 2, 2 min.
   (f) Ethanol/IMS 3, 2 min.
   (g) Distilled water, at least 2 min.

The tissue is now ready for either histological staining or immunohistochemistry.

3.5 Histological Staining

Hematoxylin and Eosin

This is a good general stain to see the overall architecture of the tissue:

1. Filter previously prepared hematoxylin (*see* **Note 12**) before use.
2. Place the rehydrated slides in a Coplin jar containing hematoxylin or lay the slides on a tray, and pipette a small amount of hematoxylin onto the tissue.
3. Allow the hematoxylin to stain for 5 min.
4. Wash excess hematoxylin with water.
5. Differentiate in acid-alcohol for a few seconds and quickly return to the water to wash off the acid-alcohol (*see* **Note 13**).
6. Place the slides in tap water for 5 min to blue the hematoxylin.
7. Wash slides in distilled water for 30 s.
8. Place the slides in a Coplin jar with eosin or lay the slides on a tray, and pipette a small amount of eosin onto the tissue. Alcoholic eosin is preferred to give bright staining. Allow staining for 2 min.
9. Wash the excess eosin from the slide by washing with water.
10. The slides are now ready for dehydration and mounting (Subheading 3.7).

Picrosirius Red

This stain highlights areas of fibrosis/tissue remodeling. Cells will be yellow, while the extracellular matrix will be pink/red. Greater staining intensity reflects denser extracellular matrix in areas of collagen deposition:

1. Prepare a Coplin jar by filling with pre-made picrosirius red solution (*see* **Note 14**).
2. Place the rehydrated slide in the Coplin jar, for 45–60 min. Optimize this based on the preferred staining intensity.
3. Remove the slide from the Coplin jar and blot the excess PSR solution from the slide with blotting paper.
4. Proceed with dehydration and mounting as described in Subheading. 3.7 (*see* **Note 15**).

3.6 Immunohistochemistry

Antigen retrieval is usually required. Perform antigen retrieval according to the antibody manufacturer's recommendations, particularly taking note of the correct buffer for antigen retrieval. We recommend using the Sequenza coverplate system or similar for manual immunohistochemistry as it saves on reagents and prevents evaporation of the reagents which can ruin the procedure.

Buffers

- Citrate buffer, pH 6 (10 mM citric acid, 0.05% Tween-20), adjust to pH 6 with 1 N acid/base solutions (hydrochloric acid and sodium hydroxide). Add the Tween-20 once the pH has been optimized.
- Tris-EDTA buffer, pH 9 (10 mM Trisma base, 1 mM EDTA, 0.05% Tween-20). Adjust to pH 9 with 1 N acid/base solutions (hydrochloric acid and sodium hydroxide). Add the Tween-20 once pH has been optimized.
- Note: Tris-EDTA can give high background staining, so citrate buffer is preferred unless the manufacturer's instructions recommend Tris-EDTA buffer. Optimization with your own equipment (e.g., for automated immunohistochemistry) is recommended, trialing both buffers for best results:

    1. *Safety: ensure no metal parts are placed in the microwave.* Pre-warm the chosen antigen-retrieval buffer in a plastic box in a microwave on the high setting until the buffer begins to boil. Immerse a plastic rack with the dewaxed and hydrated slides in the pre-warmed buffer, and microwave for 20 min at 95 °C (low setting on microwave to simmer). Keep the probe covered by buffer. This can alternatively be performed

in a pressure cooker or on an automated immunohistochemistry platform. *Safety: supervise this process at all times to ensure the container does not boil dry, if using the microwave technique* (*see* **Note 16**).

2. Using gloves, carefully remove the plastic tin, and let the solution cool down slowly (for approximately 15 min).
3. Working in a large bath of tap water, mount the Sequenza coverplates on the slides and place on the trays.
4. Align the slide with the dots at the bottom of the coverplate.
5. Make sure there are no bubbles by inspecting the mounted slides. If bubbles are present, repeat **steps 3** and **4**.
6. Insert the slide on the tray slot (until it clicks) and fill it up to the brim with distilled water; the level should reduce gradually over approximately 5 minutes (*see* **Note 17**).
7. Fill the Sequenza reservoir half full with TBST buffer (pH 7.6) to rinse the slides.
8. Apply 100 μL of peroxidase block (Novolink kit) for 5 min.
9. Wash with TBST 2 × 5 min.
10. Apply 100 μL of protein block (Novolink kit) for 5 min.
11. Wash with TBST 2 × 5 min.
12. Apply 100 μL of the appropriate dilution of the primary antibody in TBST. The Novolink kit is compatible with primary antibodies raised in either rabbit or mouse hosts (*see* **Note 18**).
13. Wash with TBST 2 × 5 min.
14. Apply 100 μL post primary (Novolink kit) for 30 min.
15. Wash with TBST 2 × 5 min.
16. Apply 100 μL Novolink polymer for 30 min.
17. While you are waiting, make up DAB solution 1:20, DAB chromogen in DAB substrate buffer. (DAB solution should be kept in the dark and used within 6 h.)
18. Wash sections with TBST 2 × 5 min.
19. Apply 100 μL DAB working solution for 5 min.
20. Wash with TBST 2 × 5 min.
21. Apply 100 μL Novolink hematoxylin for 6 min to counterstain.
22. Remove slides from Sequenza plates and place in a rack in water to remove excess hematoxylin.
23. Slides are now ready for dehydration and mounting (Subheading 3.7).

3.7 Dehydration and Mounting

It is recommended to set up a series of three alcohol (either ethanol or industrial methylated spirits) and three xylene baths in a fume hood for these steps. Note for picrosirius red staining, fewer [3] and faster dips in the alcohol baths should be used:

1. Six fast dips in alcohol I.
2. Six fast dips in alcohol II.
3. Six fast dips in alcohol III.
4. Six slow dips in xylene I.
5. Six slow dips in xylene II.
6. Place the slide rack in xylene III in readiness for mounting the coverslip with DPX.
7. Get a coverslip ready by smearing a small drop of DPX from one end of the coverslip to the other down the center (a small wooden stick or metal probe is ideal for this).
8. Take a slide from the xylene and place the slide surface gently onto the coverslip, the xylene and DPX should mix, and when the slide is lifted, the coverslip should be held on the surface. Ensure the coverslip covers the tissue.
9. Flip the slide over and gently press the coverslip onto the slide surface to push the DPX to the edges and remove air bubbles.
10. There should be a thin bead of DPX around the edge of the coverslip. Excess DPX can be removed with a cotton bud.
11. Allow the DPX to air dry in the fume hood (overnight is sufficient). The DPX will set to a resin-like consistency.

The slides can now be viewed under the microscope or digitally scanned (Fig. 2 and Table 2).

4 Notes

1. Work fairly rapidly during this step. In our experience, the collagen mixture remains liquid for a few minutes once neutralized and at low temperature. Gelling occurs rapidly at room temperature or higher.
2. Using medium with an indicator is important, as the pH is important. If the pH is too low, the gel will not set properly. If the pH is too high, cell viability will be poor.
3. It is recommended to use 1.5 mL media volume for the 12-well plate system and 750 μL for the 24-well plate system. In our experience, this ensures the growth surface is bathed in media during cell spreading. If necessary, the media volume can be reduced after the first media change to achieve an air-liquid interface, but the media may require more frequent changes.

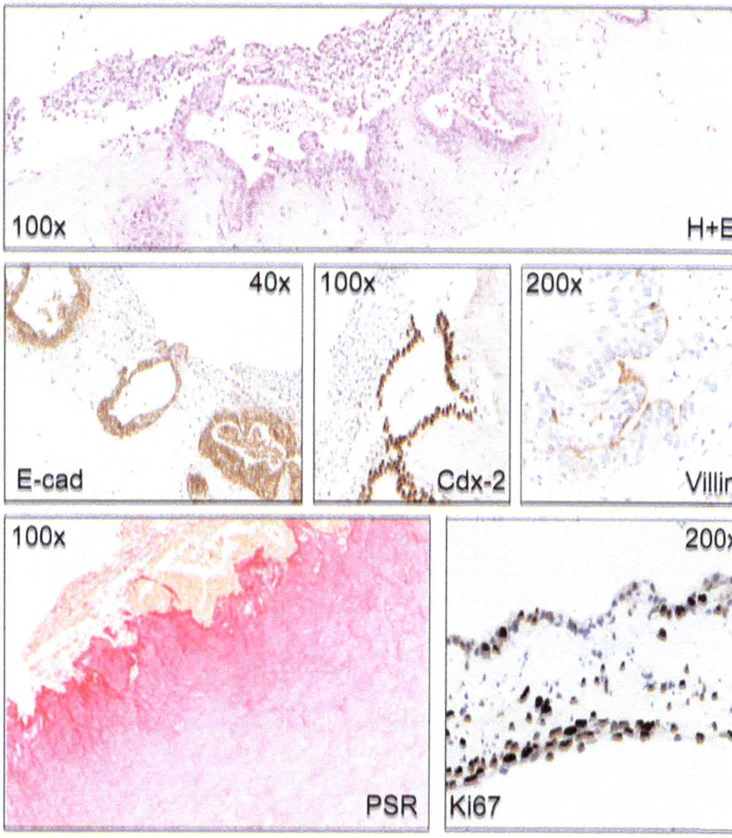

Fig. 2 Example of staining with hematoxylin and eosin (H + E), E-cadherin (E-cad), Cdx-2, villin and Ki67 immunohistochemistry, and picrosirius red (PSR) histochemical staining. All slides were prepared using the protocols described here, and images were scanned using a Roche Ventana DP200 slide scanner. Magnifications are given as 40×, 100×, and 200× corresponding to using 4×, 10×, and 20× objectives with a 10× eyepiece on a light microscope

4. A small pipette tip can be helpful to gently prize the gel out of the Transwell mold.
5. A tissue embedding mold is ideal.
6. We have tested this using the Sigma GenElute (TM) Mammalian Total RNA Miniprep Kit and successfully extracted high concentrations of RNA sufficient for qPCR analysis of multiple genes.
7. If used incorrectly, the microtome has the potential to cause serious injury by cutting. The microtome blade is extremely sharp, and additionally the stage can cause crushing and force onto the blade edge. The microtome has a range of safety features and principles which make it very safe to use if followed diligently. Ensure you have been shown the safety features and follow these meticulously each time. Proper training in microtomy is essential.

Table 2
Antibodies according to expected staining pattern and purpose. This is a list of antibodies we have found useful or may be useful for deducing various biological phenomena in 3D cultures of gastrointestinal cell populations using this system. A brief overview of their expected staining pattern and purpose is given below

Antibody	Staining pattern	Purpose
Cdx-2	Nuclear	Demonstrates intestinal differentiation within the epithelial layer; highlights morphology of the epithelial cell layer
Villin	Membranous (apical membrane)	Demonstrates intestinal epithelial differentiation. It should only be expressed on the luminal edge of the epithelial cells and therefore is useful for showing polarization of the epithelium
E-Cadherin	Membranous (basolateral membranes)	Demonstrates intestinal epithelial differentiation and if on the basolateral membranes only shows polarization of the epithelial monolayer
Chromogranin A	Cytoplasmic (within granules/vacuoles)	Shows heterogeneity of the epithelium by highlighting neuroendocrine cell differentiation
Synaptophysin	Cytoplasmic	Demonstrates neuroendocrine differentiation, often used in combination with chromogranin A
MUC2	Cytoplasmic (within a round vacuole)	Shows heterogeneity of the epithelium by highlighting goblet cell differentiation
Ki67	Nuclear	This highlights proliferative cells. Within mature intestinal epithelium, this should be restricted to the base of the crypts (within stem cells). This helps to demonstrate polarization and compartmentalization of the epithelium
SMA-α	Cytoplasmic	Demonstrates mesenchymal (fibroblast/smooth muscle) differentiation
Desmin	Cytoplasmic	Demonstrates smooth muscle differentiation
Vimentin	Cytoplasmic	Demonstrates mesenchymal differentiation
Nanog and Oct-4	Nuclear	Shows pluripotency within cell cultures. Should be negative in pure/fully differentiated intestinal cell populations
CK7	Cytoplasmic	Low molecular weight cytokeratin which is typically expressed by cells of upper gastrointestinal and other proximal endoderm origin (e.g., lung, thyroid)
CK20	Cytoplasmic	Low molecular weight cytokeratin which is expressed by cells of intestinal origin within the upper parts of the crypt. Also expressed by urothelial cells and Merkel cells
AE1/AE3	Cytoplasmic	Pan-cytokeratin, which shows epithelial differentiation. Can show aberrant staining in non-epithelial cells in malignancy
BerEP4 (EpCAM)	Membranous	Epithelial differentiation marker. As it is membranous, it is particularly useful for viable cell separation/sorting
MUC5AC	Cytoplasmic	Stains mucin particularly of upper gastrointestinal origin (stomach)

(continued)

Table 2 (continued)

Antibody	Staining pattern	Purpose
CK5/6	Cytoplasmic	High molecular cytokeratin, relatively specific for demonstrating squamous differentiation (e.g., esophagus, skin) and basal/myoepithelial cells in some tissues (e.g., salivary glands, breast)
P63	Nuclear	Transcription factor for squamous cells and basal cells
CK14	Cytoplasmic	High molecular cytokeratin, demonstrates squamous and basal/myoepithelial cell differentiation

8. Cool blocks are essential for accurate cutting and tissue preservation.
9. If one vertical edge is being cut more than the other, the horizontal alignment needs adjusting away from that edge and likewise for the horizontal edges with vertical alignment. The alignment does not need to be perfect, but ensuring good alignment at the trimming stage saves a lot of difficulty later with poor sections.
10. If the slides are to be used for H + E or special stains, standard frosted slides are sufficient. If the slides are to be used for immunohistochemistry, adherent slides (such as polysine-coated slides or TOMO slides) should be used.
11. The last section, used for lifting the ribbon/gripped by forceps, can be discarded as the quality may be poor if crushed by forceps.
12. A regressive hematoxylin such as Harris or Gill's is preferable. Hematoxylin can either be purchased pre-made or made up according to preferred formulation; consult other sources for information on this.
13. If left for too long, hematoxylin staining will be very weak in the tissue. If this is the case, the hematoxylin step can be repeated and a shorter differentiation in acid-alcohol should be used. Ideally, check the staining under a microscope to verify correct staining.
14. A ready-made solution such as Abcam's picrosirius red solution is recommended [ab246832], but this may be changed according to preference.
15. The alcohol steps should be expedited as the PSR is highly soluble in alcohol. Three quick dips in each alcohol solution are recommended. The xylene and mountant steps can be undertaken as normal. If there is residual water on the slide in the xylene step, cautiously repeat an alcohol step, again proceeding as rapidly as possible.

16. 95 °C is suitable for most antibodies, but some may require alternative temperatures or incubation times—check manufacturers' recommendations.
17. If the level reduces very quickly, it means the face of the slide is not flush with the coverplate. In this case, go back to **step 3** and repeat the following steps. If the levels reduce very slowly, it usually means there is an air bubble or other obstruction. Again, go back to **step 3** and begin the process from here.
18. It is recommended to optimize the primary antibody specifically for the Novolink kit as the antibody dilutions used for other purposes or with different kits can result in excessive background staining. We find that incubation overnight at 4 °C is optimal for most primary antibodies, but this should be optimized in consultation with the manufacturer's instructions for the antibody.

References

1. Thorne CA, Chen IW, Sanman LE, Cobb MH, Wu LF, Altschuler SJ (2018) Enteroid monolayers reveal an autonomous WNT and BMP circuit controlling intestinal epithelial growth and organization. Dev Cell 44(5):624–33.e4
2. Fujii M, Matano M, Toshimitsu K, Takano A, Mikami Y, Nishikori S et al (2018) Human intestinal organoids maintain self-renewal capacity and cellular diversity in niche-inspired culture condition. Cell Stem Cell 23(6):787–93.e6
3. Gjorevski N, Sachs N, Manfrin A, Giger S, Bragina ME, Ordóñez-Morán P et al (2016) Designer matrices for intestinal stem cell and organoid culture. Nature 539(7630):560–564
4. Hartl L, Huelsz-Prince G, van Zon J, Tans SJ (2019) Apical constriction is necessary for crypt formation in small intestinal organoids. Dev Biol 450(2):76–81
5. Qi Z, Chen YG. Efficient Culture of Intestinal Organoids with Blebbistatin. (1940–6029 (Electronic))
6. Sprangers J, Zaalberg IC, Maurice MM (2021) Organoid-based modeling of intestinal development, regeneration, and repair. Cell Death Differ 28(1):95–107
7. Zanoni M, Cortesi M, Zamagni A, Arienti C, Pignatta S, Tesei A (2020) Modeling neoplastic disease with spheroids and organoids. J Hematol Oncol 13(1):97
8. Hayden PJ, Harbell JW (2021) Special review series on 3D organotypic culture models: introduction and historical perspective. In Vitro Cell Dev Biol Anim 57(2):95–103
9. Shamir ER, Ewald AJ (2014) Three-dimensional organotypic culture: experimental models of mammalian biology and disease. Nat Rev Mol Cell Biol 15(10):647–664

Chapter 14

Fluorescence Intensity and *Fluorescence Lifetime Imaging Microscopies* (FLIM) of Cell Differentiation in the Small Intestinal Organoids Using Cholera Toxin

Irina A. Okkelman and Ruslan I. Dmitriev

Abstract

Live cell microscopies of in vitro, ex vivo, and in vivo experimental intestinal models enable visualizing cell proliferation, differentiation, and functional cellular status in response to intrinsic and extrinsic (e.g., in the presence of microbiota) factors. While the use of transgenic animal models expressing biosensor fluorescent proteins can be laborious and not compatible with clinical samples and patient-derived organoids, the use of fluorescent dye tracers is an attractive alternative. In this protocol, we describe how the differentiation-dependent intestinal cell membrane composition can be labeled using fluorescent cholera toxin subunit B (CTX) derivatives. By using the culture of mouse adult stem cell-derived small intestinal organoids, we show that CTX can bind specific plasma membrane domains in differentiation-dependent manner. Green (Alexa Fluor 488) and red (Alexa Fluor 555) fluorescent CTX derivatives also display additional contrast in a fluorescence lifetime domain, when probed by the fluorescence lifetime imaging microscopy (FLIM), and can be used together with other fluorescent dyes and cell tracers. Importantly, CTX staining remains confined to specific regions in the organoids after fixation, which enables using it in both live cell and fixed tissue immunofluorescence microscopies.

Key words Cholera toxin, Intestinal epithelium, Intestinal organoid, FLIM, Live cell imaging

1 Introduction

Live fluorescence microscopy and mesoscale imaging in 3D cell and tissue models witness a rapid growth, thanks to merging of many areas of the life sciences, such as optics, stem cell-derived organoids, biofabrication, single-cell sequencing, machine learning, and others [1–4]. This is particularly important for adoption of innovative microscopy techniques in advanced organoid, "mini-gut," and related *tissue-on-a-chip* biological models: light sheet, two-photon microscopy, macro-imaging, and fluorescence lifetime imaging microscopy (FLIM) [5].

Paloma Ordóñez-Morán (ed.), *Intestinal Differentiated Cells: Methods and Protocols*, Methods in Molecular Biology, vol. 2650, https://doi.org/10.1007/978-1-0716-3076-1_14,

To successfully analyze the distinct functional cell types (stem cells, enteroendocrine, goblet, Paneth, enterocytes, and others) within the live intestinal epithelium, a specific cell labeling is often required. This can be achieved by (i) using genetically modified and expressing fluorescent protein tags (e.g., Lgr5-GFP organoids) [6], (ii) specific staining of the live cells with fluorescent dyes and probes, and (iii) antibodies and using fixed samples [7, 8]. Typically, a number of employed fluorescence dyes ("colors") are limited to ~3–4 fluorescent spectral channels (i.e., blue, green, and red channels, typically corresponding to DAPI/Hoechst, GFP/FITC, and rhodamine), expanded to some more with laser-based excitation and spectral unmixing on the confocal and more advanced microscopes. In addition, new imaging modalities such as fluorescence lifetime imaging microscopy (FLIM) enable for improved multiplexing, i.e., based on so-called 'tau contrast' or lifetime contrast. Ideally, fluorescent labeling not only provides for cell staining but also adds functional information on the microenvironment, pH, hypoxia, and other analytes and biomarkers and can be integrated into the multiparameter live cell microscopy [5].

One of the characteristic features of differentiating intestinal epithelial cells was noted a few decades ago [9]: their ability to accumulate cholera toxin drastically changes over the course of the intestinal cell maturation and the age of the culture. Interestingly, the exact targets of the cholera toxin subunit B binding are not well defined and likely to include multivalent binding interactions with different membrane surface receptors: such high-affinity receptors as gangliosides GM1, GD1b, and fucosyl-GM1, weak receptors such as M2 [10–12], and fucosylated glycoproteins resembling the structure of histo-blood group antigen Lewis^{X} [13, 14]. Recently, a secondary binding site located on the lateral side of B-pentamer for fucosyl receptors and distinct from primary GM1 binding site ("bottom surface" of B-pentamer) was suggested to enhance cholera holotoxin attachment to the cell membranes with a few available GM1 receptors [14, 15]. The number of the fucosyl-glycoprotein CTX targets in the intestinal epithelium is still unknown. Nevertheless the decrease of fucose content along the crypts-villi axis [16] and the importance of its modification for the intestinal epithelium homeostasis are well appreciated, e.g., the role of O-fucosylation of EGF-like repeats in regulation of Notch signaling [17], required for intestinal stem cell maintenance and regulation of differentiation [18]. Another important aspect of CTX membrane binding expanding the number of CTX targets in the intestinal tissue is the cell cycle stage dependence (G0/G1 and slow decrease with S-phase) [19].

Our team has previously described the labeling of mouse intestinal organoids with Alexa Fluor 488-conjugated CTX [20]. We found that with the culture of mouse small intestinal organoids, CTX labels actively proliferating regions, co-localizing with crypts and Lgr5+ cells. When monitored by the FLIM, we also observed increased "lifetime contrast" of the CTX fluorescent conjugates. Here, we provide a detailed protocol of using CTX for labeling cell differentiation in intestinal organoid culture using widely available conventional fluorescence microscope and FLIM platforms. As a practical example, we demonstrate the utility of CTX labeling for studying vitamin D3-induced cell differentiation.

In summary, we demonstrate here that application of CTX-Alexa Fluor 488 and 555 conjugates can be a promising tool for analysis of differentiation in small intestinal organoid culture. However, some parameters of their staining still have to be evaluated: First, the exact targets of CTX have to be identified, including the nature of CTX-binding receptors on the membrane surface and the origin of cells, especially in respect to their fluorescence lifetime differences. Secondly, it will be important to reveal the binding efficiency, localization, and fluorescence lifetime characteristics of different CTX-Alexa Fluor conjugates, since the structure (e.g., charge and its hydrophobicity) of fluorescent dye would potentially affect the binding properties of CTX molecule itself. These can be done with additional analysis of CTX-Alexa Fluor conjugates binding by analyzing effects of cell surface glycosylation and lipid composition of intestinal organoids using multiparametric assays with live imaging probes and genetically encoded markers and by immunofluorescence. Intestine is a well-known target of the cholera toxin. Our experiments show that this feature is helpful for developing new live quantitative imaging method(s) for intestinal cell differentiation analysis especially using the "fluorescence lifetime contrast". In perspective, testing and discovery of additional fluorescent conjugates of bacterial toxins and lectins will help expanding the number of quantitative live microscopy intensity and FLIM imaging probes.

2 Materials

All solutions and buffers should be prepared using ultrapure Milli-Q grade (18 MΩ cm) sterile water. Store all reagents at 4 °C for no longer than 4 weeks, unless specified otherwise. Perform all cell culture procedures under the laminar flow (class II biological safety level cabinet with HEPA filters) and aseptically. Wear gloves and spray all used materials with 70% ethanol. Unless provided sterile, sterilize glass and plasticware by autoclaving (121 °C, 20 min). Media and buffer solutions must be sterilized prior to use by filtration (0.2 μm) or autoclaving where appropriate.

2.1 Small Intestinal Organoids

Mouse small intestinal organoid culture expressing Lgr5-GFP: Lgr5-EGFP-IRES-CreERT2 (Laboratory of Prof. H. Clevers, Hubrecht Institute, the Netherlands [21]) produced from the mouse available from the Jackson Laboratory (B6.129P2-*Lgr5tm1(cre/ERT2)Cle/*J https://www.jax.org/strain/008875) using conventional techniques [22].

2.2 Chemicals, Plasticware, and Equipment

1. 1 M HEPES solution pH 7.2, 100× concentrate, sterile (Sigma, H0887).
2. Penicillin-streptomycin solution (p/s) 100× concentrate, sterile (Sigma, P0781).
3. GlutaMAX solution form 100× concentrate, sterile (Gibco, 35050038).
4. 100 mM sodium pyruvate solution, 100× concentrate, sterile (Sigma, S8636).
5. Matrigel (growth factors reduced, phenol red-free, e.g., Corning 356231): thaw the original bottle at 4 °C overnight. Keep all the time on ice. Pipet well and make aliquots. Aliquots can be stored at −20 or −80 °C until the expiration date. Defrost the aliquot in advance, i.e., 1–2 days before, at 4 °C.
6. AdDF+++ (500 mL): supplement DMEM F12 Ham containing 15 mM HEPES (Sigma, D6421) with 100 U/mL penicillin-streptomycin solution (from 100× concentrate) and 2 mM GlutaMAX (from 100× concentrate). Store at 4 °C for up to 1 month.
7. B-27 media supplement, serum-free, 50× concentrate (Invitrogen, 17.504-044). Aliquot aseptically and store at −20 or −80 °C prior to use.
8. N-2 media supplement, 100× concentrate (Invitrogen, 17.502-048). Aliquot aseptically and store at −20 or −80 °C prior to use.
9. 500 mM N-acetyl-L-cysteine (NAC) 400× concentrate solution in water (Sigma, A9165). Store aliquots at −20 °C for 1 month.
10. Basal culture medium (BCM): supplement 50 mL of AdDF+++ media with 1 mL of B27 (50×), 0.5 mL of N2 (100×), and 125 μL of NAC (400×). Use for preparation of ENR, EN, or ENRVC media. Store at 4 °C for 1 month.
11. 0.1% bovine serum albumin (BSA, Sigma, A4503) solution in phosphate buffer saline (PBS), Ca^{2+}- and Mg^{2+}-free (BSA/PBS, Sigma, P4417). Sterilize by filtration (0.2 μm) prior to use.
12. Murine recombinant EGF (epidermal growth factor). Prepare 10,000× concentrate solution (500 μg/mL) in BSA/PBS from lyophilized dry powder (PeproTech, 315-09 or STEMCELL Technologies, 78016). Store aliquots at −80 °C for 1 month.

13. Murine recombinant Noggin. Prepare 1000× concentrate solution (100 μg/mL) in BSA/PBS from lyophilized dry powder (PeproTech, 250-38 or STEMCELL Technologies, 78061). Store aliquots at −80 °C for 1 month.
14. Human recombinant R-spondin. Prepare 1000× concentrate solution (1 μg/μL) in BSA/PBS from lyophilized dry powder (PeproTech, 120-38 or STEMCELL Technologies, 78213.1). Store aliquots at −80 °C for 1 month.
15. Dimethyl sulfoxide (DMSO), "Hybri-Max" grade (Sigma, D2650).
16. 3 mM CHIR99021 stock solution in DMSO (1000× concentrate). Store aliquots at −20 °C for 1 month (Sigma, SML1046).
17. 500 mM sodium valproate (VA) solution in water (100× concentrate, Sigma, P4543).
18. Intestinal organoid growth media (ENR): supplement 10 mL of basal media with 10 ng/mL Noggin, 1 μg/mL R-spondin, and 50 ng/mL EGF. In respect to the differentiation status of intestinal organoid culture required for the experimental procedure, the composition of the growth media can be changed accordingly to ENRVC (pro-proliferative conditions) or to EN (pro-differentiation conditions). Find the comparative composition of the media in Table 2.
19. 12-well and 24-well flat-bottom tissue culture (TC) grade plates, sterile.
20. Sterile plastic tubes for 1.5, 15, and 50 mL.
21. 10 mM Rock inhibitor Y-27632 stock solution in PBS (1000× concentrate, Sigma, SCM075).
22. Dulbecco's modified Eagle's medium, phenol red-, glucose-, pyruvate-, and glutamine-free (Sigma, D5030).
23. D(+)-Glucose, powder (Sigma, G8270). Prepare 1 M stock solution in sterile MQ-water and store at 4 °C.
24. Imaging medium: DMEM supplemented with sodium bicarbonate (1.2 g/L), 10 mM HEPES-Na, pH 7.2, 1 mM sodium pyruvate, 10 mM glucose, without phenol red.
25. 1α,25-Dihydroxycholecalciferol (VD3, Sigma, D1530): make a 100 μM stock solution in 99% ethanol. Seal the vial with Parafilm and store at −20 °C.
26. Humidified CO_2/37 °C incubator.
27. Swinging bucket rotor-equipped centrifuge with cooling option 4 °C.
28. Microscope setup.

For fluorescence intensity-based ("conventional") microscopy, upright live cell imaging (water-dipping 20×~60× objective) or inverted (long working distance 20×~60× objective) compatible with excitation and emission of Alexa Fluor 488 and 555 dyes, heated stage, and climate control (37 °C, optional 5% CO_2 in case of bicarbonate-supplemented growth medium or 10–20 mM HEPES, pH 7.3–7.4 buffered solution can be used instead), equipped either with light-emitting diode (LED), 488 nm laser, or multiphoton laser light sources.

Example of the fluorescence microscope for intensity-based imaging: Olympus IX81 inverted microscope (Olympus), equipped with LED excitation source covering range of 365–770 nm (CoolLED pE4000), high-speed ORCA-Flash4.0LT+ (Hamamatsu) camera, Z-axis control, T-stage control (Okolab), long working distance water immersion objective 60×/1.0, 2 mm working distance LUMPFLN60XW (Olympus), respective fluorescence filter cubes (Hoechst, FITC, rhodamine), and cellSens Dimension software (Olympus).

For FLIM, laser-based systems such as TCSPC-based from Becker & Hickl, PicoQuant, or Leica Microsystems, having appropriate excitation sources (i.e., pulsed diode, white light, or two-photon lasers), acousto-optical modulators, detectors, and respective software as described in our previous protocol [20]. The motorized XYZ stage control and possibility of performing mosaic imaging are highly desirable. A number of new custom-made solutions for low-cost FLIM upgrade also become available [23, 24].

Example of the FLIM microscope: an upright Axio Examiner.Z1 microscope (Zeiss) equipped with water-dipping 63×/1.0 W-Plan Apochromat objective, DCS-120 confocal TCSPC FLIM scanner (Becker & Hickl GmbH), custom-made heated incubator (Life Imaging Services, Basel, Switzerland) and stage T control (Zeiss/Pecon), motorized Z-axis control, pulsed diode lasers BDL-SMNI 488 nm (Becker & Hickl GmbH), and appropriate long and band-pass emission filters (495 nm longpass, 512–536 nm for Alexa Fluor 488 and 565–605 nm for Alexa Fluor 555). Software: SPCImage (Becker & Hickl), ImageJ (fiji.sc), and Microsoft Office. Additional excitation sources, i.e., for imaging Hoechst 33342, O_2 probes, and other biosensors, can be also added [20, 25–27].

2.3 Microscopy Imaging Supplies

1. *For upright microscope*: tissue culture minidish (35/10 mm). The amount of organoid growth media can be minimized by insertion of autoclave-sterilized silicon microchamber (0.1–0.2 mL growth media volume). Before the imaging, the inserted chamber can be removed, and >2.5 mL of the imaging media can be added (see the composition below).

2. *For inverted microscope*: μ-chambers, 12-well (24 × 60 mm, ibidi GmbH, cat no. 81201). The silicon part is reusable and autoclavable and can be attached to any dry plastic or glass surfaces. Use the cover glass with appropriate thickness (e.g., No. 1.5, depends on the objective specifications), and apply appropriate immersion fluid (water, oil, or glycerol) during the imaging.

2.4 Preparation of Fixed Organoid Samples

Avoid unnecessary exposure of the probes to the light, and store them in nontransparent dark plastic vials or wrap in tinfoil. Mix well after defrosting and prior to the use. The imaging probes used in this chapter are listed in Table 1:

1. 0.5 mg/mL cholera toxin, subunit B, recombinant Alexa Fluor 488 (Invitrogen, C34775) and Alexa Fluor 555 (Invitrogen, C34776) conjugate stocks in PBS. Store in small aliquots at – 20 °C prior to use or 4 °C for short term. Do not vortex.
2. 1 mM Hoechst 33342 (Sigma, B2261), stock solution in water. Store in aliquots at – 20 °C or short term at 4 °C.
3. 100 mM stock solution of 5-bromo-2′-deoxyuridine (Sigma, B5002) in DMSO. Store in aliquots at −20 °C.
4. Phosphate-buffered saline (PBS) buffer.
5. 4% paraformaldehyde in PBS.
6. DAPI or alternative nuclear stain (e.g., Hoechst 33342 dye).
7. ProLong Gold Antifade Mounting reagent (Thermo Fisher Scientific, P10144).

2.5 Data Acquisition and Analysis Software

For collection and processing of FLIM data, use the vendor-provided software (e.g., SPCImage for DCS-120 (Becker & Hickl) and LAS X for Leica SP8 Falcon and Stellaris 8 systems).

3 Methods

Aliquots of Matrigel must be defrosted on ice or in the refrigerator at 4 °C overnight before the experiment. The tissue culture dishes for intestinal organoid culture must be pre-warmed at 37 °C and the centrifuge pre-chilled to 4 °C 30 min before the procedure.

3.1 Defrosting the Intestinal Organoid Culture

1. For recovery of frozen organoids, remove the vial from the liquid nitrogen storage tank, and thaw it quickly in a 37 °C water bath, leaving a small amount of ice inside the vial.
2. Pipet and collect the organoids with a 1 mL pipette into a 15 mL centrifuge tube. To avoid osmotic shock, slowly (drop by drop) add 10 mL of ice-cold AdDF+++ medium (4 °C), and collect the organoids by centrifugation at 4 °C (500*g*, 5 min) using a swinging bucket rotor.

Table 1
Live imaging probes used in this chapter

Measured parameter/ probe	Exc./Em., nm	Staining concentration, time	Fluorescence lifetime range, application for FLIM	Comments, localization
Cholera toxin, subunit B (CTX)-Alexa Fluor 488 conjugate (**CTX-A488**)	488/510 nm **Two-photon** exc. 985 nm	1.7–10 µg / mL, 1–2 h	0.4 ~ 2.0 ns (FLIM) Overlapping with GFP fluorescence can affect observed fluorescence lifetime. Thus, ideally has to be applied to the organoid culture lacking expressed fluorescent proteins	According to the manufacturer (molecular probes, now Thermo fisher scientific), binds to ganglioside GM1 located in lipid rafts. Additional prominent targets of CTX can also be fucosylated glycoproteins [13]. In intestinal organoids stains regions with high proliferation activity very similar to the stem cell niche and amplification zone.
Cholera toxin, subunit B (CTX)-Alexa Fluor 555 conjugate (**CTX-A555**)	555/565 nm	10 µg/mL, 2 h	0.3 ~ 1.5 ns (FLIM)	According to the manufacturer (molecular probes, now Thermo fisher scientific), binds to ganglioside GM1 located in lipid rafts. Additional prominent targets of CTX can also be fucosylated glycoproteins [13]
Cell proliferation (S phase cells)/ Hoechst 33342 (± BrdU)	405–440 nm/ 430–450 nm **Two-photon** exc. 660–705 nm	1–2 µM, 2 h	0.8 ~ 2.6 ns (FLIM)	Allows live tracing of proliferating cells. The BrdU pulsing time and loading concentration may vary, depending on the model and the purpose of the experiment [20, 30]
Lgr5-GFP	488/510 nm	Endogenously expressed fluorescent marker of intestinal stem cells. See 2.1 for details	2 ~ 2.5 ns (FLIM)	Allows live imaging tracing of non-quiescent intestinal stem cells. The number of fluorescent cells in intestinal organoids depends on the growth media type applied for cultivation (*see* Table 2). GFP fluorescence is also present in proliferating daughter cells, slowly disappearing with differentiation [21]

3. Using a pipette, gently remove the supernatant and resuspend the organoids in ice-cold Matrigel. For one vial use 60 μL (total volume) of the Matrigel.
4. Dispense 10 μL of the suspension to the two to three wells of 37 °C pre-warmed 12-well plate. Incubate the plate for 3–5 min at 37 °C in a CO_2 incubator for Matrigel solidification.
5. Add 1 mL of ENR or ENRVC medium (depending on the experimental task) to each well and incubate at 37 °C. To minimize the anoikis, supplement culture medium with 10 μM Y-27632 (ROCK inhibitor) during the first 2 days of culture.
6. Medium should be refreshed 2–3 times per week. Passage organoids when they fill most of the Matrigel blob ("dome"), stop growing, or start to appear dark due to cells being shed to the organoid lumen (routinely split in 1:5 ratio every 5–7 days).

3.2 Passaging of Intestinal Organoid Culture and Seeding for Microscopy Imaging

1. Perform mechanical disruption of organoids in Matrigel with growth media by pipetting them for 20 times with the 10 μL tip on the top of 1 mL tip. Collect organoids in a 15 mL tube (*see* **Note 1**).
2. Rinse wells once with an additional volume of AdDF+++ media (room temperature or 37 °C), combine with the collected organoid culture in the tube, and adjust with AdDF+++ media to 10 mL.
3. Centrifuge at 500*g* for 5 min at 4 °C. Carefully remove supernatant and part of old Matrigel (as much as possible without disturbing the organoids on the bottom of the tube).
4. While keeping the tube on ice, resuspend the dissociated organoids in a fresh portion of cold Matrigel in a 1:5 ratio, from the previous amount of Matrigel.
5. Dispense 50 μL of organoids/Matrigel mixture per well of pre-warmed 24-well plate or 20 μL of organoids/Matrigel mixture per microscopy imaging dish. Leave to solidify at 37 °C for 5 min.
6. To proceed with organoids cultivation, add 500 μL of growth media per well of 24-well plate. Grow for 5–7 days before the next passage (*see* **Note 2**). Regularly change growth media when needed (normally every 2 days). For microscopy add 200 μL of media to the dish and proceed with imaging on the next 1–3 days after seeding.

   Optionally, imaging analysis of CTX staining can be done with paraformaldehyde-fixed organoids. In this case, we recommend seeding and staining organoids in the microwells. Alternatively, stained organoids can be collected by centrifugation, and all fixation- and staining-related procedures can be done in a microcentrifuge tube (1.5 or 2 mL) with the following sample preparation as in a regular immunofluorescence protocol.

3.3 Sample Staining and Imaging Acquisition

The microscope (lasers, camera, incubator system, computer, and other associated operating electronic blocks) has to be turned on 15–30 min before imaging to warm up and equilibrated to 37 °C, 5% CO_2, 20% O_2, or, optionally, different O_2 values (for temperature, humidity, CO_2, or hypoxia incubators). Open the microscope control software, e.g., SPCM or LAS X. Check that the appropriate filter cubes or spectral settings for required fluorescent probes are available. If several samples have to be imaged on the same day, we recommend having an appropriate gap time interval (~30–60 min per sample, depending on the probe fluorescence collection time, number of probes used for multiplexed staining, and the number of imaging replicates):

1. Dilute the required amount of fluorescent probe in the intestinal organoid growth medium (Table 2) according to recommended staining concentrations (Table 1). In case of multiparameter imaging, several probes can be mixed together in appropriate staining concentrations. For use of CTX-A488 and CTX-A555, *see* **Note 3**. For HXT/BrdU proliferation analysis, *see* **Note 4**.
2. Aseptically replace the growth medium sample with the one containing premixed probes ("loading medium"), and incubate in the incubator during staining (Table 1). Proceed with **steps 4–9** for a live imaging analysis or to **steps 3** and **5–9** for imaging analysis of fixed samples.
3. (Optional) Gently remove the loading medium and rinse 5 times with PBS pre-warmed to room temperature or 37 °C. Completely remove PBS without disturbing the Matrigel drop. Add sufficient amount of paraformaldehyde solution (e.g., 300–400 μL per 20 μL of Matrigel in a microwell of microscopy dish), and incubate for 10 min. Matrigel will be mostly solubilized by paraformaldehyde, leaving only fixed organoids on the surface of the glass. Remove the fixing solution and gently rinse 5 times with PBS. Accurately remove PBS, rinse 2 times with MQ water, and remove the water leaving organoids in microwells. Remove the silicon microwell insert. Add a drop of antifade reagent to organoids and cover with a cover glass avoiding bubbles formation. Use transparent nail polish to fix the borders of cover glass by gluing it to the sample glass surface, and wait until the nail polish will become solid (*see* **Note 5**). Proceed with **steps 5–9**.

Table 2
Composition of growth media and their effect on the number of Lgr5-GFP cells

Growth factor (concentration in media)	Function	Type of the growth media		
		ENR	ENRVC	EN
EGF (50 ng/mL)	Promotes cell proliferation, regulation of cell differentiation [36]	+	+	+
Noggin (10 ng/mL)	BMP pathway inhibitor; maintenance of $Lgr5^+$ intestinal stem cells, inducing expansion of crypt numbers [36]	+	+	+
R-spondin (1 μg/mL)	Wnt pathway agonist, potentiates Wnt/β-catenin signaling through the formation of the complex between LGR4/5/6 and RNF43, leading to the clearance of RNF43 from the plasma membrane and stabilization of frizzled receptors (Wnt receptor, ubiquitinated by RNF43 for lysosomal degradation) [35]. Maintenance of $Lgr5^+$ intestinal stem cells	+	+	–
Valproic acid (1 mM)	Histone deacetylase inactivation, combined with CHIR99021 treatment, causes the increase of $Lgr5^+$ intestinal stem cells [36]	–	+	–
CHIR99021 (3 μM)	Wnt/β-catenin signaling activation through the decreasing of β-catenin degradation by GSK3 activity inhibition [36]; increases the number of $Lgr5^+$ intestinal stem cells	–	+	–

(continued)

Table 2
(continued)

Growth factor (concentration in media)	Function	Type of the growth media		
		ENR	ENRVC	EN
Expected phenotype		*Organoid culture is highly heterogeneous; organoids contain both GFP-positive and GFP-negative cells; in general the number of Lgr5-GFP-positive cells is lower than in ENRVC organoid culture* [37]	*Organoids enriched with Lgr5-GFP-positive cells, lower number of GFP-negative cells* [37]	*The balance between proliferation and differentiation moved toward the differentiation; the number of Lgr5-GFP-positive cells slowly decreases with time; organoids contain more GFP-negative cells* [37]

4. Accurately remove the loading medium, rinse once with the imaging medium, and add 2.5–3 mL of imaging media. If required, remove the microwell insert before the imaging (*see* **Note 6**).
5. Fix the microscopy dish with stained organoids on the pre-heated microscopy table (37 °C). Bring the objective to the working position and apply correct immersion fluid, if required.
6. Preview the sample in transmission light mode. Find/select the organoids and regions of interest (ROI) for imaging, and adjust/set focus and XYZ coordinates. *See* **Note 7** on choosing the region of interest.
7. By quick preview in fluorescence intensity mode (with appropriate excitation and emission parameters; *see* Table 1), estimate the intensity and distribution of the probe staining. Choose appropriate settings for the collection of fluorescence signals: acquisition time sufficient for collection of fluorescence signal, spatial resolution/pinhole size, the desired optical section for 2D scanning or the top, the bottom and the desired range, and increment for the Z-stack for 3D scanning. It is important that the collected photon number (intensity) is sufficient for reliable calculation of fluorescence lifetime, if the FLIM analysis has to be performed (*see* **Note 8**).
8. Collect the image or the set of images (for the 3D scanning). Name the files or the work folder for appropriate saving. Proceed with the imaging of the next probe for the same optical section (multiparametric imaging) or Z-stack (3D imaging) or the next organoid.
9. Proceed to Subheading 3.4.

3.4 Processing of Microscopy Data

1. Open several SPCImage software windows, and import selected data files for different spectral channels for the same optical section of organoid, e.g., by viewing the fluorescence intensity data for CTX-A555 staining and Lgr5-GFP or CTX-A488 and HXT/BrdU.
2. Apply the appropriate fitting settings: two-component exponential fitting model by adjusting t1, t2, pixel binning, shift, and other parameters to the fluorescence decay data of CTX-Alexa Fluor conjugates, HXT, and Lgr5-GFP.
3. Choose the regions of interest (ROI) for analysis on the fluorescence intensity image by application of the ROI mask and/or by setting the threshold to exclude the areas with undesired low intensity (e.g., background intensity of the Matrigel, lumen fluorescence, or areas with weak probe staining) from the fluorescence lifetime calculations (*see* **Note 9**).

4. Calculate emission lifetime values ("decay matrix") for the chosen ROI or the whole frame (*see* **Note 10**).
5. Check the quality of fitting by estimation of the χ^2 coefficient, which ideally should be equal to 1 for all analyzed pixels of the image. If the quality is satisfactory, proceed with **step 6**. If not proceed with adjustment of the fitting settings. As soon as the fitting settings for the defined probe fluorescence are optimized, use them for the analysis of all microscopy data (*see* **Note 11**).
6. Export the phosphorescence/fluorescence intensity (photon counting), lifetime (color-coded values), and the distribution histogram (ASCII format) as well as the lifetime images as TIFF for all probes used in analysis. Note the range of used color scales. Repeat this step to obtain the sufficient number of data replicates for statistical analysis (e.g., two groups of fluorescence lifetime data of CTX-A555 for VD3-treated and untreated organoids).
7. Using Microsoft Excel or other relevant software, open the obtained numerical data on intensity, fluorescence lifetime, and fluorescence lifetime distribution histograms. For localization-based intensity and fluorescence lifetime analysis, proceed with **step 8**. Proceed with **steps 9–11** for fluorescence lifetime distribution histogram analysis (*see* **Note 12**).
8. Produce color-coded images based on numerical values for intensity and/or color-coded values using a conditional formatting function when needed for all spectral channels used for analysis of the designated imaging area. Choose the desired area in the intensity and/or fluorescence lifetime map for the first probe (e.g., GFP-positive regions, sorted by intensity data, *or* BrdU-loaded cells, sorted by combination of intensity and color-coded value data), keep/save its coordinates (the starting cell number used for selection in the table in both directions), and apply it on "intensity" or "lifetime" images for the second probe (e.g., CTX-A555 staining). Copy the chosen values to the new file and use them for the following statistical analysis. Repeat data collection to obtain the sufficient number of areas for statistical analysis (*see* **Note 13**).
9. Calculate the sum of pixel frequencies or pixel intensities (both parameters describe the detection of photon events per pixel on a microscopy image in chosen ROI) for the whole range of fluorescence lifetimes of distribution histogram. Using this number as a 100% of all events, recalculate the percentage of events for each fluorescence lifetime value of the distribution histogram. Plot these data on a graph to produce the distribution histogram curve (this can be done in Microsoft Excel, Origin, GraphPad, or related software).

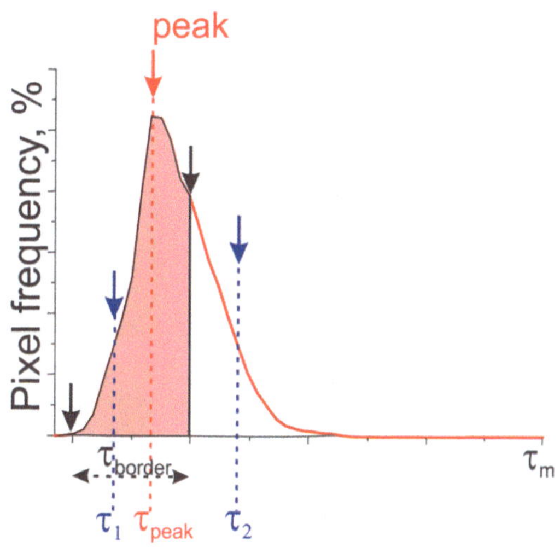

Fig. 1 Visual explanation of parameters useful for the statistical analysis of fluorescence lifetime distribution histograms

10. Collect sufficient number of distribution histograms from different microscopy images of different organoids in the culture to perform the statistical comparison (e.g., to compare an effect of VD3 treatment on fluorescence lifetime of CTX A555-stained organoids) (*see* **Note 14**).
11. The comparison of the distribution histograms can be done by the following parameters (Fig. 1): fluorescence lifetime corresponding to the maximal peak of the histogram plot (the value of fluorescence lifetime with maximal percentage of events in ROI, $_{peak}$), percentage of events for chosen lifetimes from the different sides of the plot ($_1$ and $_2$), and the value of the square under the plots curve before and/or after the chosen border lifetimes ($_{border}$), calculated by integration function, e.g., in Origin software.
12. Check the collected data to match the normal (Gaussian) distribution and apply relevant statistical methods. Present the results as bar charts, scatter (X:Y) charts, or bubble charts.

3.5 Anticipated Results

3.5.1 Multiparametric Imaging of CTX Conjugates with Live Cell Markers of Proliferation and Lgr5-GFP-Positive Small Intestinal Stem Cells

Cholera toxin subunit B (nontoxic) conjugates with Alexa Fluor 488 (CTX-A488) and Alexa Fluor 555 (CTX-A555) display robust association with plasma membranes in small intestinal organoids. This membrane staining differs between the "domains" of the organoids, being higher at the end of "the protrusions" and most completely disappearing close to the center of the organoid "body" with well-developed lumen (Fig. 2). The CTX staining of the protrusions correlated with the high cell proliferation of these regions, clearly visible on 3D reconstructions of live organoids with the FLIM-visualized cell proliferation (Figs. 2a and 3a). The

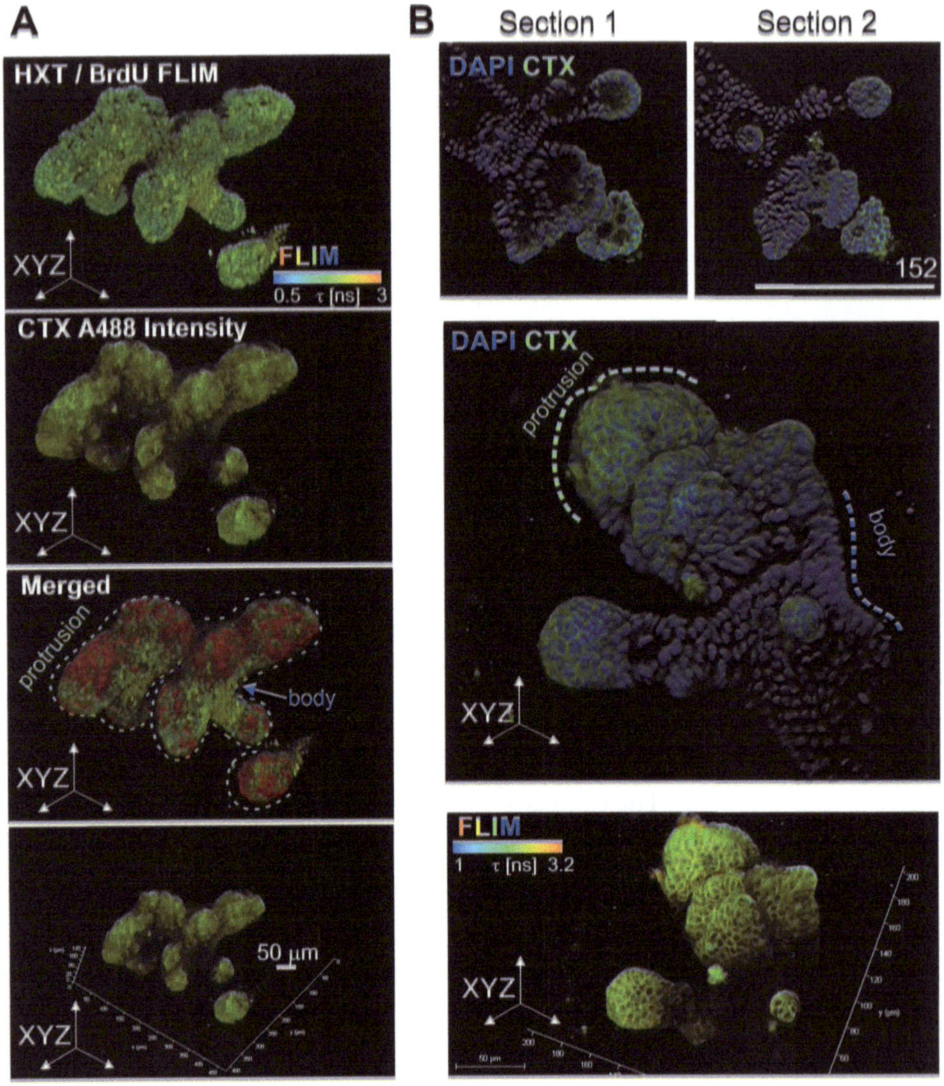

Fig. 2 Expected distribution of the CTX staining in live (a) and paraformaldehyde-fixed mouse small intestinal organoids (B) visualized using 3D microscopy. (**a**) Multiphoton 3D FLIM analysis of live small intestinal organoid (grown in ENR media) stained with Hoechst 33342 (HXT) and CTX-A488. FLIM of BrdU-treated and HXT-stained organoids (18 h) was used to analyze the areas with high proliferation and correlated with the intensity of CTX A488 staining. The intensive CTX A488 probe staining was observed at high proliferative protrusion ends of the intestinal organoids. (**b**) CTX-A488 staining in paraformaldehyde-fixed organoids. Two optical sections (top) and 3D reconstructions (below) of CTX-A488 fluorescence intensity (confocal microscopy) and confocal FLIM are shown. Scale bar and fluorescence lifetime maps are indicated

regions lacking incorporation of BrdU label had almost nil staining with CTX. Thus, CTX subunit B binding can be used as a valuable parameter for assessment of the amount of stem cell niches in relation to differentiated cell areas. CTX staining was well preserved after paraformaldehyde fixation, demonstrating the same localization characteristics (Fig. 2b), making possible the detailed analysis

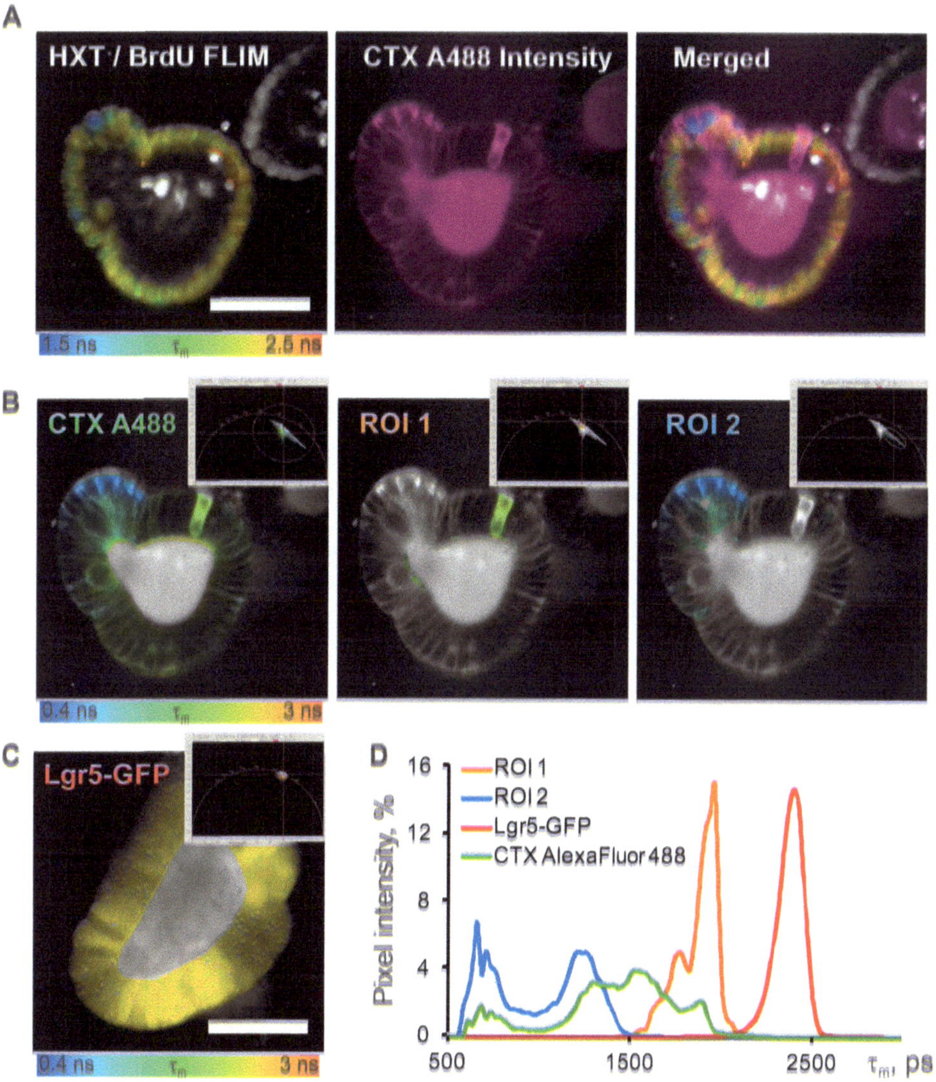

Fig. 3 (**a**) Example of multiparametric imaging of proliferation (HXT/BrdU FLIM, 3-h loading with 100 μM BrdU) and cholera toxin subunit B binding (CTX-A488 intensity, purple) in small intestinal organoids (ENR media 10 days). (**b**) Example of FLIM of CTX-A488 with the phasor plot diagrams. Left to right: overall imaging area (green), FLIM image of ROIs with high (orange, ROI 1) and low (blue, ROI 2) fluorescence lifetimes of CTX-A488 chosen by the selection on a phasor diagram. (**c**) Lgr5-GFP FLIM image of small intestinal organoids (ENRVC media) and phasor plot diagram obtained with the CTX-A488 fitting settings. (**d**) Fluorescence lifetime distribution histograms of overall CTX-A488 FLIM image (green), chosen ROIs (ROI 1, orange and ROI 2, blue), and Lgr5-GFP fluorescence (red line). Scale bar is 50 μm

of cell composition using immunofluorescence. Interestingly, in addition to the heterogeneous intensity profile of CTX-A488 staining, we noticed its prominent fluorescence lifetimes range being between 0.5 and 2 ns (Fig. 3b); this was distinct from the range of fluorescence lifetimes of Lgr5-GFP, the fluorescent marker of intestinal stem cells (Fig. 3c, d). Such "fluorescence lifetime contrast"

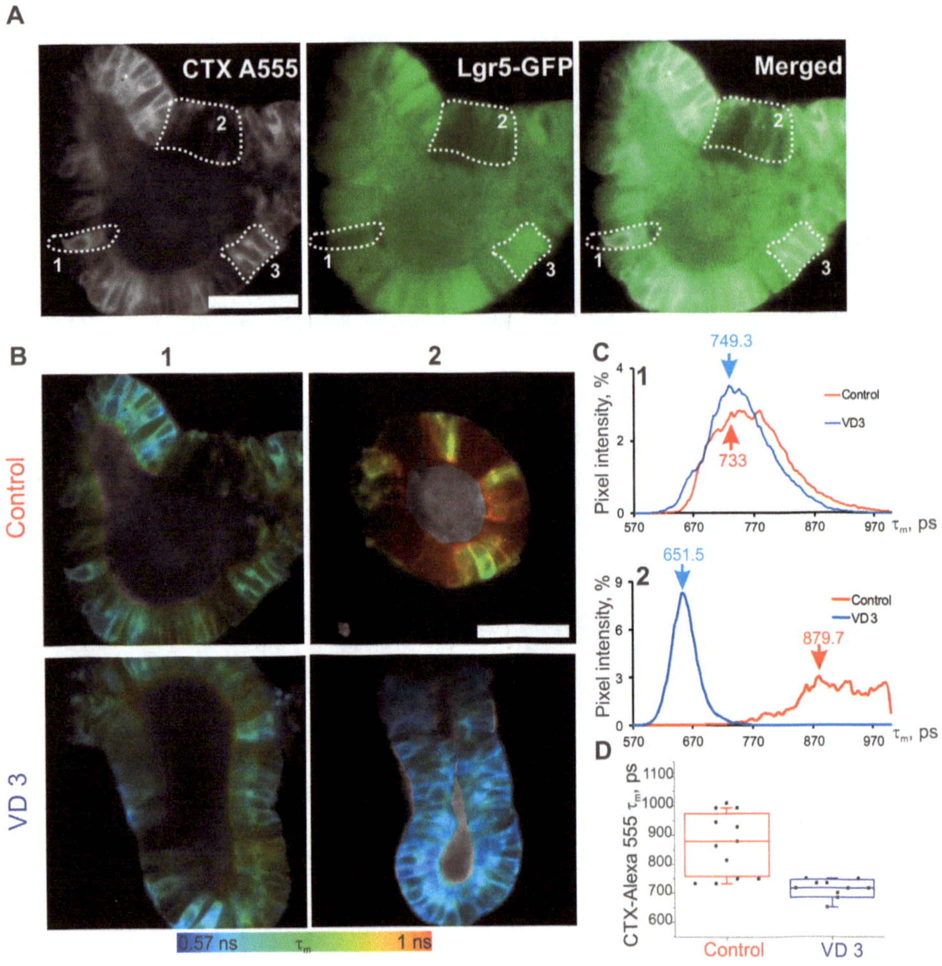

Fig. 4 (**a**) Example of multiparametric imaging of Lgr5-GFP small intestinal organoids (ENRVC-grown) with CTX-A555 staining (10 μg/mL, 2 h). (1) GFP $^{-}$/CTX-A555 $^{+}$ area, (2) GFP $^{-}$/CTX A555^{-} area, (3) GFP $^{+}$/CTX A555 $^{+}$ area. (**b**) Examples of CTX-A555 FLIM images of organoids (ENRVC) from control and VD3-treated (100 nM, 4 days) groups. Scale bar is 50 μm. (**c**) Examples of CTX-A555 fluorescence lifetime distribution histograms of corresponding FLIM images from (b). Arrows indicate the histogram peak values chosen for the statistical comparison of control and VD3-treated groups in (d). (**d**) Comparison of CTX-A555 fluorescence lifetime distribution histograms of control and VD3-treated organoid groups made by analysis of fluorescence lifetime values with the maximal percentage of pixel-by-pixel photon count events (corresponding to max peak on the plots). Control group $n = 11$, VD3 group $n = 10$. Boxes show standard deviation, and whiskers show 5 and 95 percentiles. Statistical difference was observed using *t*-test ($p < 0.05$), one experimental replicate

remained after fixation of CTX-A488-stained organoids with paraformaldehyde (Fig. 2b, bottom panel). Another CTX fluorescent conjugate, CTX-A555, also demonstrated a response in a fluorescence lifetime domain with a time range between 0.3 and 1.5 ns (Fig. 4).

Preliminary analysis of fluorescence lifetime distribution histograms of individual organoids in cultures pre-stained (1 h) with a different concentration of CTX-A488 (1.7, 5, and 10 μg/mL) showed no visible changes in fluorescence lifetime range or staining intensity in these organoid cultures. Instead, we noticed overall highly heterogeneous intensity staining of individual organoids in cultures with fluorescence lifetimes varying in a similar range (unpublished data). This suggests that at this range of the probe concentrations, the availability and spatial localization of CTX-binding receptors on the cell surfaces were the decisive factors of CTX-A488 staining, reflected in the fluorescence lifetime characteristics.

A closer look at the brightly CTX-stained protrusion organoid compartments revealed further complexity of cell types labeled with CTX-A488 and CTX-A555. Figure 3 indicates the existence of cell types with different intensity and characteristic fluorescence lifetimes. Interestingly some staining patterns with the comparable levels of intensity could have completely different lifetime values (ROI 1 and ROI 2, Fig. 3b). These patterns belong to cells at different cell cycle stages (see multiparametric imaging of CTX-A488 with HXT/BrdU FLIM on Fig. 3a): ROI 1 cell was at G0 or G1 phase, while ROI 2 area most likely did correspond to the S or G2 phases of cell cycle. These findings point at quite complex relationship between the intensity and fluorescence lifetime parameters of CTX-Alexa Fluor conjugates, which cannot be explained by simply self-quenching of the fluorescence and are probably caused by the CTX-binding site composition. This includes such factors as lipid composition, glycosylation characteristics, or availability of CTX receptors and the ability to bind primary and/or secondary sites at the CTX surface.

We compared CTX staining with those of stem cells (Lgr5-GFP) and CTX-A555 (Fig. 4a). This analysis showed that even though overall CTX staining correlated with highly proliferative zones in organoids, some GFP^- (minus) cells were also able to efficiently bind CTX A555. We defined three types of cells in respect to their CTX-A555 binding ability: rare GFP^- differentiated or quiescent cells (ROI 1, Fig. 4a) commonly present GFP^+ cells (correspond to intestinal stem cells or their daughter cells, ROI 3, Fig. 4a) with a prominent CTX-A555 staining and GFP^- cells (most likely representing fully differentiated cells, ROI 2, Fig. 4a) with no CTX staining.

3.5.2 Effect of Vitamin D3 Treatment on CTX-A555 Fluorescence Lifetime

To see if the fluorescence lifetimes of tested CTX conjugates could be affected by the cell membrane composition, we treated ENRVC-grown mouse small intestinal organoids with 100 nM VD3 for 4 days and compared the fluorescence lifetime distribution histograms of VD3- treated organoids with the vehicle (ethanol)-treated organoid culture. VD3 is a multi-level repressor of Wnt/β-catenin pathway, shown to be an important environmental determinant factor of Lgr5-stem cell function in vivo and inducer of intestinal cell maturation [28]. It can also affect lipid membrane composition and increase the number of lipid droplets on the basal membrane of ENRVC small intestinal organoids [25]. The comparison of fluorescence lifetime distribution histograms of organoids from VD3-treated and control groups showed that even with high deviations of fluorescence lifetime values of individual organoids in culture, VD3 decreased the fluorescence lifetime values of membrane-bind CTX-A555 probe (Fig. 4b–d) with the negligible effect on its fluorescence intensity. Interestingly the deviation of CTX-A555 fluorescence lifetimes was also decreased in VD3-treated organoid group, pointing at the strong effect of VD3 on overall intestinal tissue homeostasis and ability of CTX-A555 to report these changes.

4 Notes

1. Application of mechanical disruption for organoids during the passage allows them to quickly restore their growth and has a lower negative impact on their viability. At the same time, pipetting technique seriously affects the size of organoid units, which regenerate to new organoids, and thus impacts on the overall organoid culture heterogeneity (size, cell composition, and related metabolic characteristics of individual organoids). It is important that each individual experimental replicate is performed on the culture of organoids from the same passage and several independent replicates are done to provide data for adequate research conclusions.
2. Depending on the experimental task, grow your intestinal organoids for a desired period of time in one of the suggested growth media composition variants: ENR, ENRVC, or EN. In our experience organoid culture has to be maintained in new growth media conditions for at least 1 week to reliably change proliferation to differentiation balance (based on Lgr5-GFP fluorescence and HXT/BrdU live proliferation analysis). Transfer of the culture from ENR media to ENRVC or EN conditions will take less time than adaptation of the culture constantly cultivated on ENRVC.

   Optionally, the long-term treatment (more than 3 days) with additional agents (e.g., 100 nM VD3) added to the chosen growth media can be performed during the initial culturing period. For the imaging, organoids have to be transferred to the microscopy dishes as described before. Short-term treatments (less than 3 days) can be also performed prior to imaging procedure on imaging dishes.

3. Frequently, the staining concentration and probe loading time require optimization due to different microscope detector sensitivity, laser power, microscopy settings, or the "new" experimental parameters. With high molecular protein probes such as CTX-Alexa Fluor conjugates (>50 kDa), the diffusion through Matrigel matrix has to be taken into account. Please ensure that all optimizations are done in preliminary imaging experiments. Check that the CTX-Alexa Fluor conjugate chosen for analysis will not have a spectral cross talk with the internal organoid fluorescence (e.g., Lgr5-GFP fluorescent marker of intestinal stem cells). For FLIM-based studies, standardization of probe staining conditions can be very important due to potential effects of the probe self-quenching [27, 29]. In our experience, loading of intestinal organoids with CTX-A488 probe in the range of concentrations 1.7–10 μg/mL had no visible impact on fluorescence lifetime distribution in CTX-A488-stained targets, but rather affected number of organoids with high brightness. In all staining conditions, the CTX-A488-stained organoids demonstrated heterogeneity of the fluorescence intensity and lifetime values, allowing speculation that the ability to bind CTX is intrinsic and probably depends on cell composition of individual organoids in culture.

   For CTX-A488 staining, Lgr5-GFP organoid culture has to be propagated either in ENR or EN culture conditions (to decrease GFP fluorescence) prior to imaging and ideally with the CTX probe having spectral characteristics different from GFP (e.g., CTX-A555). In case of intestinal organoids derived from *Lgr5-EGFP-ires-CreERT2* mice, CTX conjugates can be easily distinguished from Lgr5-GFP signal by their specific cellular membrane localization (Fig. 3). Time domain-based separation of CTX-A488 and GFP fluorescence is also possible: see fluorescence lifetime distribution histograms in Fig. 3d.

4. Short-term BrdU loading pulse (up to 3 h) can be combined with Hoechst 33342 (HXT) staining [30]. For long-term BrdU loading experiments, staining with HXT can be done prior to imaging. To avoid BrdU toxicity, lower BrdU concentrations down to 5 μM can be applied for long-term (~18 h) pulse. For short-term pulse, 100 μM BrdU concentration should be used due to its faster effect on HXT fluorescence lifetime. No-BrdU (negative control) samples are important since the Hoechst fluorescence lifetime can depend on staining concentration and loading time.

5. Fixed CTX-Alexa Fluor-stained organoids are fully compatible with immunofluorescence and can be additionally co-stained with corresponding primary and secondary Alexa Fluor-conjugated antibodies. In this case the additional membrane

permeabilization step has to be incorporated after paraformaldehyde fixation [31, 32]. In our experience paraformaldehyde fixation demonstrates good preservation of CTX-stained membrane structures. Prior to use in experimental design, alternative fixation methods have to be validated for preserving the membrane-associated CTX-stained structures.

6. If some agents have to be present in the media during the analysis, their addition to the imaging media is recommended. It is also possible to perform imaging in the culturing media, while using PBS or its analogues can affect organoids viability and is not recommended.
7. Due to overall organoids heterogeneity, the choice of tissue regions for imaging analysis of cell proliferation and differentiation is highly important [31]. However, the three-dimensional nature of intestinal organoids, presence of lumen with strong autofluorescence, heterogeneity of their size and shape, and the complexity of their cell composition make the targeted choice of imaging regions difficult without preliminary testing. In this case focusing on regions with high fluorescence intensity of the chosen analytical parameter (e.g., fluorescence lifetime of the CTX-Alexa conjugates) and the increase of the collected microscopy data replicates for statistical analysis is highly recommended.
8. For multiplexing and further microscopy data processing (application of ROI mask or data export with following analysis in Excel or other programs), all images including different spectral channels should be collected with the same spatial resolution (e.g., 512 × 512 pixels) and the pixel size.
9. The same ROI mask can be applied simultaneously for the different spectral channels obtained from the same imaged area. Using several ROI masks for the analysis of the same imaged region allows their easy initial preselection, e.g., comparison of the fluorescence lifetime and intensity of CTX-Alexa Fluor conjugates between proliferating and nonproliferating cells (Fig. 3a, b) or Lgr5-GFP-positive and Lgr5-GFP-negative regions (Fig. 4a). Choosing appropriate ROI or setting the intensity threshold limits the initial heterogeneity of microscopy data collected for analysis, decreasing their variations and justifying the application of fluorescence lifetime distribution histogram analysis of ROIs.
10. After calculation of fluorescence lifetime values, the SPCImage software shows the false-color lifetime distribution map and the distribution histogram for the whole image frame or the chosen ROI. It also produces numerical datasets for lifetime (color-coded values), distribution histogram, and intensities (photons counting) on pixel-by-pixel basis which can be

exported and used for analysis in Excel software. However, exporting data for the selected ROIs is not always straightforward and has to be carefully controlled. At this stage calculation of phasor plot diagram [33–35] combined with fluorescence lifetime distribution analysis is also possible. Choosing the areas within a designated range of fluorescence lifetimes on a phasor plot diagram and their allocation on the microscopy imaging area can simplify the accurate choice of the regions of interest for fluorescence lifetime analysis (Fig. 3b).

11. The application of 2D correlation analysis of χ^2 coefficient and mean fluorescence lifetime (e.g., in SPCImage software, Becker & Hickl GmbH) provides a simple visual way of analysis of the quality of fitting for all pixels in the ROI.
12. Localization-based intensity and fluorescence lifetime analysis is laborious, but it allows to study the relation between different measurement parameters (e.g., the relation between stemness or differentiation by Lgr5-GFP marker assessment and ability of cellular membranes to bind cholera toxin). In comparison, fluorescence lifetime histogram distribution analysis is faster and allows to analyze the chosen ROI at once, but its usage for correlation analysis can be complicated [20].
13. Using this multiparametric analysis can be difficult when probes have different subcellular localization (e.g., nuclear staining of HXT vs. membrane staining of CTX-Alexa Fluor conjugates). In this case the numerical color-coded "intensity map" can simplify the selection of the areas of interest.
14. It is important that all distribution histograms are made and exported for the same range of fluorescence lifetimes and cover the whole range of photon events detected for the chosen ROI. Calculated in percentage of total events per ROI, they will always have equal square under the plotted curve, which can be easily determined by integration function (e.g., in Origin software).

Acknowledgments

This work was supported by the BOF Universiteit Gent grant BOF/STA/202009/003 and in part by the Science Foundation Ireland (SFI) grants 18/IF/6238 and 13/SIRG/2144. We thank Dr. H. Glauner and Dr. L. Alvarez for support with imaging at Leica Falcon SP8 microscopes (Leica Training Centre, Mannheim, Germany) and Prof. D.B. Papkovsky for support with imaging at the Biophysics and Bioanalysis Lab (University College Cork, Cork, Ireland).

References

1. Murphy KJ, Reed DA, Trpceski M, Herrmann D, Timpson P (2021) Quantifying and visualising the nuances of cellular dynamics in vivo using intravital imaging. Curr Opin Cell Biol 72:41–53
2. Perez-Ramirez CA, Christofk HR (2021) Challenges in studying stem cell metabolism. Cell Stem Cell 28(3):409–423
3. Puschhof J, Pleguezuelos-Manzano C, Clevers H (2021) Organoids and organs-on-chips: insights into human gut-microbe interactions. Cell Host Microbe 29(6):867–878
4. Hofer M, Lutolf MP (2021) Engineering organoids. Nat Rev Mater 6(5):402–420. https://doi.org/10.1038/s41578-021-00279-y
5. Dmitriev RI, Intes X, Barroso MM (2021) Luminescence lifetime imaging of three-dimensional biological objects. J Cell Sci 134(9):1–17
6. Snippert HJ, van der Flier LG, Sato T, van Es JH, van den Born M, Kroon-Veenboer C, Barker N, Klein AM, van Rheenen J, Simons BD, Clevers H (2010) Intestinal crypt homeostasis results from neutral competition between symmetrically dividing Lgr5 stem cells. Cell 143(1):134–144. https://doi.org/10.1016/j.cell.2010.09.016
7. Rios AC, Clevers H (2018) Imaging organoids: a bright future ahead. Nat Methods 15(1):24
8. Dmitriev RI (ed) (2017) Multi-parametric live cell microscopy of 3D tissue models, vol 1035. Advances in experimental medicine and biology. Springer International Publishing AG. https://doi.org/10.1007/978-3-319-67358-5
9. Bresson JL, Pang KY, Walker WA (1984) Microvillus membrane differentiation: quantitative difference in cholera toxin binding to the intestinal surface of newborn and adult rabbits. Pediatr Res 18(10):984–987. https://doi.org/10.1203/00006450-198410000-00015
10. Majoul I, Schmidt T, Pomasanova M, Boutkevich E, Kozlov Y, Söling H-D (2002) Differential expression of receptors for Shiga and cholera toxin is regulated by the cell cycle. J Cell Sci 115(4):817–826. https://doi.org/10.1242/jcs.115.4.817
11. Lauer S, Goldstein B, Nolan RL, Nolan JP (2002) Analysis of cholera toxin–ganglioside interactions by flow cytometry. Biochemistry 41(6):1742–1751. https://doi.org/10.1021/bi0112816
12. Krishnan P, Singla A, Lee C-A, Weatherston JD, Worstell NC, Wu H-J (2017) Hetero-multivalent binding of cholera toxin subunit B with glycolipid mixtures. Colloids Surf B: Biointerfaces 160:281–288. https://doi.org/10.1016/j.colsurfb.2017.09.035
13. Wands AM, Fujita A, McCombs JE, Cervin J, Dedic B, Rodriguez AC, Nischan N, Bond MR, Mettlen M, Trudgian DC (2015) Fucosylation and protein glycosylation create functional receptors for cholera toxin. eLife 4:e09545
14. Heim JB, Hodnik V, Heggelund JE, Anderluh G, Krengel U (2019) Crystal structures of cholera toxin in complex with fucosylated receptors point to importance of secondary binding site. Sci Rep 9(1):12243. https://doi.org/10.1038/s41598-019-48579-2
15. Heggelund JE, Burschowsky D, Bjørnestad VA, Hodnik V, Anderluh G, Krengel U (2016) High-resolution crystal structures elucidate the molecular basis of cholera blood group dependence. PLoS Pathog 12(4):e1005567. https://doi.org/10.1371/journal.ppat.1005567
16. Jaswal VMS, Babbar HS, Mahmood A (1988) Changes in sialic acid and fucose contents of enterocytes across the crypt-villus axis in developing rat intestine. Biochem Med Metab Biol 39(1):105–110. https://doi.org/10.1016/0885-4505(88)90064-3
17. Schneider M, Kumar V, Nordstrøm LU, Feng L, Takeuchi H, Hao H, Luca VC, Garcia KC, Stanley P, Wu P (2018) Inhibition of Delta-induced Notch signaling using fucose analogs. Nat Chem Biol 14(1):65
18. Demitrack ES, Samuelson LC (2016) Notch regulation of gastrointestinal stem cells. J Physiol 594(17):4791–4803
19. Sethi A, Wands AM, Mettlen M, Krishnamurthy S, Wu H, Kohler JJ (2019) Cell type and receptor identity regulate cholera toxin subunit B (CTB) internalization. J R Soc Interface Focus 9(2):20180076
20. Okkelman IA, Puschhof J, Papkovsky DB, Dmitriev RI (2020) Visualization of stem cell niche by fluorescence lifetime imaging microscopy. In: Ordóñez-Morán P (ed) Intestinal stem cells: methods and protocols. Springer US, New York, pp 65–97. https://doi.org/10.1007/978-1-0716-0747-3_5
21. Sato T, Vries RG, Snippert HJ, Van De Wetering M, Barker N, Stange DE, Van Es JH, Abo A, Kujala P, Peters PJ (2009) Single Lgr5 stem cells build crypt-villus structures

in vitro without a mesenchymal niche. Nature 459(7244):262–265

22. Mizutani T, Clevers H (2020) Primary intestinal epithelial organoid culture. In: Ordóñez-Morán P (eds) Intestinal stem cells. Methods in Molecular Biology, vol 2171. Humana, New York, NY. https://doi.org/10.1007/978-1-0716-0747-3_11
23. Levitt JA, Poland SP, Krstajic N, Pfisterer K, Erdogan A, Barber PR, Parsons M, Henderson RK, Ameer-Beg SM (2020) Quantitative real-time imaging of intracellular FRET biosensor dynamics using rapid multi-beam confocal FLIM. Sci Rep 10(1):5146. https://doi.org/10.1038/s41598-020-61478-1
24. Zhang Y, Guldner IH, Nichols EL, Benirschke D, Smith CJ, Zhang S, Howard SS (2021) Instant FLIM enables 4D in vivo lifetime imaging of intact and injured zebrafish and mouse brains. Optica 8(6):885–897
25. Okkelman IA, McGarrigle R, O'Carroll S, Berrio DC, Schenke-Layland K, Hynes J, Dmitriev RI (2020) Extracellular Ca2+-sensing fluorescent protein biosensor based on a collagen-binding domain. ACS Appl Bio Mater 3(8): 5310–5321. https://doi.org/10.1021/acsabm.0c00649
26. Okkelman IA, Neto N, Papkovsky DB, Monaghan MG, Dmitriev RI (2020) A deeper understanding of intestinal organoid metabolism revealed by combining fluorescence lifetime imaging microscopy (FLIM) and extracellular flux analyses. Redox Biol 30: 101420. https://doi.org/10.1016/j.redox.2019.101420
27. Okkelman IA, Papkovsky DB, Dmitriev RI (2020) Estimation of the mitochondrial membrane potential using fluorescence lifetime imaging microscopy. Cytometry A 97(5): 471–482. https://doi.org/10.1002/cyto.a.23886
28. Peregrina K, Houston M, Daroqui C, Dhima E, Sellers RS, Augenlicht LH (2014) Vitamin D is a determinant of mouse intestinal Lgr5 stem cell functions. Carcinogenesis 36(1):25–31. https://doi.org/10.1093/carcin/bgu221
29. Gehlen MH (2020) The centenary of the Stern-Volmer equation of fluorescence quenching: from the single line plot to the SV quenching map. J Photochem Photobiol C: Photochem Rev 42:100338. https://doi.org/10.1016/j.jphotochemrev.2019.100338
30. Okkelman IA, Dmitriev RI, Foley T, Papkovsky DB (2016) Use of fluorescence lifetime imaging microscopy (FLIM) as a timer of cell cycle S phase. PLoS One 11(12):e0167385. https://doi.org/10.1371/journal.pone.0167385
31. Okkelman IA, Foley T, Papkovsky DB, Dmitriev RI (2017) Live cell imaging of mouse intestinal organoids reveals heterogeneity in their oxygenation. Biomaterials 146:86–96. https://doi.org/10.1016/j.biomaterials.2017.08.043
32. Okkelman IA, Foley T, Papkovsky DB, Dmitriev RI (2017) Multi-parametric imaging of hypoxia and cell cycle in intestinal organoid culture. In: Dmitriev R (eds) Multi-parametric live cell microscopy of 3D tissue models. Advances in Experimental Medicine and Biology, vol 1035. Springer, Cham. https://doi.org/10.1007/978-3-319-67358-5_6
33. Digman MA, Caiolfa VR, Zamai M, Gratton E (2008) The phasor approach to fluorescence lifetime imaging analysis. Biophys J 94(2): L14–L16
34. Lakner PH, Monaghan MG, Möller Y, Olayioye MA, Schenke-Layland K (2017) Applying phasor approach analysis of multiphoton FLIM measurements to probe the metabolic activity of three-dimensional in vitro cell culture models. Sci Rep 7:42730. https://doi.org/10.1038/srep42730
35. Stringari C, Cinquin A, Cinquin O, Digman MA, Donovan PJ, Gratton E (2011) Phasor approach to fluorescence lifetime microscopy distinguishes different metabolic states of germ cells in a live tissue. Proc Natl Acad Sci 108(33):13582–13587
36. Merenda A, Fenderico N, Maurice MM (2020) Wnt signaling in 3D: recent advances in the applications of intestinal organoids. Trends Cell Biol 30(1):60–73
37. Lindeboom RG, van Voorthuijsen L, Oost KC, Rodríguez-Colman MJ, Luna-Velez MV, Furlan C, Baraille F, Jansen PW, Ribeiro A, Burgering BM, Snippert HJ, Vermeulen M (2018) Integrative multi-omics analysis of intestinal organoid differentiation. Mol Syst Biol 14(6):e8227. https://doi.org/10.15252/msb.20188227

Chapter 15

In Vitro Morphogenesis and Differentiation of Human Intestinal Epithelium in a Gut-on-a-Chip

Woojung Shin and Hyun Jung Kim

Abstract

The establishment of a three-dimensional (3D) epithelial structure and cytodifferentiation in vitro is necessary to recapitulate in vivo-relevant structure and function of the human intestine. Here, we describe an experimental protocol to build an organomimetic gut-on-a-chip microdevice that allows inducing 3D morphogenesis of human intestinal epithelium using Caco-2 cells or intestinal organoid cells. Under physiological flow and physical motions, intestinal epithelium spontaneously recreates 3D epithelial morphology in a gut-on-a-chip that offers enhanced mucus production, epithelial barrier, and longitudinal host-microbe co-culture. This protocol may provide implementable strategies to advance traditional in vitro static cultures, human microbiome studies, and pharmacological testing.

Key words Human intestinal epithelium, Organoid, Gut-on-a-chip, Morphogenesis, In vitro model

1 Introduction

Intestinal epithelial cells such as the Caco-2 cell line and human intestinal organoids have been extensively used to recreate an intestinal mucosal interface in vitro for studying epithelial permeability [1–3], molecular transport [4, 5], and host-microbe interaction [6–8]. When these human intestinal epithelial cells are cultured in a gut-on-a-chip [9–14] under physiological flow and mechanodynamic cues, cells spontaneously undergo 3D morphogenesis in vitro, establishing an in vivo-relevant epithelial structure. The underlying mechanism was discovered [15] in which the removal of morphogen antagonists that are basolaterally secreted (e.g., Dickkopf-1, Wnt inhibitory factor 1, secreted frizzled-related protein 1, or Soggy-1) is critical to induce 3D epithelial morphogenesis. Based on this mechanistic understanding, we have demonstrated reproducible induction of 3D morphogenesis, in vivo-relevant differentiation of the established epithelium, and advanced intestinal epithelial functions [10, 16].

Paloma Ordóñez-Morán (ed.), *Intestinal Differentiated Cells: Methods and Protocols*, Methods in Molecular Biology, vol. 2650, https://doi.org/10.1007/978-1-0716-3076-1_15,

In this protocol, we provide a methodology for fabricating a microfluidic gut-on-a-chip, growing intestinal epithelium (Caco-2 or intestinal organoid cells) in a gut-on-a-chip, inducing 3D morphogenesis, and analyzing intestinal functions. We also describe cellular and molecular characteristics of 3D epithelial microarchitecture that illustrate tissue-specific histogenesis and lineage-dependent cytodifferentiation via multimodal imaging techniques. This protocol can be implemented to study intestinal mucosal biology, host-microbe interaction, and the mechanism of intestinal diseases such as inflammatory bowel disease (IBD) and colorectal cancer (CRC). Incorporating patient-derived samples will also enable to model patient-specific intestinal microenvironment.

2 Materials

2.1 Culture of Human Intestinal Epithelium

1. Human intestinal epithelium: Caco-2 human adenocarcinoma or tissue-derived human colonic organoids.
2. Cell dissociating solution: 0.25% trypsin/1 mM ethylenediaminetetraacetic acid (EDTA; trypsin/EDTA) for dissociating Caco-2 cells; TrypLE Express for dissociating colonoid cells.
3. Phosphate-buffered saline (PBS; pH 7.4, Ca^{2+}- Mg^{2+}-free).
4. Culture medium for growing Caco-2 cells: Dulbecco's modified Eagle medium (DMEM) containing 20% (v/v) fetal bovine serum (FBS; heat-inactivated) and antibiotics (penicillin and streptomycin).
5. Culture medium for growing organoids: Organoid basal medium (Table 1) containing organoid growth supplements (Table 2).
6. Matrigel, growth factor reduced (83% protein that gels).
7. 24-well culture plate (surface treated).
8. Cell strainer (cutoff size, 100 μM).

2.2 Fabrication of a Gut-on-a-Chip

1. Pre-patterned silicon wafers for preparing an upper and a lower microchannel layer as well as a porous membrane [16].
2. Polydimethylsiloxane (PDMS) and curing agent.
3. Ethyl alcohol (ethanol), 200 proof and 70% (v/v).
4. Fluoropolymer-coated polyester film.
5. Coverslip (No. 1, 60 × 48 mm).
6. Surgical scalpel.
7. Dry oven.
8. Corona treater.

Table 1
Composition of an organoid basal medium

Component	Final concentration	Amount required
Advanced DMEM/F12		485 mL
GlutaMAX (100×)	1×	5 mL
4-(2-hydroxyethyl)-1-piperazineethanesulfonic acid (HEPES) (1 M)	10 mM	5 mL
Penicillin-streptomycin (100×)	1×	5 mL

Table 2
Composition of an organoid culture medium

Component	Final concentration	Amount required
Organoid basal medium		337.7 mL
Wnt3a (100 μg/mL)[a]	100 ng/mL	500 μL
R-spondin-conditioned medium		100 mL
Noggin-conditioned medium		50 mL
N-2 MAX media supplement (100×)[a]	0.5×	2.5 mL
B-27 supplement (50×)[a]	0.5×	5 mL
Nicotinamide (1 M)[a]	5 mM	2.5 mL
N-acetylcysteine (500 mM)[a]	0.5 mM	500 μL
Mouse recombinant epidermal growth factor (100 μg/mL)[a]	50 ng/mL	250 μL
SB202190 (100 mM)[a]	10 μM	50 μL
A-8301 (5 mg/mL)[a]	500 nM	2.11 μL
Gastrin (0.5 mg/mL)[a]	10 nM	20.98 μL
Y-27632, dihydrochloride (10 mM)[a]	10 μM	500 μL
Primocin (50 mg/mL)[a]	100 μg/mL	1 mL

[a]Numbers inside the parenthesis indicate the concentration of a stock solution or a product

2.3 Culture of Human Intestinal Epithelium in a Gut-on-a-Chip

1. Silicone tubing (0.8 mm ID, 2.4 mm OD).
2. Y-connector (1.57 mm ID).
3. Blunt-end needle (18 gauge).
4. 1 mL syringes, slip tip.
5. Luer lock syringe (3 mL, sterile).
6. UV/ozone generator.

7. Flexcell tension system (vacuum-assisted mechanical motion generator).
8. CO_2 incubator (37 °C, humidified, 5% CO_2).
9. Extracellular matrix (ECM) solution: a mixture of Matrigel (100-fold dilution) and collagen I (30 μg/mL).
10. Polyethylenimine (PEI) solution (average molecular weight ~ 2,000 g/mol; final concentration, 1%, w/v; filter-sterilized at 0.2 μm cutoff).
11. Glutaraldehyde solution (final concentration, 0.1%, w/v; filter-sterilized at 0.2 μm cutoff).
12. Syringe pump.

2.4 Morphological Assessment

1. Inverted phase-contrast microscope.
2. Laser-scanning confocal microscope.
3. Image analysis software.
4. Triton X-100 (0.3%, v/v).
5. Bovine serum albumin (2%, w/v).
6. 4′,6-Diamidino-2-phenylindole (DAPI; 1 mg/mL).
7. Fluorophore-conjugated phalloidin.
8. Hexamethyldisilazane.
9. Conductive carbon tape.

3 Methods

3.1 Culture of Human Intestinal Epithelium: Caco-2 Cells

1. Thaw a frozen vial of Caco-2 cells at 37 °C.
2. Transfer the thawed cells to a T75 flask containing 10 mL of pre-warmed Caco-2 culture medium.
3. Incubate the flask in a humidified 5% CO_2 incubator at 37 °C.
4. Change culture medium every 2–3 days.
5. When cells are ~90% confluent, detach the cells using 1 mL of trypsin/EDTA and pass 1×10^6 cells to a new T75 flask.
6. Repeat the whole step for the maintenance of Caco-2 cells.

3.2 Culture of Human Intestinal Epithelium: Intestinal Organoids

1. Thaw a frozen vial of human intestinal organoids at 37 °C.
2. Prepare Matrigel and organoid culture medium (Table 2; *see* **Note 1**) at 4 °C.
3. Resuspend the thawed organoid cells with 10 mL of organoid culture medium, spin down the organoid cells (100 × *g*, 4 °C, 5 min), aspirate the supernatant using a 10 mL serological

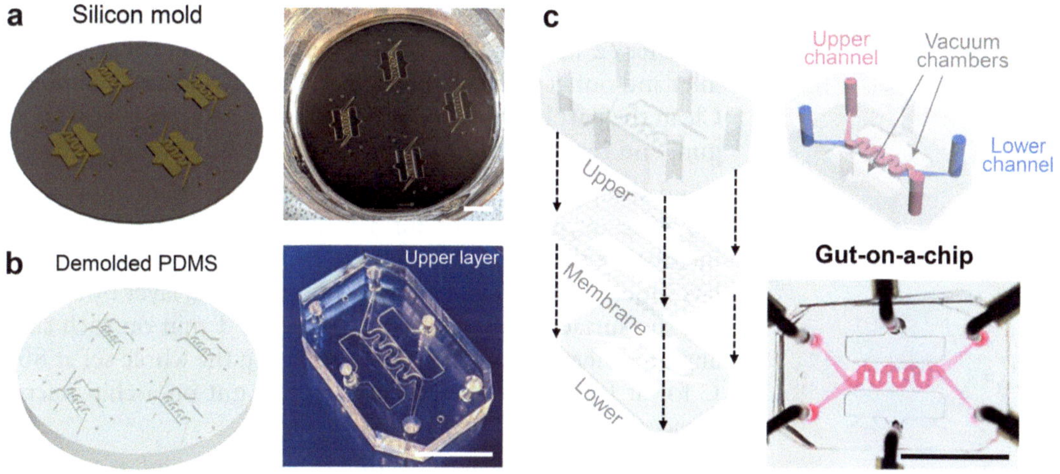

Fig. 1 Configuration of a gut-on-a-chip microdevice. (**a**) A schematic (left) and a photograph (right) of a silicon mold that has epoxy patterns of four gut-on-a-chips. (**b**) A schematic (left) and a photograph (right) of a PDMS layer that contains the patterns of microchannels. (**c**) A schematic that describes the layer-by-layer bonding (left) for creating a gut-on-a-chip microdevice (right top). The upper and lower microchannel is highlighted in pink and blue, respectively. A photograph of a complete gut-on-a-chip device (right bottom). Bars, 1 cm. Reproduced from [16]

pipette, and then mix the organoid cell pellet with liquid Matrigel on ice (100–200 organoids per 30 μL of Matrigel).

4. Place 30 μL of the organoid-Matrigel suspension in a well of a 24-well plate at room temperature, and incubate the well plate for gelation at 37 °C in a humidified 5% CO_2 incubator for 10 min.
5. Add 500 μL of organoid culture medium to each well and incubate the plate in a humidified CO_2 incubator at 37 °C.
6. Change medium every other day until organoids fully grow for up to 7–10 days (*see* **Note 2**).

3.3 Fabrication of a Gut-on-a-Chip

1. Pour a degassed silicone polymer mix (PDMS/curing agent = 15:1, w/w) on a clean silicon mold that has the patterns of an upper or a lower microchannel (Fig. 1a).
2. Drop the same uncured PDMS mix (~1 mL) on a clean silicon mold that has an array of micropillars (diameter, 10 μm; center-to-center distance, 25 μm); overlay a fluoropolymer-coated polyester film on the PDMS-covered silicon mold without generating air bubbles, 1 cm thick flat PDMS slab, frosted glass slide, sequentially; and place a weight (~3 kg) on top.
3. Incubate the silicon molds containing uncured PDMS in a 60 ° C dry oven for >4 h to cure PDMS, and then cut out the solidified PDMS parts from the molds using a surgical scalpel (Fig. 1b).

4. Prepare an upper layer by punching holes using a biopsy puncher (2 mm in diameter) at the connecting ports of both inlet and outlet microchannels as well as the vacuum chambers. Clean the part by spraying 100% ethanol and drying with an air gun, and remove lint using tape prior to the bonding process.
5. Bond an upper microchannel layer and a porous membrane by exposing oxygen plasma for 1.5 min in a plasma cleaner and incubating the setup at 80 °C for 12 h. Then, bond the upper layer-membrane set with a lower microchannel layer by activating the surface using a corona treater for 1 min on each side, aligning them within 30 s, and incubating the whole set at 80 °C for at least 12 h to make a complete gut-on-a-chip microdevice (Fig. 1c).
6. Assemble gas-permeable silicone tubing to the inlets and outlets of each microchannel using cut blunt-end needles, sterilize the channels and tubing using 70% ethanol, and keep the fabricated gut-on-a-chip device in a dry oven until use.

3.4 Induction of 3D Intestinal Morphogenesis in a Gut-on-a-Chip

1. To grow Caco-2 cells, activate the surface of the microchannels in a gut-on-a-chip by treating the chip setup in a UV/ozone generator for 40 min, and then incubate the chip at 37 °C in a humidified 5% CO_2 incubator for 1 h after introducing a chilled ECM solution (Matrigel and collagen I) into microchannels using a 1 mL sterile syringe. After coating, replace the ECM solution with Caco-2 culture medium by slowly infusing the medium using 3 mL Luer lock syringes placed in a syringe pump.
2. To culture intestinal organoid cells, activate the surface of the microchannels in a gut-on-a-chip by treating the setup in a UV/ozone generator for 40 min. Quickly introduce 30 μL of 1% PEI solution, incubate at room temperature for 10 min, introduce 0.1% glutaraldehyde solution, incubate at room temperature for 20 min, wash the channels with deionized water, and then dry the whole setup at 60 °C overnight. Next, add a chilled ECM mixture containing collagen I (final concentration, 60 μg/mL) and Matrigel (50× dilution) into the microchannels, incubate the device setup at 37 °C in a humidified CO_2 incubator for 1 h, and then replace the ECM mixture with organoid culture medium using 3 mL Luer lock syringes placed in a syringe pump.
3. Introduce dissociated human intestinal epithelial cells (*see* **Note 3**) into the upper microchannel, let the cells attach on the porous membrane at 37 °C in a humidified 5% CO_2 incubator for 1–3 h (*see* **Note 4**), and then flow suitable culture medium (for Caco-2 or organoid cells) into the upper microchannel alone at 50 μL/h (*see* **Note 5**) for 24–36 h by attaching the syringes to a syringe pump.

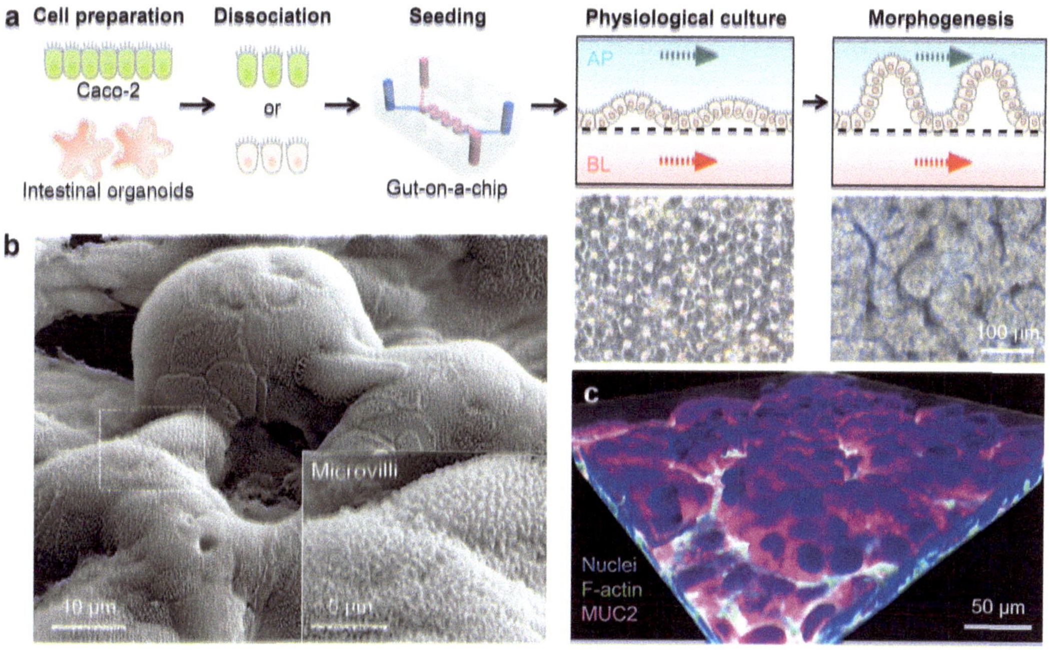

Fig. 2 Regeneration of 3D intestinal epithelial layer in a gut-on-a-chip. (**a**) A schematic illustrates the procedure of 3D morphogenesis in a gut-on-a-chip using either Caco-2 cells or intestinal organoid epithelium. Arrows in "Physiological culture" and "Morphogenesis" show the flow of a cell culture medium inside the microchannels. Black dashed lines indicate a porous basement membrane in a gut-on-a-chip. Phase-contrast images corresponding to each schematic display representative morphology of Caco-2 cells grown in a gut-on-a-chip on days 2 (left) and 5 (right), respectively. (**b**) An SEM image highlights the 3D microarchitecture of a Caco-2 cell layer recreated in a gut-on-a-chip. A dashed box and a corresponding inset show a high-power magnification of the part of an apical brush border regenerated on a 3D Caco-2 layer. (**c**) An overlaid immunofluorescence micrograph that visualizes the expression of mucin 2 (MUC2, magenta), F-actin (green), and nuclei (blue) in a 3D Caco-2 epithelial layer formed in a gut-on-a-chip. Reproduced from [16]

4. Perfuse a corresponding culture medium (for Caco-2 or organoid cells) into both upper and lower microchannels at 50 μL/h using a syringe pump once the cell monolayer is uniformly formed with 100% confluency. Maintain this dual flow to the gut-on-a-chip for additional 4–6 days until the cells undergo spontaneous 3D morphogenesis (Fig. 2a; *see* **Note 6**).
5. Apply stretching motions to the cells growing in a gut-on-a-chip using a Flexcell module with 10% of cell strain and 0.15 Hz of frequency.

3.5 Morphological Assessment of Differentiated Epithelium in a Gut-on-a-Chip

1. To monitor the morphogenesis process of an intestinal epithelium in a gut-on-a-chip, use a phase-contrast or differential interference-contrast (DIC) microscope (Fig. 2a, micrographs). As soon as the imaging is finished, put the gut-on-a-chip back to a humidified CO_2 incubator.

2. To analyze the surface topology of a 3D intestinal epithelium formed in a gut-on-a-chip, perform scanning electron microscopy (SEM) imaging at a high-power magnification. To prepare a sample, fix the epithelial cells cultured in a gut-on-a-chip (4% PFA, for 30 min), remove the upper PDMS layer, dehydrate the cell layer in ethanol (200 proof) and hexamethyldisilazane, dry the setup in a desiccator overnight, mount the sample on a conductive carbon tape, coat with gold, and then image using an SEM (Fig. 2b).
3. To characterize the expression of proteins (e.g., differentiation markers) or analyze the cellular components (e.g., F-actin and nuclei), perform an immunofluorescence microscopic analysis using a laser-scanning confocal microscope (Fig. 2c). To stain protein markers on the cells grown in a gut-on-a-chip, run a fixation (30 μL of 4% PFA solution, 15 min), permeabilization (0.3% Triton X-100 in PBS, 30 min), and blocking process (2% BSA in PBS, 1 h) sequentially at room temperature. Next, apply a primary antibody solution (appropriate concentration upon manufacturer's recommendation, at room temperature for 3 h or 4 °C for overnight) and a secondary antibody solution (appropriate concentration upon manufacturer's recommendation, at room temperature 3 h under light protection) in sequence. Rinse the cells between steps with PBS. Perform counterstaining for visualizing F-actin and nuclei by applying fluorophore-labeled phalloidin and DAPI, respectively, by incubating the staining solution at room temperature for 30 min.

4 Notes

1. R-spondin- and Noggin-conditioned media are prepared in-house by culturing R-spondin and Noggin producing HEK293T cells, respectively, and harvesting their supernatant [16].
2. To pass organoids, remove the culture medium, add 500 μL of cold Cell Recovery Solution, and incubate the plate at 4 °C for 30–40 min. Collect the organoids in a 15 mL conical tube, spin them down at 100 × *g* for 5 min, add 1 mL of TrypLE Express solution, add 5 mL of organoid basal medium, pipette 10 times with a 1 mL micropipette, and spin down at 100 × *g* for 5 min. Remove the supernatant, and resuspend the pellet with a desired amount of Matrigel. A typical split ratio is 1:3. Place 30 μL of the dissociated organoid suspension in a new 24-well plate, and culture the organoids in a humidified 37 °C CO_2 incubator.

3. To prepare a single-cell suspension of Caco-2 cells, wash a T75 flask that contains ~90% confluent Caco-2 cells with PBS (Ca^{2+}- and Mg^{2+}-free) twice, add 1 mL of pre-warmed trypsin/EDTA solution, incubate at 37 °C for 3–5 min, resuspend the dissociated cells with 9 mL of serum-containing Caco-2 medium, centrifuge the cell suspension (300 × *g* at 4 °C for 3 min), and then resuspend the cell pellet with the culture medium at 1.0×10^7 cells/mL. To prepare a dissociated organoid cell suspension, remove the culture medium in the plate with fully grown organoids, incubate the plate with 500 μL of cold Cell Recovery Solution in each well at 4 °C for 30–40 min, collect the organoids in a 15 mL conical tube, and spin down at 100 × *g* at 4 °C for 5 min. After removing the supernatant, add 1 mL of pre-warmed TrypLE Express solution containing Y-27632 (final concentration,10 μM), incubate the cell suspension in a 37 °C water bath for 5 min, add 5 mL of warm PBS and resuspend aggressively, filter the cell suspension using a cell strainer (cutoff, 100 μm), centrifuge the suspension at 300 × *g* at 4 °C for 3 min, resuspend in organoid culture medium, and then adjust the cell density to 1.0×10^7 cells/mL.
4. Inspect the cell attachment before flowing culture medium. If the cell attachment is poor, the attachment time can be extended in a humidified 37 °C CO_2 incubator overnight.
5. Flow a culture medium into both upper and lower microchannels using a syringe pump at 50 μL/h volumetric flow rate to establish 0.02 dyne/cm^2 of shear stress in a 500 μm height microchannel.
6. Inspect the generation of air bubbles in the microchannels during the culture. Air bubbles may compromise cell viability if staying in the channel for a while. To remove air bubbles, carefully flow a culture medium through the microchannels.

Acknowledgments

This work was supported in part by the National Cancer Institute of the National Institutes of Health (K00CA245801 to WS; R21CA236690 to HJK) and the Leona M. & Harry B. Helmsley Charitable Trust (Grant #1912-03604 to HJK).

References

1. Hubatsch I, Ragnarsson EG, Artursson P (2007) Determination of drug permeability and prediction of drug absorption in Caco-2-monolayers. Nat Protoc 2(9):2111–2119. https://doi.org/10.1038/nprot.2007.303
2. den Daas SA, Soffientini U, Chokshi S, Mehta G (2022) A permeability assay for mouse intestinal organoids. STAR Protoc 3(2):101365. https://doi.org/10.1016/j.xpro.2022.101365

3. Bardenbacher M, Ruder B, Britzen-Laurent N, Schmid B, Waldner M, Naschberger E, Scharl M, Muller W, Gunther C, Becker C, Sturzl M, Tripal P (2019) Permeability analyses and three dimensional imaging of interferon gamma-induced barrier disintegration in intestinal organoids. Stem Cell Res 35:101383. https://doi.org/10.1016/j.scr.2019.101383
4. Artursson P, Palm K, Luthman K (2001) Caco-2 monolayers in experimental and theoretical predictions of drug transport. Adv Drug Deliv Rev 46(1–3):27–43. https://doi.org/10.1016/s0169-409x(00)00128-9
5. Bijvelds MJC, Roos FJM, Meijsen KF, Roest HP, Verstegen MMA, Janssens HM, van der Laan LJW, de Jonge HR (2021) Rescue of chloride and bicarbonate transport by elexacaftor-ivacaftor-tezacaftor in organoid-derived CF intestinal and cholangiocyte monolayers. J Cyst Fibros 21:537. https://doi.org/10.1016/j.jcf.2021.12.006
6. Lee YK, Puong KY, Ouwehand AC, Salminen S (2003) Displacement of bacterial pathogens from mucus and Caco-2 cell surface by lactobacilli. J Med Microbiol 52(Pt 10):925–930. https://doi.org/10.1099/jmm.0.05009-0
7. Puschhof J, Pleguezuelos-Manzano C, Martinez-Silgado A, Akkerman N, Saftien A, Boot C, de Waal A, Beumer J, Dutta D, Heo I, Clevers H (2021) Intestinal organoid cocultures with microbes. Nat Protoc 16(10):4633–4649. https://doi.org/10.1038/s41596-021-00589-z
8. Han X, Mslati MA, Davies E, Chen Y, Allaire JM, Vallance BA (2021) Creating a more perfect union: modeling intestinal bacteria-epithelial interactions using organoids. Cell Mol Gastroenterol Hepatol 12(2):769–782. https://doi.org/10.1016/j.jcmgh.2021.04.010
9. Kim HJ, Huh D, Hamilton G, Ingber DE (2012) Human gut-on-a-chip inhabited by microbial flora that experiences intestinal peristalsis-like motions and flow. Lab Chip 12(12):2165–2174. https://doi.org/10.1039/c2lc40074j
10. Kim HJ, Ingber DE (2013) Gut-on-a-Chip microenvironment induces human intestinal cells to undergo villus differentiation. Integr Biol 5(9):1130–1140. https://doi.org/10.1039/c3ib40126j
11. Kim HJ, Li H, Collins JJ, Ingber DE (2016) Contributions of microbiome and mechanical deformation to intestinal bacterial overgrowth and inflammation in a human gut-on-a-chip. Proc Natl Acad Sci USA 113(1):E7–E15. https://doi.org/10.1073/pnas.1522193112
12. Shin W, Ambrosini YM, Shin YC, Wu A, Min S, Koh D, Park S, Kim S, Koh H, Kim HJ (2020) Robust formation of an epithelial layer of human intestinal organoids in a polydimethylsiloxane-based gut-on-a-Chip microdevice. Front Med Technol 2:2. https://doi.org/10.3389/fmedt.2020.00002
13. Sontheimer-Phelps A, Chou DB, Tovaglieri A, Ferrante TC, Duckworth T, Fadel C, Frismantas V, Sutherland AD, Jalili-Firoozinezhad S, Kasendra M, Stas E, Weaver JC, Richmond CA, Levy O, Prantil-Baun R, Breault DT, Ingber DE (2020) Human colon-on-a-Chip enables continuous in vitro analysis of colon mucus layer accumulation and physiology. Cell Mol Gastroenterol Hepatol 9(3):507–526. https://doi.org/10.1016/j.jcmgh.2019.11.008
14. Shin YC, Shin W, Koh D, Wu A, Ambrosini YM, Min S, Eckhardt SG, Fleming RYD, Kim S, Park S, Koh H, Yoo TK, Kim HJ (2020) Three-dimensional regeneration of patient-derived intestinal organoid epithelium in a Physiodynamic mucosal Interface-on-a-Chip. Micromachines 11(7):663. https://doi.org/10.3390/mi11070663
15. Shin W, Hinojosa CD, Ingber DE, Kim HJ (2019) Human intestinal morphogenesis controlled by Transepithelial morphogen gradient and flow-dependent physical cues in a microengineered gut-on-a-Chip. iScience 15:391–406. https://doi.org/10.1016/j.isci.2019.04.037
16. Shin W, Kim HJ (2022) 3D in vitro morphogenesis of human intestinal epithelium in a gut-on-a-chip or a hybrid chip with a cell culture insert. Nat Protoc 17(3):910–939. https://doi.org/10.1038/s41596-021-00674-3

Chapter 16

Co-culturing Human Intestinal Enteroid Monolayers with Innate Immune Cells

Janet F. Staab, Jose M. Lemme-Dumit, Rachel Latanich, Marcela F. Pasetti, and Nicholas C. Zachos

Abstract

The coordinated interaction between the intestinal epithelium and immune cells is required to maintain proper barrier function and mucosal host defenses to the harsh external environment of the gut lumen. Complementary to in vivo models, there is a need for practical and reproducible in vitro models that employ primary human cells to confirm and advance our understanding of mucosal immune responses under physiologic and pathophysiologic conditions. Here we describe the methods to co-culture human intestinal stem cell-derived enteroids grown as confluent monolayers on permeable supports with primary human innate immune cells (e.g., monocyte-derived macrophages and polymorphonuclear neutrophils). This co-culture model reconstructs the cellular framework of the human intestinal epithelial-immune niche with distinct apical and basolateral compartments to recreate host responses to luminal and submucosal challenges, respectively. Enteroid-immune co-cultures enable multiple outcome measures to interrogate important biological processes such as epithelial barrier integrity, stem cell biology, cellular plasticity, epithelial-immune cells crosstalk, immune cell effector functions, changes in gene expression (i.e., transcriptomic, proteomic, epigenetic), and host-microbiome interactions.

Key words Intestinal organoids, Enteroids, Monolayer, Macrophages, Neutrophils, Co-culture

1 Introduction

Human intestinal enteroids are derived from actively dividing LGR5$^+$ intestinal stem cells obtained from patient biopsies or surgical resections [1]. These are epithelial-only cultures that faithfully recapitulate the functional phenotype of the intestinal segment from which they are derived [2, 3]. Human enteroids (duodenum, jejunum, ileum) and colonoids (ascending-, transverse-, descending-, and sigmoid colon, and rectum) can be maintained indefinitely as three-dimensional (3D) cultures in a basement

Janet F. Staab and Jose M. Lemme-Dumit contributed equally to this work.

Paloma Ordóñez-Morán (ed.), *Intestinal Differentiated Cells: Methods and Protocols*, Methods in Molecular Biology, vol. 2650, https://doi.org/10.1007/978-1-0716-3076-1_16,

membrane-enriched matrix (e.g., Matrigel). Culture media containing essential growth factors potentiate Wnt signaling to maintain and propagate intestinal stem cells. Removal of Wnt signaling allows for terminal differentiation of enteroid/colonoid cultures to express, in addition to absorptive enterocytes, the other major epithelial cell lineages including goblet, enteroendocrine, Paneth, tuft, and M cells (with supplementation of RANKL) representing the intestinal segments from which they were obtained [4–8]. In addition, human enteroids/colonoids also preserve disease-specific phenotypes and thus have been studied to enhance our understanding of the pathophysiology of infectious diseases, genetic disorders, metaplasia and cancer [9–12]. Together, these properties make human enteroids a relevant, translational ex vivo model of the human intestinal epithelium that has the potential to serve as preclinical model for drug target discovery and development.

Intestinal epithelial barrier function depends upon the combined participation of the epithelium and underlying stromal cells including immune, lymphatic, nerve, and mesenchymal cells. Traditional in vitro modeling of the human gut mucosa has mainly considered the epithelial cells in the absence of other cell populations. The biological processes by which stromal cells influence the function of the human intestinal epithelium and their coordinated actions have been difficult to explore due to the lack of a relevant in vitro model that recapitulates the spatial proximity of primary cells. Most enteroid studies are conducted as 3D structures in Matrigel with the apical surface facing inward toward an enclosed lumen. This conformation has been used to co-culture stromal cells within Matrigel; however, their spatial proximity to the epithelium cannot be controlled, and exposure of luminal content must be introduced into each structure via microinjection. Alternatively, "apical-out" enteroids overcome the technical obstacles for luminal exposure to gut antigens/microbes but do not allow for basolateral incorporation of stromal cells [13]. A convenient feature of the enteroid model is the ability to generate a confluent monolayer on permeable supports (i.e., Transwell inserts) with controllable access to both apical and basolateral compartments [14]. To further improve this model, we developed a co-culture system consisting of human enteroids harboring primary immune cells; these immune enteroids are sought to bridge some of the tissue culture gaps in modeling the human intestinal epithelium [15–17].

Here, we outline methods for establishing human enteroid monolayers and their co-culture with innate immune cells (e.g., monocytes, macrophages, and neutrophils) (*see* Fig. 1).

Three-dimensional enteroids are fragmented and subsequently seeded onto the upper chamber of human collagen IV-coated cell culture inserts. Primary human polymorphonuclear neutrophils (PMN) or monocytes are isolated from fresh human whole blood or frozen peripheral blood mononuclear cells (PBMC) obtained

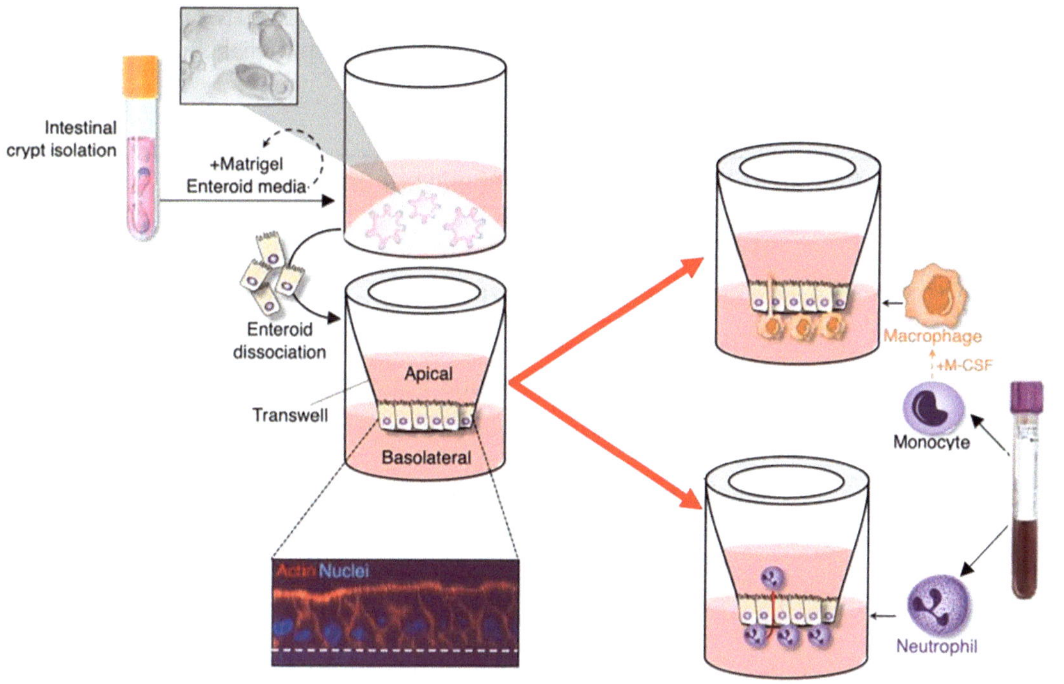

Fig. 1 Generating human intestinal enteroid monolayers co-cultured with primary human innate immune cells obtained from peripheral blood. Enteroid/colonoid monolayers are produced from Lgr5+ stem cells isolated from intestinal crypts and grown on Transwell inserts. Phagocytic cells, i.e., polymorphonuclear neutrophils and monocyte-derived macrophages, are obtained from human peripheral blood and added to the basolateral side of confluent enteroid/colonoid monolayers

from healthy donors. Monocytes are enriched from PBMC and subsequently differentiated into macrophages by supplementation of macrophage colony-stimulating factor (M-CSF) in the culture media for 6 days. Monocyte-derived macrophages exhibit a $CD14^{+}$ $CD16^{low}$ $CD64^{low}$ $CX3CR1^{-}$ phenotype and have the capacity to phagocytose bacteria [16]. Co-cultures containing macrophages can be maintained for 48 h. PMN exhibiting the phenotype $CD15^{+}$ $CD16^{+}$ $CD14^{-}$ are used the same day of isolation [15] due to the limited lifespan [18] and remain viable in co-culture for 4 h. Successful assembly of the co-culture requires careful coordination of timing to obtain confluent, differentiated enteroid monolayers that can be combined with fully differentiated macrophages or freshly isolated PMN. The required preincubation and planning times are built in the protocol described. The methods are divided into the two main components of the model: (a) generation of enteroid/colonoid monolayers and isolation of immune cells and (b) assembly of immune enteroids. Enteroid-immune co-cultures enable multiple outcome measures, including transepithelial electrical resistance (TEER), production of cytokines/chemokines and other biomarkers, phenotypic adaptation of immune cells, tissue immunofluorescence imaging, protein and mRNA expression, antigen or microbe uptake, and other epithelial and immune cell functions.

2 Generating Confluent Human Enteroid Monolayers on Transwells

2.1 Three-Dimensional Enteroid Cultures

2.1.1 Materials

Base Media (CMGF-)

1. Advanced DMEM/F12 (Gibco/Thermo Fisher 12634028).
2. 10 mM HEPES buffer (Gibco/Thermo Fisher 15630080).
3. 1% GlutaMAX (Thermo Fisher 35050061).
4. 1% penicillin-streptomycin (Quality Biological 120-095-721).

Non-differentiation/Propagation Medium (NDM)

1. CMGF-.
2. 50 ng/mL EGF (R&D Systems/Tocris 236-GMP).
3. 2% B27 Supplement (50×; Gibco/ThermoFisher 17504044)
4. 10 nM Gastrin (Anaspec AS-64149)
5. 500 nM A83-01 (R&D Systems/Tocris 2939).
6. 10 μM SB 202190 (Sigma-Aldrich S7067); prepare stock solution in DMSO.
7. 125 μg/mL Primocin (InvivoGen ant-pm-2); antimicrobial reagent for primary cells.
8. 50% Wnt3A-conditioned medium.
9. 15% R-Spondin 1-conditioned medium.
10. 10% Noggin-conditioned medium.

NDM will keep at 4 °C for 1 week.

Differentiation Medium (DFM)

1. CMGF- (without pen/strep for bacterial infection of co-cultures).
2. 50 ng/mL EGF (R&D Systems/Tocris 236-GMP).
3. 2% B27 supplement (50×; Gibco/Thermo Fisher 17504044).
4. 10 nM gastrin (Anaspec AS-64149).
5. 500 nM A83-01 (R&D Systems/Tocris 2939).
6. 125 μg/mL Primocin (InvivoGen ant-pm-2); antimicrobial reagent for primary cells (omit for bacterial infections of co-culture).
7. 10% Noggin-conditioned medium.

DFM will keep at 4 °C for 1 week.

Other Reagents

1. Y-27632 dihydrochloride (R&D Systems/Tocris 1254); prepare 10 mM in sterile ddH_2O. Store aliquots at −20 °C.
2. CHIR99021 (R&D Systems/Tocris 4423); prepare 10 mM in DMSO. Store aliquots at −20 °C.
3. Matrigel phenol red-free (Corning 356231).

Monocyte-Derived Macrophages (MoDM) Complete Medium

1. RPMI (Thermo Fisher 11875093).
2. 10% FBS, heat inactivated (Sigma F4135).
3. 1× MEM nonessential amino acids (Sigma M7145).
4. 1 mM sodium pyruvate (100 mM solution; Sigma S8636).
5. 55 μM 2-mecaptoethanol (Gibco, Thermo Fisher 21985023).
6. 1% penicillin-streptomycin (Quality Biological 120-095-721).
7. 50 ng/mL macrophage colony-stimulating factor (M-CSF) (PeproTech 300-25).

MoDM will keep at 4 °C for up to 2 weeks. Omit M-CSF for culturing monocytes.

2.2 Seeding Enteroid Fragments onto Transwell Inserts

2.2.1 Materials

1. Enteroids/colonoids embedded in Matrigel in NDM for cell propagation.
2. 1.0 μm pore-size Transwell (TW)/culture inserts (PET membrane; MilliporeSigma MCSP24H48) for 24-well plates for macrophage or monocyte co-cultures.
3. 3.0 μm pore-size Transwell (TW)/culture inserts (PET membrane; Corning 3472) for 24-well plates for PMN co-cultures (PET membranes are translucent and permit visualization of the underlying immune cells through the monolayer for immunofluorescence).
4. Collagen IV, from human placenta (Sigma-Aldrich C5533). Prepare 1 mg/mL in 100 mM (or 0.5 M) acetic acid; store in single-use aliquots at −20 °C.
5. TrypLE Express Enzyme 1× no phenol red (Thermo Fisher 12604013): aliquot and store protected from light at room temperature.
6. Organoid harvesting solution (Cultrex/R&D Systems 3700-100-01).
7. Sterile mini cell scrapers (United Biosystems MCS-200).
8. Enteroid propagation (NDM) and differentiation (DFM) media; see below.
9. CMGF- (see below).

10. Centrifuge (swinging bucket).
11. 37 °C water bath.
12. P200 and P1000 micropipettes.
13. Sterile Pasteur pipettes.
14. Orbital shaker at 4 °C.
15. Tissue culture incubator (37 °C, 5% CO_2).
16. Voltohmmeter to monitor TEER to establish confluence and differentiation status (World Precision Instruments EVOM^2 with STX2 electrode).
17. Inverted light microscope capable of 5× and 10× magnification.

2.2.2 Enteroid Monolayer Formation

1. Dilute 1.0 mg/mL collagen IV 1:30 in sterile PBS to obtain a 33 μg/mL solution.
2. Coat the desired number of TWs with 100 μL of diluted human collagen IV (either overnight at 4 °C or ≥ 2 h at 37 °C). Discard unused diluted collagen IV.

   Include TWs for enteroid monolayers only, no immune cells (culture controls).
3. Have a minimum of one dense well of enteroids (approximately 100+ enteroids in a 25–30 μL dome of Matrigel) in a 24-well plate for every two TWs.

   The seeding enteroids should be as large as possible so that when they are fragmented, the fragments will lay and attach to the insert TWs. Use enteroids that have been in propagation medium for 6–8 days. A ratio higher or lower than one well for two TWs may be more appropriate depending on the well densities.
4. Aspirate the medium from the wells containing enteroids, and add 1.0 mL of cold cell recovery/dissociation solution. Dislodge the Matrigel and enteroids from the bottom of the well using a mini cell scraper.
5. Shake the plate on an orbital shaker at 4 °C as per the cell recovery/dissociation solution manufacturer's recommendations, usually 30–40 min at 250 rpm.

   *Matrigel removal is critical (see **Note in step 8**).*
6. Recover the dislodged enteroids using a P1000 micropipette and transfer the suspended enteroids to a 15 mL conical tube.

   As much as possible, maintain enteroids in an intact state, avoiding small cellular clumps.
7. Add equal volume of CMGF- (see recipe above), and pellet the enteroids by centrifugation at 400 × *g* for 10 min at 4 °C.

8. Aspirate/remove as much of the supernatant as possible without disturbing the enteroid.pellet. The enteroid pellet should be free of Matrigel.

   *Any remaining Matrigel will form a translucent layer above the enteroid pellet. To remove residual Matrigel, add 2 mL of cell recovery/dissociation solution, and gently suspend the pellet using a P1000 micropipette. Place the 15 mL conical tube on its side on the orbital shaker at 4 °C and shake for 10 min. Repeat **step** 7 to recover the enteroids.*

9. Add 50 μL/well of TrypLE Express to the enteroids; mix gently 5× by pipetting with a P1000 micropipette.

   Once the TrypLE is added to the enteroid pellet, work quickly to limit exposure to the enzyme.

10. Place the suspended enteroids in a 37 °C water bath for 90 s. Less time for smaller enteroids (~75 s). Swirl the tube occasionally during the incubation to suspend the enteroids.

11. Immediately, add 5–8 mL of cold CMGF- in sterile conditions. Pellet the enteroid fragments by centrifugation as above.

12. Aspirate the supernatant, and gently suspend the enteroid fragments at 100 μL NDM per TW (with 10 μM each Y-27632 and CHIR99021 inhibitors for the ileum and colon) using a P1000 micropipette.

    For example, enteroid fragments from one well will be suspended in 200 μL of propagation medium to seed two insert TWs. Set aside at room temperature.

13. Aspirate the collagen solution from the TWs, and wash with 100–200 μL of CMGF-. Repeat (two washes in total). Aspirate wash.

14. Gently suspend the enteroid fragments using a P1000 micropipette, and plate 100 μL per TW using a P200 micropipette.

    Gently resuspend the enteroid fragments with a P200 micropipette each time prior to adding 100 μL into TWs. The enteroid fragments will settle and need resuspension between plating.

15. Add 600 μL of NDM to the well of the receiver 24-well plate (with 10 μM each Y-27632 and CHIR99021 inhibitors for the ileum or colon).

16. Incubate at 37 °C, 5% CO_2.

17. Change the media at 48 h to propagation media (NDM) without Y-27632 and CHIR99021 inhibitors, and continue to incubate at 37 °C with media changes every other day.

18. Observe for patches formation and eventual closed monolayers. Monolayer confluency is usually reached in 7–14 days.

*Measure TEER (refer to voltohmmeter manufacturer for instructions on how to perform TEER measurements), or observe cultured inserts under an inverted microscope to monitor confluence (TEER range depends on intestinal segment). Raw TEER values of < 200 ohms indicate incomplete monolayer coverage of the TW. Newly confluent monolayers will have raw TEER values in the range of 300–500 ohms (100–165 ohms*cm²). TEER values will continue to increase over time, while the monolayers are maintained in propagation medium (NDM).*

19. Once confluent, change the monolayer media to DFM for 5 days. Change DFM on days 2 and 4. Monitor for an increase in TEER as a sign of monolayer maturation. Start monolayer differentiation after 1–3 days of confluency. For experiments interrogating crypt-like epithelium, maintain the monolayers in propagation medium (NDM) and proceed to co-culture setup.

    Undifferentiated (crypt-like) monolayers in propagation medium will remain viable and confluent for up to 7–10 days; however it is best to start the differentiation process soon after the monolayers become confluent as older monolayers tend to fall apart. TEER values will increase rapidly over the 5-day period ending at > 2–3× the initial ohms after start of monolayer differentiation [16].

2.2.3 Alternate Monolayer Protocol

Some enteroid or colonoid lines will not form monolayers when fragmented with TrypLE Express. For these instances, substitute TrypLE for mechanical trituration:

1. After recovery of the enteroids from Matrigel (**step 7**), suspend the enteroids in propagation medium (NDM) (with 10 μM each Y-27632 and CHIR99021 inhibitors for the ileum or colon) at 100 μL per TW using a P1000 micropipette.
2. Using a P200 micropipette, triturate the enteroids by vigorously pipetting up and down 25–30× as for fragmenting enteroids for propagation.
3. Transfer 100 μL of mechanically fragmented enteroids into washed TWs (**step 13**).
4. Proceed with the remaining steps above.

3 Isolation of Immune Cells from Human Peripheral Blood

3.1 Human Monocytes

3.1.1 Materials

1. Access to human peripheral blood: 60 mL of whole blood should suffice for isolating the number of monocytes needed to set up to >10 co-cultures. Collect blood into ETDA tubes (BD Vacutainer® 366643).

2. Ficoll-Paque PREMIUM (Cytiva 17544202).
3. Pan Monocyte Isolation Kit (Miltenyi Biotec 130-096-537).
4. Centrifuge (swinging bucket).
5. Macrophage colony-stimulating factor (M-CSF) (PeproTech 300-25).
6. Six-well tissue culture plate.
7. Monocyte-derived macrophage medium (MoDM; see below).
8. 15-mL conical tubes.
9. Tissue culture incubator (37 °C, 5% CO_2).

3.1.2 Protocol

1. Six days prior to setting up the enteroid-macrophage co-culture, isolate monocytes from 60 mL human peripheral blood.
2. Isolate PBMC by diluting whole blood with PBS (1:1) and centrifugation gradient over Ficoll-Paque PREMIUM per the manufacturer's recommendations.
3. Monocytes are isolated from PBMC using a Pan Monocyte Isolation Kit (Miltenyi Biotec, USA), according to the manufacturer's instructions.

   The PBMC and monocyte isolation will take 4–5 h.
4. Plate the monocytes in monocyte-derived macrophages (MoDM) medium with 50 ng/mL macrophage colony-stimulating factor (M-CSF) into six-well plates at 1×10^6/mL. Plate 2 mL per well.

   For monocyte culture, omit M-CSF from the medium.
5. Differentiate the monocytes into macrophages for 6 days by replacing the culture media (MoDM + M-CSF) every other day.

   Observe for cell adherence to the plate and production of cell membrane extensions. Rounding and loss of adhesion are signs of cell death.
6. Five days before assembly of the co-culture, change the propagation medium (NDM) of the confluent enteroid/colonoid monolayers to DFM. Change DFM on days 2 and 4. The co-culture is assembled on day 5 of enteroid/colonoid monolayer differentiation.

3.2 Human PMN

3.2.1 Materials

1. Access to human peripheral blood: 40 mL of whole blood should suffice for isolating the number of PMN needed to set up >10 co-cultures. Collect blood into ETDA tubes (BD Vacutainer® 366643). Start PMN isolation early in the day of co-culture that represents day 5 of differentiated enteroid/colonoid monolayers.

2. Ficoll-Paque PREMIUM (Cytiva 17544202).
3. 6% dextran (Alfa Aesar J63702); see Unit 7.23 for preparation.
4. PBS (1×), pH 7.4 (Quality Biological 114-058-101).
5. PBS (10×), pH 7.4 (Quality Biological 119-069-131).
6. Cold sterile Milli-Q water.
7. 50 mL conical tubes.
8. Centrifuge.
9. Hemacytometer or other means for counting cells.
10. Inverted light microscope.

3.2.2 Protocol

1. In a 50 mL conical tube, transfer 15 mL of whole blood and add PBS (1×) to a final volume of 50 mL.
2. Spin for 10 min at 400 × *g*, 21 °C with the acceleration/deacceleration brake set at 5/5.
3. Aspirate the supernatant (mix of PBS and plasma) without disturbing the pellet, and resuspend the cells in PBS (1×) to a final volume of 35 mL.
4. Dispense 15 mL of Ficoll-Paque Premium in a new 50 mL conical tube, and carefully layer on top the 35 mL of washed blood cells from **step 3**.
5. Spin for 35 min at 300 × *g*, 21 °C without break.

   It is important to avoid the use of the break during spin-down process to maintain the gradient.
6. Collect the PBMC layer and remove the Ficoll-Paque without disturbing the PMN/erythrocytes pellet.
7. Aspirate the PMN/erythrocytes pellet and transfer to a new 50 mL conical tube.

   It is important to transfer to a fresh conical tube to avoid PBMC contamination that can remain on the wall of the tube.
8. Add PBS (1×) to a final volume of 22.5 mL.
9. Add 7.5 mL of 6% dextran (final dilution 1:4). Mix the tube contents by gentle inversion (10 times).
10. Set aside to allow the erythrocytes to sediment by gravity for 15–20 min at room temperature.
11. Transfer the supernatant to a new 50 mL conical tube and bring the volume to 50 mL with PBS (1×).
12. Spin for 10 min 300 × *g*, 21 °C with the acceleration/deacceleration break set at 5/5.
13. Aspirate the supernatant and suspend the pellet in the remaining liquid.

14. Add 18 mL of cold sterile water to lyse the erythrocytes. Mix the cell suspension by inversion for 20 s, and immediately add 2 mL of cold 10× PBS and 20 mL of cold 1× PBS.
15. Spin for 10 min at 300 × *g* at 4 °C with the acceleration/deacceleration break set at 5/5.
16. Repeat **steps 14** and **15** if necessary, to lyse any remaining erythrocytes.
17. Resuspend the PMN in 1–2 mL of DFM without antibiotics and count using a hemocytometer or an automated cell counter.

    Keep the PMN suspension at 4 °C while counting.
18. Adjust the PMN concentration to 1×10^7 viable cells/mL in DFM without antibiotics, and immediately proceed to co-culture setup.

4 Co-culture Assembly

4.1 Materials

1. 15 mL sterile conical tubes.
2. Sterile cell scrapers (United Biosystems MCS-200).
3. 12-well flat-bottom tissue culture plate(s).
4. Small metal forceps.
5. Sterile Pasteur pipettes.
6. FIREBOY Safety Bunsen burner (Integra Biosciences Corp 144010).
7. Centrifuge.
8. Hemacytometer or other means to count cells.
9. Light microscope.
10. Sterile PBS (1×) pH 7.4.
11. Tissue culture incubator (37 °C, 5% CO_2).
12. NDM or DFM (without antibiotics for bacterial infections).

4.2 Assembly of Immune-Enteroid Co-cultures

1. The morning of the experiment, measure TEER if this is a readout parameter of the assay. Return the plate to the tissue culture incubator until the macrophages are ready or when PMN isolation is complete (approximately 4 h). Once PMN are in DFM without antibiotics, proceed to **step 7** without delay.
2. Using a sterile cell scraper, gently remove the attached macrophages from the six-well culture plate, and collect in a 15 mL centrifuge tube.
3. Pellet the macrophages by centrifugation (swinging bucket; 400 × *g* for 5 min, at 4 °C).

4. Suspend the macrophages in sterile PBS and count using Trypan blue to determine viability.
5. Pellet the macrophages by centrifugation, and suspend in enteroid medium (propagation or DFM) with 10 ng/mL M-CSF at a concentration of 2×10^6 viable cells/mL. Set aside on ice.
6. Using sterile forceps, pick up a monolayer-containing TW and gently invert into an empty 12-well plate. Retain the 12-well plate lid.

   Some of the apical medium will drain; however, the monolayer will retain some media. Set up the number of inverted monolayers needed.
7. Gently aspirate any remaining medium on the bottom (now facing up) of culture inserts.
8. Using a P200 micropipette, gently add 50 μL of the macrophage suspension (1×10^5 cells) in enteroid medium (propagation or DFM) + M-CSF onto the bottom of the insert. For PMN, add 50 μL (5×10^5 cells) in DFM without antibiotics.

   For monocytes, add 50 μL (1×10^5 cells) in enteroid medium (propagation or DFM).
9. Repeat until the immune cells have been deposited onto the inverted inserts.
10. Gently place the lid onto the 12-well culture plate.

    The lid will contact the 50 μL drop containing the immune cells and form a bevel.
11. Return the plate to the 37 °C, 5% CO_2 incubator for 2 h.

    Longer incubations of the inverted inserts do not improve adherence of the immune cells. Keep the original insert TW 24-well culture plate with media at 37°C, 5% CO_2. Alternatively, a new 24-well plate can be used to upright the co-culture insert TWs.
12. Move plate from the incubator with the inverted TWs and place under the biosafety cabinet.
13. Using sterile forceps, pick up the inverted TWs, and flip back into the original or a new 24-well culture plate.
14. Add 100 μL of enteroid medium (propagation or DFM) to the apical side and 600 μL to the well of the tissue culture plate + M-CSF at 10 ng/mL for macrophages (omit for PMN or monocytes) if using a new 24-well plate.

    If the original 24-well culture plate is used, add M-CSF to a final concentration of 10 ng/mL to the remaining media if co-culturing monolayers with macrophages.

15. Return the plate to the incubator and run a co-culture time course. The standard co-culture time is 24 h at the time of seeding the macrophages onto the bottom of the inserts (22 h after placing the uprighted TWs at 37 °C, 5% CO_2).

    PMN co-cultures run for up to 2 h after returning the inserts/TWs to their upright position. These experiments run for a total of 4 h. Monocyte co-cultures can be extended up to 48 h as for macrophage co-cultures. See below.

16. Measure TEER at 24 h of co-culture and again at the end of any treatment (i.e., bacterial infection) if this is a read-out.

    Some of the macrophages will drop from the bottom of the insert/TW onto the receiver plate > 48 h. Any treatment is for up to 24 h after co-culture establishment (e.g., *24-h co-culture is treated/interrogated for up to* $\leq$ *24 h* = *48 h total).*

5 Notes

1. Universal safety precautions should be followed when handling human blood and tissue samples. Institutional Review Board-approved protocols are required for collection and use of human tissue and for conduct of any research involving human subjects.
2. All solutions and material used need to be sterile, and proper sterile technique maintained throughout. PBMC and cell isolation/manipulations should be performed in a biosafety level 2 (BSL2) cabinet using careful sterile technique to prevent contamination.
3. Confluent monolayer formation is essential for proper assembly of immune enteroids. Proficiency seeding enteroids/colonoids onto inserts/TWs and achieving properly formed monolayers is necessary prior to undertaking establishment of immune enteroids [15, 16]. Technical experience on enteroid/colonoid monolayer formation and PBMC isolation/cell differentiation is necessary to adequately time co-culture experiments. We recommend using 1 well (from a 24-well culture plate) with approximately 100 enteroids to seed 2 TWs or equivalent inserts (24-well plates, 0.33 cm^2 growth area). The number of enteroid wells needed for seeding in addition to wells needed for enteroid propagation needs to be considered. Once TWs are seeded, a 7–14-days growth period is needed for monolayers to form, followed by 5 days of differentiation (12–19 days in total). Enzymatic treatment facilitates generation of enteroid/colonoid fragments to form monolayers. However, some lines will need mechanical trituration (*alternate protocol*); this can only be determined empirically. TEER

will gradually increase as monolayers become confluent in propagation medium. Alternatively, monolayer confluence may be monitored by light microscopy under low magnification (e.g., inverted light microscope).

4. Co-culture of monolayers and macrophages are useful to model intestinal homeostasis (resident cells) and, under activating conditions, immunity to pathogens. PMN in enteroid co-cultures model inflammation. The day of co-culture assembly will coincide with 5 days of monolayer maturation and 6 days of monocyte-derived macrophage differentiation. PMN are recovered from whole blood the same day the co-culture is established. Perform all steps under a biosafety cabinet.

5. Co-cultures for bacterial infection require monolayer differentiation in media without antibiotics. All other co-cultures for noninfection experiments are established in media with antibiotics [15, 16].

6. Media manipulation permits modeling of the enteroids to reflect a crypt-like or villus-like cellular compartment. A crypt-like epithelium is induced by maintaining enteroids in propagation medium (NDM), and removal of Wnt signaling (DFM) promotes epithelial differentiation to mimic a villus-like epithelium of the small intestine or surface cells of the colon. Other media manipulations can further drive enrichment of specific epithelial cell types [19, 20] to suit experimental needs. For example, the rare M cell type found in the follicle-associated epithelium of Peyer's patches or isolated lymphoid follicles can be induced in ileal enteroids grown as spheroids or monolayers by addition of TNF-α and RANKL [5–7] or retinoic acid, lymphotoxin, and RANKL [4]. Small intestinal enteroids were recently engineered to increase the differentiation and numbers of hormone-secreting enteroendocrine cells by overexpression of neurogenin 3 [21].

7. *Critical Parameters and Troubleshooting*: A key variable for generating monolayers is to have well-established, healthy enteroid cultures propagating at a rate that requires splitting every 6–7 days. When planning the culture of monolayers, it is important to consider the number of wells needed for seeding in addition to those needed to maintain enteroid propagation and maintenance. Often, it takes 2–3 weeks of enteroid expansion to generate enough wells for seeding monolayers and continued propagation. Monolayer formation across culture inserts is not always complete. It is therefore advisable to seed at least two additional monolayers per experiment. Monolayer formation in culture inserts/TWs of larger pore size ($\geq$3.0 μm) usually takes longer, but these inserts are needed for basolateral

bacterial cell invasion assays [6, 22]. Here we use the 3.0 μm-pore culture TWs for PMN-enteroid co-cultures to allow PMN migration through the monolayer [15]. We have not noticed differences in phenotypic features of the monolayers (e.g., TEER, differentiation) grown on 0.4, 1.0, or 3.0 μm-pore culture TWs ([16] and unpublished data). It is important to monitor confluence by light microscopy and/or TEER to plan for the start of differentiation, which will determine the day of co-culture assembly. Monolayers in NDM with TEER values >70 ohms*cm^2 have reached confluence and are ready for differentiation. Start the differentiation process 1–3 days upon monolayer confluence because older monolayers sometimes lose integrity after culture in DFM for 5 days. It is common to find that some monolayers lag in reaching confluence when it is necessary to begin differentiation in order to keep with immune cell maturation or isolation plans. Under these circumstances, we have found that monolayers can complete their closing in DFM if they are near confluence (approximately 90% closed). This process takes 1–2 days in DFM, and thus day 5 of differentiation should now be considered as the fifth day post changing the media to DFM. There is no appreciable difference in day 5 TEER values or other physiological parameters between monolayers differentiated from fully closed or nearly closed monolayers. Media changes affect TEER values; therefore, it is recommended to perform measurements before replacing the media. The times to measure TEER should be during monolayer growth, before assembly of the co-cultures, 24 h later (pretreatment), and 0–24 h posttreatment.

8. A key step that affects enteroid fragment adhesion to culture inserts/TWs is the human collagen IV coating. The stock 1 mg/mL solution in acetic acid is prepared in the shipping vial (5 mg) and allowed to reconstitute at 4 °C for several hours with occasional mixing (refer to Sigma-Aldrich recommendations). The collagen must be completely dissolved to allow for the enteroid fragments to properly adhere and spread onto the culture insert. The stock collagen suspension should be stored in single-use aliquots at −20 °C and can be used for several months. Replace the collagen IV stock solution at the first sign that enteroid fragments are failing to adhere to culture inserts. Culture inserts need to be coated with a saturating concentration of collagen IV (10 μg/cm^2) for efficient adhesion and spreading of enteroid fragments. Collagen coating can be done the day before seeding by placing the culture inserts/TWs with 33 μg/mL collagen IV in a Parafilm-sealed receiver plate at 4 °C. On the day of seeding, move the plate from 4 °C storage, and place it in the biosafety cabinet at the start of the

protocol to allow the plate to come to room temperature. Culture inserts/TWs with 33 μg/mL collagen IV can be left at 4 °C for up to 2 days without detriment. This is a convenient method to save time (2 h) on the day of seeding.

Acknowledgments

The protocols developed for co-culture studies were funded by the National Institute of Allergy and Infectious Diseases (P01-AI125181 to MP and NZ). The authors also wish to acknowledge the Integrated Physiology and Imaging Cores of the Hopkins Conte Digestive Disease Basic and Translational Research Core Center (DK-089502 to NZ) for resources used to develop the human immune-enteroid co-culture models.

References

1. Toshiro S, Hans C (2013) Growing self-organizing mini-guts from a single intestinal stem cell: mechanism and applications. Science 340(6137):1190–1194. https://doi.org/10.1126/science.1234852
2. Zachos NC et al (2016) Human enteroids/colonoids and intestinal organoids functionally recapitulate normal intestinal physiology and pathophysiology. J Biol Chem 291(8):3759–3766. https://doi.org/10.1074/jbc.R114.635995
3. Sabine M, Kerstin S, Wiegerinck CL, Michal M, Akkerman RDL, Simone W, Hans C, Nieuwenhuis EES (2014) Adult stem cells in the small intestine are intrinsically programmed with their location-specific function. Stem Cells 32(5):1083–1091. https://doi.org/10.1002/stem.1655. https://doi.org/10.1002/stem.v32.5
4. Ding S, Song Y, Brulois KF, Pan J, Co JY, Ren L et al (2020) Retinoic acid and Lymphotoxin signaling promote differentiation of human intestinal M cells. Gastroenterology 159:214. https://doi.org/10.1053/j.gastro.2020.03.053
5. Fasciano AC, Blutt SE, Estes MK, Mecsas J (2019) Induced differentiation of M cell-like cells in human stem cell-derived Ileal Enteroid monolayers. J Vis Exp 149. https://doi.org/10.3791/59894
6. Ranganathan S, Doucet M, Grassel CL, Delaine-Elias B, Zachos NC, Barry EM (2019) Evaluating Shigella flexneri pathogenesis in the human Enteroid model. Infect Immun 87(4). https://doi.org/10.1128/IAI.00740-18
7. Wood MB, Rios D, Williams IR (2016) TNF-alpha augments RANKL-dependent intestinal M cell differentiation in Enteroid cultures. Am J Physiol Cell Physiol 311(3):C498–C507. https://doi.org/10.1152/ajpcell.00108.2016
8. Sato T, Vries RG, Snippert HJ, van de Wetering M, Barker N, Stange DE, van Es JH, Abo A, Kujala P, Peters PJ, Clevers H (2009) Single Lgr5 stem cells build crypt-villus structures in vitro without a mesenchymal niche. Nature 459(7244):262–265. https://doi.org/10.1038/nature07935
9. Dotti I, Mora-Buch R, Ferrer-Picón E, Planell N, Jung P, Masamunt MC, Leal RF, de Carpi JM, Llach J, Ordás I, Batlle E, Panés J, Salas A (2017) Alterations in the epithelial stem cell compartment could contribute to permanent changes in the mucosa of patients with ulcerative colitis. Gut 66(12):2069–2079. https://doi.org/10.1136/gutjnl-2016-312609
10. Howell KJ et al (2018) DNA methylation and transcription patterns in intestinal epithelial cells from pediatric patients with inflammatory bowel diseases differentiate disease subtypes and associate with outcome. Gastroenterology 154(3):585–598. https://doi.org/10.1053/j.gastro.2017.10.007
11. Nanki K et al (2018) Divergent routes toward Wnt and R-spondin niche independency during human gastric carcinogenesis. Cell 174(4):

856–869.e17. https://doi.org/10.1016/j.cell.2018.07.027

12. Lehle AS et al (2019) Intestinal inflammation and dysregulated immunity in patients with inherited caspase-8 deficiency. Gastroenterology 156(1):275–278. https://doi.org/10.1053/j.gastro.2018.09.041
13. Co JY et al (2019) Controlling epithelial polarity: a human enteroid model for host-pathogen interactions. Cell Rep 26(9):2509–2520.e4. https://doi.org/10.1016/j.celrep.2019.01.108
14. VanDussen KL, Marinshaw JM, Shaikh N, Miyoshi H, Moon C, Tarr PI, Ciorba MA, Stappenbeck TS (2015) Development of an enhanced human gastrointestinal epithelial culture system to facilitate patient-based assays. Gut 64(6):911–920. https://doi.org/10.1136/gutjnl-2013-306651
15. Lemme-Dumit JM, Doucet M, Zachos NC, Pasetti MF (2022) Epithelial and neutrophil interactions and coordinated response to Shigella in a human intestinal enteroid-neutrophil co-culture model. mBio 13(3):e0094422. https://doi.org/10.1128/mbio.00944-22
16. Noel G, Baetz NW, Staab JF, Donowitz M, Kovbasnjuk O, Pasetti MF, Zachos NC (2017) A primary human macrophage-enteroid co-culture model to investigate mucosal gut physiology and host-pathogen interactions. Sci Rep 7:45270. https://doi.org/10.1038/srep45270
17. Staab JF, Lemme-Dumit JM, Latanich R, Pasetti MF, Zachos NC (2020) Co-culture system of human enteroids/colonoids with innate immune cells. Curr Protoc Immunol 131(1):10.1002/cpim.v131.1. https://doi.org/10.1002/cpim.113
18. Kolaczkowska E, Kubes P (2013) Neutrophil recruitment and function in health and inflammation. Nat Rev Immunol 13(3):159–175. https://doi.org/10.1038/nri3399
19. Beumer J, Artegiani B, Post Y, Reimann F, Gribble F, Nguyen TN et al (2018) Enteroendocrine cells switch hormone expression along the crypt-to-villus BMP signalling gradient. Nat Cell Biol 20(8):909–916. https://doi.org/10.1038/s41556-018-0143-y
20. Beumer J, Puschhof J, Bauza-Martinez J, Martinez-Silgado A, Elmentaite R, James KR et al (2020) High-resolution mRNA and secretome atlas of human enteroendocrine cells. Cell 181(6):1291–1306. e1219. https://doi.org/10.1016/j.cell.2020.04.036
21. Chang-Graham AL, Danhof HA, Engevik MA, Tomaro-Duchesneau C, Karandikar UC, Estes MK et al (2019) Human intestinal enteroids with inducible neurogenin-3 expression as a novel model of gut hormone secretion. Cell Mol Gastroenterol Hepatol 8(2):209–229. https://doi.org/10.1016/j.jcmgh.2019.04.010
22. Koestler BJ, Ward CM, Fisher CR, Rajan A, Maresso AW, Payne SM (2019) Human intestinal Enteroids as a model system of Shigella pathogenesis. Infect Immun 87(4). https://doi.org/10.1128/IAI.00733-18

Part IV

Differentiation in Colon Cancer

Chapter 17

Identifying Cell Differentiation in Colorectal Cancer

Isabel Puig, Irene Chicote, and Héctor G. Pálmer

Abstract

The intestinal epithelium is a rapid self-renewing tissue. Stem cells at the bottom of the crypts first give rise to a proliferative progeny that finally differentiates to a variety of cell types. These terminally differentiated intestinal cells are mostly present in the villi of the intestinal wall and serve as functional units to sustain the main purpose of the organ: food absorption. But for a balance homeostasis, the intestine is composed not only by absorptive enterocytes but also by other cell types such as goblet cells that secrete mucus to lubricate the intestinal lumen, Paneth cells that secrete antimicrobial peptides to control microbiome, and others. Many relevant conditions affecting the intestine including chronic inflammation, Crohn's disease, or cancer can alter the composition of these different functional cell types. As a consequence, they can lose their specialized activity as functional units and further contribute to disease progression and malignancy. Measuring the amount of these different cell populations in the intestine is essential to understand the bases of these diseases and their specific contribution to their malignancy. Interestingly, patient-derived xenograft (PDX) models faithfully recapitulate patients' tumors including the proportion of the different cell lineages present in the original tumor. Here we expose some protocols for evaluating the differentiation of intestinal cells in colorectal tumors.

Key words Colorectal cancer, Differentiated intestinal cells, Patient-derived xenografts, Imaging, Immunofluorescence

1 Introduction

The intestine is composed by a monolayer of cells presenting complementary differentiation lineages that play distinctive roles for sustaining a balanced homeostasis of the self-renewing tissue. Colon or rectal cancer can occur as a consequence of acquiring initiating mutations mostly in Wnt pathway components. These initiating mutations promote the expansion of undifferentiated intestinal cells giving rise to early adenomas. Later acquisition of other mutations promotes the transition to carcinomas and patients enter a disease phase that can compromise their lives. During this oncogenic process, the proportion and functionality of the different cell populations can drastically change. Although a general

Paloma Ordóñez-Morán (ed.), *Intestinal Differentiated Cells: Methods and Protocols*, Methods in Molecular Biology, vol. 2650, https://doi.org/10.1007/978-1-0716-3076-1_17,

expansion of undifferentiated cells is observed during cancer progression, not all carcinomas retain the same proportion of differentiated cell lineages. This is, for instance, relevant for distinguishing between mucinous adenocarcinomas, full of goblet-like secretory cells, and conventional adenocarcinomas mostly built by enterocytes. Such distinctive cellular qualities can be related with patient's outcome, and therefore a precise histological evaluation of differentiated cells is clinically relevant [1–3] (Figs. 1 and 2).

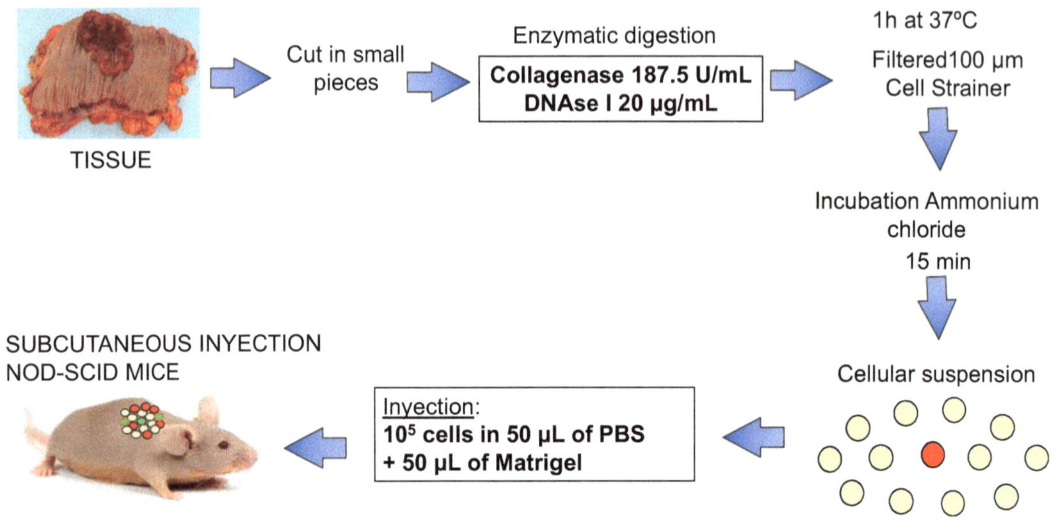

Fig. 1 Scheme showing patient-derived xenografts protocol

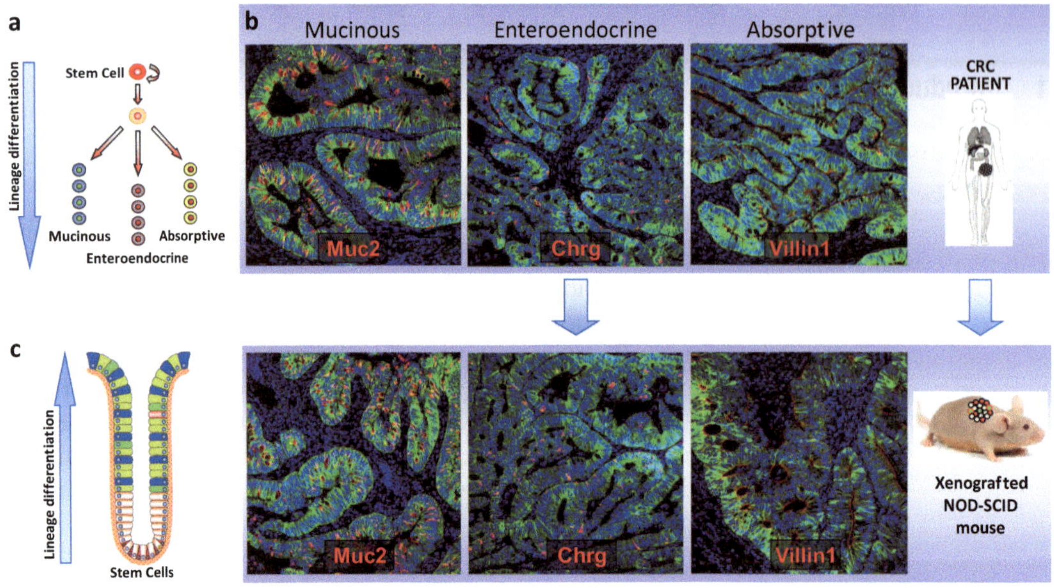

Fig. 2 Patient-derived xenograft (PDX) models faithfully recapitulate patients' tumors including the proportion of the different cell lineages present in the original tumor

2 Materials

2.1 Patient-Derived Xenografts

1. Four-week-old NOD.CB17-Prkdcscid/NcrCrl mice (Charles River Laboratories).
2. Phosphate-buffered saline (PBS), sterile.
3. Trypan blue (Life Technologies).
4. 29G U-100 insulin syringes (BD, catalog number: 320926).
5. 15 mL tubes.
6. 50 mL tubes.
7. 10 cm Petri dish.
8. Filter strainer 100 microns (BD Biosciences # 352360).
9. Ammonium chloride solution RBC lysis buffer (Labclinics # 07800 (00-4333-57)).
10. Matrigel, basement membrane matrix, 5 mL (BD Biosciences # 356234).
11. DNase I. Stock concentration 2 mg/mL (100×) (Sigma D4263-5VL).
12. Collagenase. Stock concentration 150 mg/mL (100×) (Sigma C0130-500MG).
13. Desmed medium:

    DMEM /F12 Liq (Life Technologies 21331020).

    250 μg/mL pen/strep (Life Technologies 5140122).

    10 μg/mL Fungizone (Life Technologies 15290026).

    10 μg/mL kanamycin (Life Technologies 15160047).

    50 μg/mL gentamycin (Life Technologies 15750037).

    5 μg/mL Nistatin (Sigma N4014-50MG).
14. CoCSCM 6Ab medium:

    DMEM /F12 Liq (Life Technologies 21331020).

    6% D-glucose (Sigma G6152).

    1 mg/mL apotransferrin (Sigma T1147).

    250 μg/mL insulin (Sigma I9278).

    96 μg/mL putrescin (Sigma P5780-5G).

    52 μg/mL sodium selenite (SS) (Sigma S5261-25G).

    63 ng/mL progesterone (Sigma P0130-25G).

    100 μg/ml pen/strep (Life Technologies 5140122).

    10 μg/mL Fungizone (Life Technologies 15290026).

    10 μg/mL kanamycin (Life Technologies 15160047).

    50 μg/mL gentamycin (Life Technologies 15750037).

5 μg/mL Nistatin (Sigma N4014-50MG).
1× supplement B27 (Life Technologies 17504044).
4 μg/mL heparin sodium salt (HSS) (Sigma H4784-250MG).
1% nonessential amino acids (Life Technologies 11140035).
1% sodium pyruvate (Life Technologies 11360039).
2 mM L-glutamine (Life Technologies 25030024).
20 ng/mL EGF (PEPROTECH 100-15). Add fresh.
10 ng/mL FGF2 (PEPROTECH 100-18B). Add fresh.

15. RBC lysis buffer (eBioscience #00-4333-57).

2.2 Equipment and Tools for Patient-Derived Xenografts Methodology

1. Forceps (sterilize before use) (Fine Science Tools).
2. Surgical scissors (sterilize before use) (Fine Science Tools).
3. Blade #24, sterile (BRAUN).
4. Cell culture hood.
5. Incubator.
6. Microscope.
7. Centrifuges.

2.3 Immunostaining

1. Fixation buffer: 4% paraformaldehyde (PFA) diluted in PBS.
2. Permeabilization buffer: 0.1% Triton-X-100 diluted in PBS.
3. Tween-20.
4. Washing buffer: 0.1% Tween-20 diluted in PBS (PBS-T).
5. Blocking buffer: 3% bovine serum albumin (BSA) diluted in PBS-T.
6. Primary antibodies of interest to detect differentiated cells:

   Anti-MUC2 (Clone CCP58; BD Bioscience; Cat n° 555926) 1/100.

   Anti-VILL (Clone 12C; Lifespan Biosciences; Cat n° 381526-USB) 1/100.

   Anti-CHGA (Clone LK2H10; Bio-Rad; Cat n° MCA5696) 1/100.
7. Dye-conjugated secondary antibodies (e.g., Alexa-conjugated antibodies).

   Anti-mouse-Alexa 488, diluted 1/200 in the case of Invitrogen Ab.
8. Nuclear staining (Hoechst 33342 1/10,000).
9. Mounting media (VECTASHIELD).

2.4 Equipment and Tools for the Immunostaining Methodology

1. Water bath.
2. Oven.
3. Microwave.
4. Slides.
5. Cover slips.
6. Pan pen.
7. Confocal microscope.

3 Methods

3.1 Patient-Derived Xenografts

3.1.1 Fresh Patient Tissue

1. Wash 3 times with PBS the fresh sample of the patient tumor.
2. Maintain in Desmed medium at 4ª C o/n.
3. Tissues should be digested as soon as possible the next day.

3.1.2 Mice Tumor Tissue

1. A mouse with a previously established subcutaneous colorectal tumor is euthanized.
2. The subcutaneous tumor is removed using sterile technique. Extract the tumor from the body (free from the skin), carefully removing as much excess tissue surrounding the tumor.
3. Store the harvested tumors in PBS on ice until the digestion procedure (*see* **Note 1**).

3.1.3 Cell Preparation

This part should be performed in a biological cabinet at room temperature:

1. In a 10 cm Petri dish with 1 mL of complete CoCSCM 6Ab medium (to make mincing easier), dissect the tumor with a scalpel blade until you get a homogeneous sample, and place into a 15 mL conical tube.
2. Add up to 5 mL of complete CoCSCM 6Ab medium (*see* **Note 2**).
3. Add 50 μL of DNase I and 50 μL of collagenase. Vortex the sample.
4. Incubate the tube 1 h at 37 °C in the cell culture incubator in an inclined position. Pipette the sample every 15 min with a 5 mL pipette.
5. Dilute the 5 mL digested mixture with complete CoCSCM 6Ab medium at a 1:1 ratio.
6. Filter the mixture with a 100 μM cell strainer into another sterile 50 mL conical tube.

7. Centrifuge the filtered cells at 1600 rpm 10 min at room temperature.
8. Remove supernatant.
9. Resuspend in 3 mL of RCB lysis buffer.
10. Incubate 10 min at RT.
11. Add 3 mL of complete CoCSCM 6Ab medium and centrifuge at 1600 rpm 10 min at RT.
12. Cell counting: Remove supernatant and resuspend the pellet with 5–10 mL of complete CoCSCM 6Ab medium (depending on how big the pellet is) (*see* **Note 3**).
13. Resuspend cells in fresh PBS to a concentration 1×10^5 cells/50 μL.
14. Slowly put 50 μL of Matrigel in an insulin syringe.
15. Slowly put 50 μL of cells (1×10^5 cells/50 μL) in the Matrigel of an insulin syringe (*see* **Note 4**).
16. The insulin syringe with the tumor cell suspension should be kept on ice until injection.

3.1.4 Subcutaneous Injection Procedure

1. The insulin syringe with the tumor cell suspension should be kept on ice until injection.
2. Remove hair from the mice flanks region.
3. Pinch the skin of the mouse and pull the skin away from the body of the mouse.
4. Inject slowly and evenly, creating a single bubble of cells beneath the skin and avoiding too much spread of the cells.

3.2 Immunostaining on Formalin-Fixed Paraffin-Embedded Sections of Intestinal Tissue

1. Dry the slides in an oven, overnight at 60 °C.
2. Dewax: Xylene-EtOH 100%-EtOH 96%-EtOH 70%-H2O MQ (5 min each).
3. Antigen retrieval: Immerse slides in 10 mmol/L sodium citrate buffer (pH = 6) and introduce them in a microwave.
4. Boil for 7 min at a reduced power.
5. Cool at room temperature (RT) for 30 min.
6. Wash twice in PBS 1×, 10 min/wash.
7. Permeabilization: incubate the slides in PBS-T 15 min at RT.
8. Wash twice in PBS 1×, 10 min/wash.
9. Blocking: Incubate the slides with blocking solution, 1 h at RT.
10. Incubate the slides with the primary antibodies diluted in blocking solution, in a moist chamber, o/n at 4 °C.
11. Wash twice in PBS 1×, 10 min/wash.

12. Incubate the slides with the secondary antibody diluted in blocking solution, in a moist chamber, 1 h at RT (protect from the light).
13. Wash twice in PBS 1×, 10 min/wash.
14. Mounting slides in mounting medium VECTASHIELD (Vector Laboratories, Ref. H-1000).
15. Add coverslip on top.

4 Notes

1. Tumors should be digested within 24 h after surgery. After this time, the success of implantation decreases considerably.
2. It is important that there is no more than 3 mL of disaggregated tissue in the same tube. Then, fill to 5 mL with CoCSCM 6Ab medium.
3. Make sure you have a homogeneous cell suspension.
4. Pipette the sample frequently to homogenize the cell suspension.

Acknowledgments

This work was supported by Instituto de Salud Carlos III (ISCIII), CIBERONC, Cellex and Fero Foundations, and Asociación Española Contral el Cáncer (AECC).

References

1. Negri FV, Wotherspoon A, Cunningham D, Norman AR, Chong G, Ross PJ (2005) Mucinous histology predicts for reduced fluorouracil responsiveness and survival in advanced colorectal cancer. Ann Oncol 16(8):1305–1310
2. Catalano V, Loupakis F, Graziano F et al (2009) Mucinous histology predicts for poor response rate and overall survival of patients with colorectal cancer and treated with first-line oxaliplatin- and/or irinotecan-based chemotherapy. Br J Cancer 100(6):881–887
3. Verhulst J, Ferdinande L, Demetter P, Ceelen W (2012) Mucinous subtype as prognostic factor in colorectal cancer: a systematic review and meta-analysis. J Clin Pathol 65(5):381–388

Chapter 18

Intestinal Cell Differentiation and Phenotype in 2D and 3D Cell Culture Models

Magda Martínez-Espuga, Alvaro Mata, and Paloma Ordóñez-Morán

Abstract

Three-dimensional (3D) culture models are more physiologically relevant than two-dimensional (2D) cell culture models. 2D approaches cannot reproduce the complexity of the tumor microenvironment and are less able to translate biological insights; and drug response studies have many limitations to be extrapolated to the clinics. Here, we use the Caco-2 colon cancer cell line, which is an immortalized human epithelial cell line that under specific conditions can polarize and differentiate into a villus-like phenotype. We describe cell differentiation and cell growth in both 2D and 3D culture conditions, concluding that cell morphology, polarity, proliferation and differentiation are highly dependent on the type of cell culture system.

Key words Caco-2 cells, 2D, 3D, Differentiation, Gene expression, Markers

1 Introduction

In two-dimensional (2D) culture models, the cells are grown as monolayers attached to a tissue-culture-treated plastic. Although this model has been widely used and has increased our understanding of the mechanism of drugs action, there are many limitations associated with it. 2D cell culture models do not replicate the intra-tumor heterogeneity, they lack a complex extracellular microenvironment, and in consequence, this model cannot reproduce the type of cancer which the cells are derived from [1]. The implementation of three-dimensional (3D) cell culture models has addressed the limitations of 2D models. 3D cell culture systems are improved models because they are able to reproduce mechanical and biochemical features of the tumor, such as tissue stiffness, specific gradients, and cell-cell/cell-extracellular matrix (ECM) interactions [2]. To obtain these 3D cell culture models is essential to have a scaffold to promote cell adhesion, proliferation/differentiation, and migration to recapitulate cell-ECM interactions. Matrigel, a basement membrane-like matrix, derived from a mouse

Paloma Ordóñez-Morán (ed.), *Intestinal Differentiated Cells: Methods and Protocols*, Methods in Molecular Biology, vol. 2650,
https://doi.org/10.1007/978-1-0716-3076-1_18,

sarcoma tumor, has been extensively used as the gold standard scaffold material to provide a 3D environment to a wide range of healthy and cancer cell types.

Caco-2 cells, derived from a patient's colon adenocarcinoma, are an immortalized human epithelial cell line used as a model of the intestinal epithelial barrier. This cell line is one of the most relevant in vitro models for the study of differentiation and regulation of intestinal homeostasis. In this process of differentiation, epithelial cells acquire the functional properties of mature enterocytes [3]. This differentiation leads to the formation of a monolayer of polarized cells that form a physical and biochemical barrier which express features of these differentiated cells [4]. In this chapter, we compare 2D vs 3D models and how the different approaches affect colon cancer cell phenotype, proliferation, and differentiation (into both lineages: absorptive and secretory). For this reason, Caco-2 cells were seeded in tissue-culture-treated plastic plates (2D), and in parallel, the same cells were embedded in Matrigel matrix (3D) and afterward seeded in tissue-culture-treated plastic plates. For oncoming CRISPR experiments, we generated by lentiviral infection a modified Caco-2 Cas9-RFP cell line.

Overall, Caco-2 cells have the ability to proliferate and differentiate in vitro. However, to obtain a model that closely resembles and behaves as the intestinal crypts is essential to maintain the ratio of stem cells. These cells are continuously self-renewing to give rise to progenitors' cells that form the small intestine. Therefore, long-term experiments should be done to better understand the effect of culture conditions on cells. Matrigel presents many limitations that need to be considered when using this material to support the expansion of cells in a 3D environment. Other synthetic materials are being investigated for their ability to mimic the dynamic native microenvironment that play a key role in differentiation, proliferation, and stemness. These synthetic materials may not only replace animal-derived gels but also may be tuned with specific molecular moieties and defined mechanical properties to better recreate the tumor microenvironment.

2 Materials

2.1 Caco-2 Cas-9 RFP Culture (2D)

1. Dulbecco's Modified Eagle Medium (DMEM) supplemented with 1% penicillin/streptomycin and 10% of fetal bovine serum (FBS).
2. Trypsin.
3. PBS 1×.
4. 24-well plate (tissue-culture-treated plastic).
5. T-flasks (75 cm^2) (tissue-culture-treated plastic).

2.2 Caco-2 Cas-9 RFP Culture (3D)

1. Matrigel™, basement membrane matrix.
2. Dulbecco's Modified Eagle Medium (DMEM) supplemented with 1% penicillin/streptomycin and 10% of fetal bovine serum (FBS).
3. Trypsin.
4. PBS 1×.
5. 24-well plate (tissue-culture-treated plastic).

2.3 Total RNA Purification

1. RNeasy Mini Spin Columns.
2. Collection tubes (1.5 mL).
3. Collection tubes (2 mL).
4. RLT buffer.
5. RW1 buffer.
6. RPE buffer.
7. RNase-free water.
8. β-Mercaptoethanol.
9. Ethanol 70%.

2.4 cDNA Synthesis

1. RNA template (total RNA).
2. 100 μM oligo dT master mix (18 and 24).
3. dNTPs (10 mM each).
4. 0.1 M DTT.
5. 250 mM Tris–HCl pH 8.3; 375 mM KCl, 15 mM $MgCl_2$ 5× RT buffer.
6. 200 U/μL reverse transcriptase (SuperScript II).
7. 40 U/μL ribonuclease inhibitor (RNasin).
8. Nuclease-free water.
9. Thermal cycler.

2.5 cDNA Amplification by qPCR

1. cDNA pool: samples and negative controls (samples without SuperScript enzyme).
2. PCR primers for the genes of interest and housekeeping genes (Table 1).
3. SYBR green which includes dNTPs, Taq polymerase, and $MgCl_2$.
4. MicroAmp Fast Optical 96-well plate.
5. MicroAmp Optimal Adhesive Film (strip lids).
6. Quantitative real-time thermal cycler.

Table 1
Primers used for qPCR analysis of intestinal gene expression

Human gene	Forward primer	Reverse primer
GAPDH	GAGTCAACGGATTTGGTCGT	TTGATTTTGGAGGGATCTCG
MKI67	ATGCAGAATCAGAAAGGGAAAGG	TTGTCTTTCTTGATCTCAGGCAC
PCNA	AGGCACTCAAGGACCTCATCA	GAGTCCATGCTCTGCAGGTTT
ALDOB	TCCAGAATACCCACCCAAGAAA	TTCTTTGGATGAGGAGCCGATA
ANPEP	CATTATGACACACCCTACCCACT	CTCATGAGCAATCACAGTGACC
MUC2	ACCCGCACTATGTCACCTTC	GGACAGGACACCTTGTCGTT

3 Methods

3.1 Caco-2 Cas9-RFP Cell Culture

1. Culture Caco-2 Cas9-RFP cells in T-flasks (75 cm^2) in a CO_2 incubator at 37 °C and 5% CO_2. The culture medium is Dulbecco's Modified Eagle Medium (DMEM) supplemented with 1% penicillin/streptomycin and 10% of fetal bovine serum (*see* **Note 1**).
2. Split the Caco-2 Cas9-RFP at a ratio of 1:5 and at a confluency of 40–50% for cell maintenance.
3. Change the culture medium every 48–72 h.
4. For 2D culture conditions: seed the cells in 24-well plates at a confluency of 20%.
5. For 3D culture conditions: mix the same number with 50 μL of Matrigel previously thawed in ice, and seed the cells in 24-well plates at a confluency of 20%.
6. To check viability and growth of the cells, you can visualize them by using the microscope (brightfield and fluorescence for RFP) (Figs. 1 and 2) (*see* **Note 2**).

3.2 RNA Isolation and cDNA Synthesis

1. Extract total RNA of Caco-2 Cas9-RFP using the RNeasy MicroKit following manufacturer protocol to obtain high quality of RNA (use buffers RLT, RW1, RPE and elute total RNA with RNase-free water).
2. Check the RNA quality with the NanoDrop.
3. To obtain cDNA, first do the primer annealing step. Mix 1 μg of the RNA sample, 1 μL of oligo dT master mix, and 1 μL of dNTPs master mix.
4. Heat the tubes to 65 °C for 5 min to melt the secondary structure within the template.

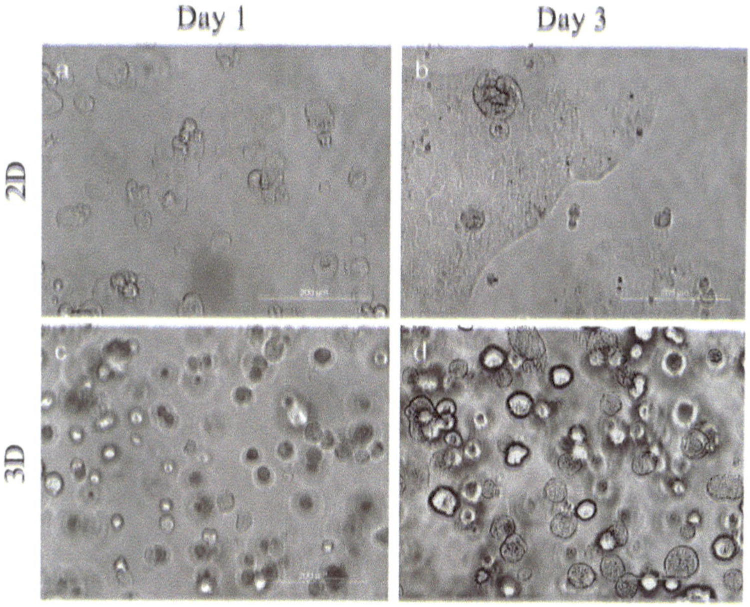

Fig. 1 Caco-2 Cas9-RFP cells seeded in 2D and embedded in Matrigel (3D) for 1 day (**a**, **b**) and 3 days (**b**, **d**)

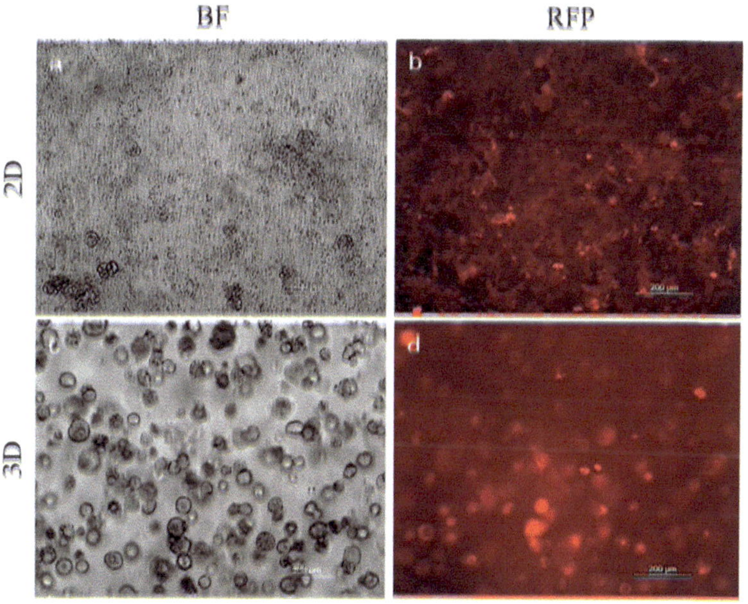

Fig. 2 Caco-2 Cas9-RFP seeded in 2D and embedded in Matrigel (3D) for 6 days, brightfield (BF: **a**, **b**) and fluorescence (RFP: **c**, **d**)

5. Cool the tubes immediately on ice to prevent secondary structure from reforming.
6. Centrifuge briefly to collect the solution at the bottom of the tube.

7. Reverse transcription step: mix 12 μL of the annealed primer/RNA/sample, 2 μL of DTT 0.1 M, 4 μL of 5× reaction buffer, 1 μL ribonuclease inhibitor (RNasin), and 1 μL of SuperScript II Reverse Transcriptase.
8. Incubate at 40 °C for 50 min.
9. Incubate at 70 °C for 15 min to inactivate the reverse transcriptase enzyme (SuperScript).
10. Incubate at 4 °C to cool down.
11. Store the obtained cDNA at −20 °C or continue the protocol.

3.3 Quantitative Real-Time PCR (qPCR)

1. Prepare a 1/20 dilution of cDNA mixture in RNase-free water.
2. Choose the set of primers that will amplify your gene of interest (Table 1). We selected the ones below to determine differentiation into both lineages, proliferation, and stem/Wnt pathway activation:

Process/pathway	Markers	Analysis (Fig. 3)
Absorptive lineage	*ANPEP, ALDOB*	*see* **Note 3**
Secretory lineage	*MUC2*	*see* **Note 4**
Wnt and stemness	*LGR5*	*see* **Note 5**
Proliferation	*MKi67, PCNA*	*see* **Note 6**

3. For gene expression analysis, in a 96-well plate, prepare a PCR master mix containing gene primers mix (forward and reverse), SYBR green, and RNase-free water. Find an example below for the mix done for the housekeeping gene GAPDH:

Master mix 1 (*GAPDH*)	Sample (1×)	Total
GAPDH primers mix	1.5 μL	1.5 μL × number of wells
SYBR green	7.5 μL	7.5 μL × number of wells
H_2O	3 μL	3 μL × number of wells
Total volume	12 μL	

4. Dispense 12 μL of the master mix into each well of the 96-well plates.
5. Dispense 3 μL of RT and RTneg cDNA into each well of the 96-well plates. Take care to pipette accurately into the wells as small variations will affect the assay.

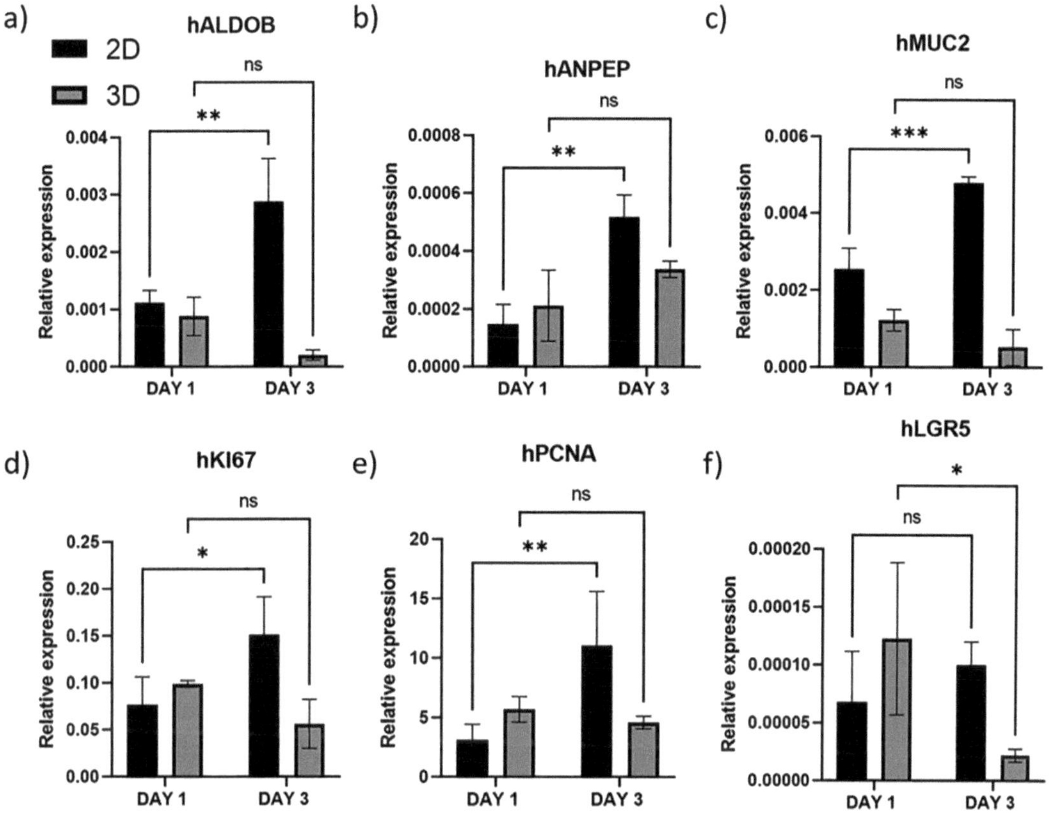

Fig. 3 qRT-PCR measurement of indicated genes studied in Caco-2 Cas9-RFP cells seeded in plastic culture plates (2D) and, in parallel, cells embedded in Matrigel (3D) for 1 and 3 days. (**a**, **b**) Enterocyte markers, ALDOB and ANPEP; (**c**) goblet cell marker, MUC2; (**d**, **e**) proliferation markers, MKi67 and PCNA; and (**f**) stemness and Wnt pathway marker, LGR5

6. Place the optical strip lids on the wells.
7. The thermal cycler should be programmed as described below:

98 °C – 30 s (initial denaturation)	
98 °C – 10 s (denaturation) 57 °C – 30 s (primer annealing) 72 °C – 60 s (extension)	40 cycles (it can be modified)
72 °C – 10 min (to fill in the protruding ends of newly synthesized PCR products).	

8. Ct values (threshold cycle) were obtained after qPCR.
9. qPCR data based on the $2^{-\Delta\Delta Ct}$ method were analyzed (Fig. 3).

4 Notes

1. We have generated Caco-2 Cas9-RFP by lentivirus infection that expresses a red fluorescent tag (RFP) that allows cell tracing to visualize cell proliferation among other parameters by microscopy (Fig. 2).
2. When comparing Caco-2 cells seeded in the tissue-culture plate (2D) and the cells embedded in Matrigel (3D), we observed that cells in 2D at day 3 start to polarize acquiring a characteristic apical brush border. The cells embedded in Matrigel formed clusters; however there were no significant changes in the morphology of these cells after 3 days (Fig. 1). Other studies have shown that shifting Caco-2 cells from a 2D to 3D environment can induce changes in gene expression, differentiation, and metabolism [5]. After 6 days in culture, Caco-2 cells in 2D covered the entire culture flask and they even grew on top of each other. Opposed to this behavior, Caco-2 cells in 3D also proliferated but not at the same extent as cells in 2D (Fig. 2). We state that Caco-2 cells on plastic culture flask proliferate quicker than in 3D. The same cells embedded in Matrigel organize a multicellular complex structure.
3. To further demonstrate that Caco-2 cells differentiate into intestinal enterocytes depending on the culture conditions, we analyzed the expression of two gene-specific markers (Fig. 3). *ALDOB* and *ANPEP* were selected as they encode for genes enriched in enterocytes [6, 7]. The relative expression of both genes was statistically significant comparing day 1 vs day 3 in 2D. However, in 3D cell culture, we did not detect significant changes. Enterocyte expression increased with time in 2D cell culture in contrast to 3D cell culture where we found lower expression of enterocytes (Fig. 3).
4. MUC2 was selected as a marker of goblet cells of the intestine. MUC2 is the major intestinal mucin expressed in goblet cells and is resistant to endogenous proteases [8]. We observed that MUC2 show similar behavior as the enterocytes' markers previously tested. We found that there is a significant change in the relative expression between day 1 and day 3 in 2D cell culture that is not observed in 3D conditions (Fig. 3).
5. LGR5 was selected as is a target of Wnt signaling and a marker of stem cells in healthy and colorectal tissues [9]. As in 3D conditions we found less differentiation, we were expecting that this system would sustain the expression of stem cells. However, we observed the opposite. After some days of adaptation, it is likely that the relative expression of LGR5 would increase in 3D cell culture models (Fig. 3).

6. MKi67 and PCNA were chosen as they are proliferation markers. Ki67 is a nucleolar protein specifically expressed in G1-S-G2-M phase of the cell cycle, so this marker is an indicator of cell cycle progression [10]. Besides, PCNA is a proliferating cell nuclear antigen directly involved in DNA synthesis [11]. We observed that the proliferation is increased in 2D in contrast to 3D. Further experiments would need to be done to confirm cell proliferation at longer periods of time (Fig. 3).

References

1. Rodrigues J, Heinrich MA, Teixeira LM, Prakash J (2021) 3D in vitro model (r)evolution: unveiling tumor–stroma interactions. Trends Cancer 7: 249–264. Preprint at https://doi.org/10.1016/j.trecan.2020.10.009
2. Katt ME, Placone AL, Wong AD, Xu ZS, Searson PC (2016) In vitro tumor models: Advantages, disadvantages, variables, and selecting the right platform. Front Bioeng Biotechnol 4. Preprint at https://doi.org/10.3389/fbioe.2016.00012
3. Jumarie C, Malo C Caco-2 cells cultured in serum-free medium as a model for the study of enterocytic differentiation in vitro
4. Verhoeckx K, Cotter P, López-Expósito I, Kleiveland C et al The impact of food bioactives on health in vitro and ex vivo models
5. Samy KE et al (2019) Human intestinal spheroids cultured using sacrificial micromolding as a model system for studying drug transport. Sci Rep 9:9936
6. Gassler N et al (2006) Molecular characterisation of non-absorptive and absorptive enterocytes in human small intestine. Gut 55: 1084–1089
7. Ozawa T et al (2015) Generation of enterocyte-like cells from human induced pluripotent stem cells for drug absorption and metabolism studies in human small intestine. Sci Rep 5
8. Johansson MEV, Sjövall H, Hansson GC (2013) The gastrointestinal mucus system in health and disease. Nat Rev Gastroenterol Hepatol 10: 352–361. Preprint at https://doi.org/10.1038/nrgastro.2013.35
9. Barker N et al (2007) Identification of stem cells in small intestine and colon by marker gene Lgr5. Nature 449:1003–1007
10. Basak O et al (2014) Mapping early fate determination in L gr5 + crypt stem cells using a novel K i67- RFP allele. EMBO J 33:2057–2068
11. Krol M, Benson WH (1994) Detection of proliferating cell nuclear antigen in tissues of three small fish species

Chapter 19

Confocal Laser Scanning Imaging of Cell Junctions in Human Colon Cancer Cells

Peixun Zhou and M. Angeles Juanes

Abstract

The intestinal epithelium is formed by a single layer of cells. These cells originate from self-renewal stem cells that give rise to various lineages of cells: Paneth, transit-amplifying, and fully differentiated cells (as enteroendocrine, goblet cells, and enterocytes). Enterocytes, also known as absorptive epithelial cells, are the most abundant cell type in the gut. Enterocytes have the potential to polarize as well as form tight junctions with neighbor cells which altogether serve to ensure both the absorption of "good" substances into the body and the blockage of "bad" substances, among other functions. Culture cell models such as the Caco-2 cell line have been proved to be valuable tools to study the fascinating functions of the intestine. In this chapter we outline some experimental procedures to grow, differentiate, and stain intestinal Caco-2 cells, as well as image them using two modes of confocal laser scanning microscopy.

Key words Differentiated intestinal cells, Caco-2, Confocal, Imaging, Cell junctions, Immunofluorescence

1 Introduction

The intestinal epithelium is the first line of defense against ingested potential pathogens while interplaying with the bacteria and the immune system [1–3]. The intestinal epithelium is also critical for food absorption, digestion, and secretion of mucus [3–5]. In addition, the intestine is under constant renewal [5–9]. All these functions are critical to sustain gut homeostasis and therefore a healthy gut. Accordingly, disruption of homeostasis leads to human diseases such as inflammatory bowel disease and cancers (reviewed in [10]).

Intestinal functions are achieved by specific cell types that compose the intestinal epithelium and which organize into highly proliferative invaginations called crypts of Lieberkühn and differentiated cells in fingerlike villus structures [9, 11, 12]. Self-renewing stem cells are located at the bottom of the crypts. These

Paloma Ordóñez-Morán (ed.), *Intestinal Differentiated Cells: Methods and Protocols*, Methods in Molecular Biology, vol. 2650,
https://doi.org/10.1007/978-1-0716-3076-1_19,

stem cells will give rise to a lineage of cells: (i) Paneth cells which will remain in the crypt, (ii) partially differentiated cells (also known as transit-amplifying cells), and (iii) fully differentiated and migratory cells that when reaching the top of the villus undergo apoptosis and eventually shed off [6, 13–15]. This cell shedding is counterbalanced by the continuous proliferation of the stem cells that reside at the bottom of the crypts [6]. Fully differentiated cells are enteroendocrine (that release hormones), goblet cells (that secrete mucus), and enterocytes (or absorptive cells). Enterocytes are the most abundant cells in the intestinal epithelium. Enterocytes are polarized/differentiated cells which present distinctive domains: apical and basolateral [16, 17]. The apical domain faces to the lumen of the gut and consists of an array of actin-based membrane protrusions (or microvilli brush border) that increases the capacity of absorbing molecules, and it is rich in digestive enzymes. The basolateral domain mediates intercellular connections with neighbor cells, including adherens junctions (AJ) which provide strength and tight junctions (TJ) which serve as a barrier to solutes and as a fence to prevent diffusion of components.

Many human colon cancer cell lines derived from colon carcinomas have been used to investigate the large panel of cell biology functions of the gut. However, only a few cell lines differentiate and develop enterocyte features in in vitro culture conditions [18, 19]. The Caco-2 cell line has been extensively used as model because it exhibits spontaneous differentiation into enterocytes which varies in function of their growth as well as presents similar features than the intestine, including microvilli structure, brush border, and tight cell junctions [18, 20–27]. At early days after seeding (0–4 days, subconfluent), these cells present a homogenous undifferentiated morphology with an underdeveloped brush border. As they grow (5–20 days, confluent), they start to polarize and differentiate, developing dense microvilli and tight junctions. After 21 days, Caco-2 cells are more homogenous, fully polarized and differentiated, exhibiting a typical enterocytes-like morphology (Fig. 1). Note that despite 3D cell culture models mimic closer physiological conditions than 2D cell culture systems, the latter are cheaper and very simple to manipulate, grow, and analyze; therefore they are still heavily used to ask fundamental gut questions [28–30].

In this chapter we describe several protocols in order to successfully (i) grow and differentiate Caco-2 cells (Subheading 3.1), (ii) stain cell junctions (Subheading 3.2), and (iii) image them using confocal laser scanning microscopy in conventional (Subheading 3.3.1) or lightning mode (Subheading 3.3.2), along with some representative images. At the end of this chapter, we have included a section named "Notes" to help the research community to troubleshoot potential experimental problems and/or provide advice about alternative reagents/solutions to use.

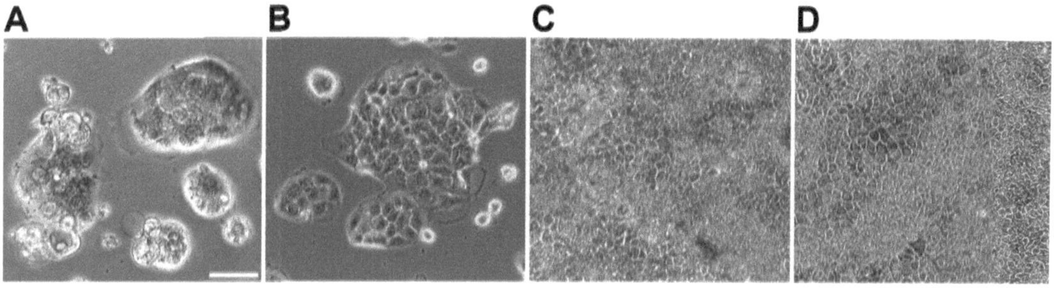

Fig. 1 Illustrates the Caco-2 cells at various stages of growth. Representative phase-contrast microscopy images of Caco-2 cells at day 2 (**a**) and day 3 (**b**) subconfluent and not differentiated; day 12 (**c**) confluent; and day 21 (**d**) confluent and fully differentiated using a light microscope. Scale bar 100 μM

2 Materials

2.1 Cell Growth and Differentiation of Caco-2 Cells

1. Caco-2 cells.
2. Complete DMEM: Dulbecco's Modified Eagle Medium (DMEM) containing 4.5 g/L high glucose, 0.5 mM sodium pyruvate, and 2.5 mM L-glutamine and supplemented with 100 U/mL penicillin and 100 U/mL streptomycin.
3. Fetal bovine serum (FBS; *see* **Notes 1** and **2**).
4. Phosphate-buffered saline (PBS): 137 mM NaCl, 2.7 mM KCl,10 mM Na2HPO4, 1.8 mM KH2PO4 (without calcium and magnesium).
5. Trypsin-ethylenediaminetetraacetic acid (EDTA): 2.5 g/L trypsin and 1 mM EDTA.

2.2 Immuno-fluorescence

1. Fixation buffer: 4% paraformaldehyde (PFA) diluted in PBS (*see* **Note 3**).
2. Permeabilization buffer: 0.1% Triton-X-100 diluted in PBS (*see* **Note 4**).
3. Tween-20.
4. Washing buffer: 0.1% Tween-20 diluted in PBS (PBS-T; *see* **Note 5**).
5. Blocking buffer: 3% bovine serum albumin (BSA) diluted in PBS-T (*see* **Note 6**).
6. Primary antibodies of interest (e.g., rabbit alpha-occludin, mouse alpha-cadherin).
7. Dye-conjugated secondary antibodies (e.g., Alexa-conjugated antibodies: goat anti-rabbit Alexa Fluor-555 to stain occludin and anti-mouse Alexa Fluor-633 to stain cadherin).
8. Nuclear staining (e.g., 4,6-diamidino-2-phenylindole dihydrochloride, DAPI) (*see* **Note 7**).

9. Mounting media (*see* **Note 8**).
10. Sealing reagent (e.g., Fixogum rubber cement; *see* **Note 9**).

2.3 Equipment and Others

1. Water bath or sand-bead incubator at 37 °C.
2. Cell culture incubator at 37 °C, 5% CO_2.
3. 25 cm^2 flasks.
4. Sterile pipettes.
5. Sterile tips.
6. Polycarbonate Cell Culture Inserts in Multi-Well Plates.
7. Six-well plates.
8. Forceps.
9. Slides and cover slips.
10. Falcon tubes.
11. Eppendorfs.
12. Centrifuge.
13. Light microscope, e.g., Leica DMi1 microscope (*see* **Note 10**).
14. Confocal microscope, e.g., Leica TCS SP8 SMD laser scanning microscope (*see* **Note 11**).

3 Methods

3.1 Growth and Differentiation of Caco-2 Cells

1. Thaw one frozen vial of Caco-2 cells in a 37 °C water bath or sand-bead incubator for 3–4 min.
2. Add 1 mL of complete DMEM medium containing 10% FBS into the cells, pipette up and down, and add into a 15 mL falcon tube.
3. Add another 1 mL of complete DMEM medium containing 10% FBS into the frozen vial to take any leftover cells, and add into the same 15 mL falcon containing cells.
4. Optional to add extra 2–3 mL of complete DMEM medium containing 10% FBS to the falcon tube with cells.
5. Spin down the falcon tube with cells in a centrifuge at 600 × *g* for 1 min at room temperature.
6. Aspirate the medium and resuspend the cell pellet with 2–3 mL of complete DMEM medium containing 10% FBS.
7. Repeat **steps 5** and **6**. Since cells are usually stocked in DMSO, this extra wash is recommended to remove any DMSO which would be toxic for the cells.

8. Transfer cells with medium into a 25 cm^2 flask, and refill with enough media to cover the bottom of the flask (usually 5–6 mL of complete DMEM containing 10% FBS).
9. Keep the cells growing in an incubator at 37 °C, 5% CO2, and replace the medium every 2–3 days until differentiated (*see* **Note 12**).
10. After 5–6 days from seeding cells, confluent cells could be split and seeded into the upper part of "Cell Culture Inserts" containing polycarbonate filters fitted in Multi-Well Plates (as explained in **step 18**). For the purpose of this experiment, filters should have 4 μM porous size to allow cells to differentiate and acquire gut-like structure without migrating through the filter (*see* **Note 13**).
11. Wash cells with 1–2 mL of PBS.
12. Add 1–2 mL of trypsin-EDTA to detach cells from the flask and then leave the cells in the incubator at 37 °C for 5 min (*see* **Note 14**).
13. Add 1–2 mL of complete DMEM medium with 10% FBS to stop the trypsin reaction.
14. Transfer cells resuspended in the DMEM medium from the flask to a falcon tube.
15. Spin down cells from the falcon tube in a centrifuge at 600 g for 1 min at room temperature.
16. Aspirate media and resuspend cell pellet with 2–3 mL of complete DMEM medium containing 10% FBS.
17. Count cells and seed 50,000 cells into the upper part of a six-well plate with inserts containing polycarbonate filters 4 μM porous size (*see* **Note 15**).
18. Add 500 μL of complete DMEM medium supplemented with 10% FBS in the upper part of the insert and 1250–1500 μL in the bottom part. Those DMEM volumes are recommended for a six-well plate but may be modified for other type of plates (*see* **Note 2**) [26].
19. Optional. The day after the cells have been seeded on the filter, the medium can be carefully exchanged to avoid cells from pilling.
20. Replace medium each 2–3 days or when it starts to be yellowish during the 21 days until cells fully polarize/differentiate.
21. Optional: Visualize and/or image cells in a light microscope to follow their development to differentiated cells until day 21 (Fig. 1). In this case imaging was performed in an inverted Leica DMil microscope equipped with a 5× 0.12 air Plan I objective lens, 5 W LED illumination coupled to a Leica

MC170 HD camera. Pictures were captured after 30 ms exposure, gain1 in color mode using a Leica applications suite X 3.6.0.20104 (LAS X) software.

3.2 Immunofluorescence

The immunofluorescence protocol detailed here is based on previous methodologies [26, 27, 31–33]. Volumes below are appropriate for a six-well plate and may be adjusted for other plates according to their size. It is important to not touch the filters from the inserts with the tip of the pipette in any step to avoid disrupting the cell monolayer:

1. Fix cells by adding 3.7% of PFA diluted in PBS onto the cells/filters during 15–20 minutes at room temperature (*see* **Notes 3** and **16**).
2. Remove the fixative reagent and wash the filter with 1 mL of PBS.
3. Remove the filter from the insert by carefully punching it with forceps.
4. Place the filter in a clean plate (e.g., regular six-well plate without inserts) containing 1 mL of PBS, and ensure that the filter is covered by the buffer, no floating on it, to avoid it dries at any time.
5. Permeabilize the cells for 10–15 min by adding 1 mL of PBS containing 0.5% Triton X-100 onto the cells/filter.
6. Remove the permeabilization buffer and wash the filter with 1 mL of PBS-T (*see* **Note 5**).
7. Add 1 mL of 3% BSA dissolved in PBS-T (blocking buffer) per well, and incubate for 1 h at room temperature under gentle agitation, e.g., in a rocket platform (*see* **Note 6**).
8. Remove the blocking buffer and wash the filter with 1 mL of PBS-T.
9. Incubate filters (from 12 h to overnight) at 4 °C with primary antibodies (1:500 rabbit anti-occludin and 1:500 mouse anti-cadherin) diluted with PBS-T.
10. Remove the antibodies and wash cells with PBS-T during 15 min.
11. Repeat washes (three times in total).
12. From this point onward, protect the filters from the light by wrapping the plate with aluminum foil.
13. Incubate filters for 1 h at room temperature with secondary antibodies (1:1000 goat anti-rabbit Alexa Fluor-555 and 1: 1000 donkey anti-mouse Alexa Fluor-633).
14. Wash three times with PBS-T, 15 min each wash.
15. Wash once with PBS for 5 min.

16. Dilute 1 mg/mL DAPI stock solution to 0.5 μg/mL in PBS (i.e., 1:2000 dilution) (*see* **Note 7**).
17. Add 1 mL fresh diluted DAPI on the filters to stain the cell nucleus and keep at room temperature under agitation during 5 min.
18. Wash once with PBS for 5 min.
19. Add some drops of mounting medium to cover the surface of the filter (approximately 30 μL) on a clean slide (*see* **Note 8**).
20. Place filters facing down on the mounting medium. Press the filter with the tweezers carefully to allow a homogenous distribution of the mounting medium around the sample, without squeezing the sample.
21. Add a cover slip on top of the filter.
22. Seal filters with gum to avoid from drying (*see* **Note 9**).
23. Allow mounting medium and rubber cement to solidify. This could take from 4 h to overnight at 4 °C.
24. At this point, slides can be immediately imaged or kept protected from the light at 4 °C for some weeks prior to imaging.

3.3 Imaging Using Confocal Laser Scanning Microscopy

Confocal laser scanning microscopy (CLSM) blocks out-of-focus and scattered light from images with laser point scanning and a pinhole aperture, ensuring that detected light only comes from the excited location and a single focal plane [34–36]. CLSM focuses laser beam to illuminate a small spot on the specimen. This is different than conventional imaging processes such as widefield fluorescence microscopy where the entire specimen is illuminated [34–36]. In this study, a Leica TCS SP8 SMD laser scanning microscope coupled to a LAS X software (v3.5.5.19976) was used for imaging (Fig. 2). It is worth noting that considerations for successful multicolor imaging include spectral cross talk between channels and pixel saturation. Sequential scan can be set up manually or with the *Dye assistant* tool in LAS X software to minimize spectral cross talk (*see* **Note 17**). In addition, laser power or detector gain can be decreased until saturated blue-color pixels disappear in the LAS X Over-/Underexposure toggle mode (*see* **Note 18**). Here we describe the settings that best suit to specifically image stained Caco-2 cells—processed as described in Subheading 3.2—in both conventional confocal and lightning mode (Fig. 3).

3.3.1 Conventional Confocal Mode (Fig. 2)

1. From the top left part of the LAS X software, select "TCS SP8."
2. From the "Acquisition" settings on the left side of the LAS X software, select the below parameters as follows:

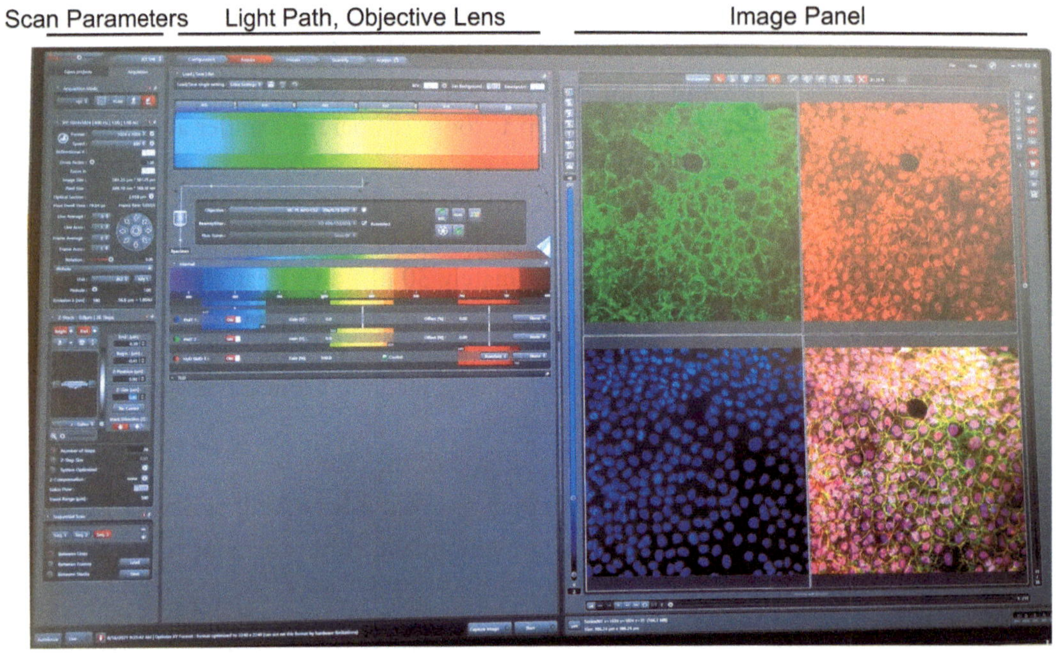

Fig. 2 Shows the interface displayed in the Leica Application Suite X (LAS X) software for acquiring images (as described in this protocol) in the TCS SP8 microscope using the conventional confocal mode

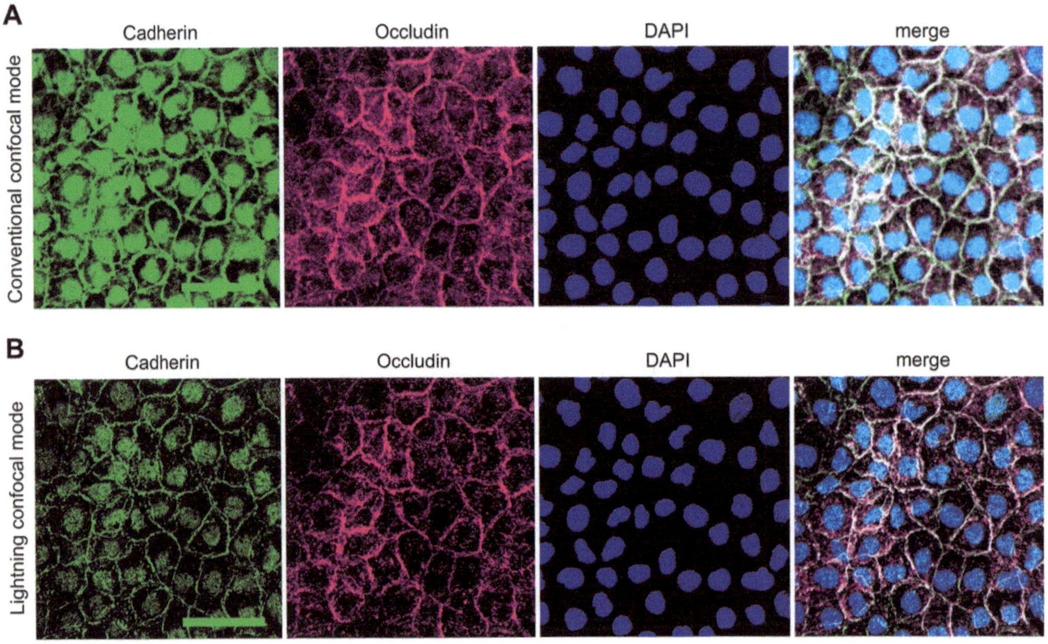

Fig. 3 Shows fully differentiated Caco-2 cells stained and then imaged with a confocal laser point scanning microscope. Representative images of Caco-2 cells fixed at day 21 showing (from the left to the right) various stainings: cadherin (adherens junction marker), occludin (tight junction marker), and DAPI (nuclear marker), as well as the merge of the three signals, captured in either conventional confocal mode (a) or lightning confocal mode to minimize background. Scale bar 50 μM

(a) Acquisition mode: XYZ.

(b) XY (*see* **Note 19**):

Format: 1024 × 1024 pixels.

Speed: 600 Hz.

Line average: 3.

Frame average: 2.

Pinhole: 1.00 AU Airy.

(c) Z-stack setting (*see* **Note 20**):

Z-Size: 20 μM.

Number of steps: 25.

Z-Step size: 0.8 μM.

(d) Sequential scan settings and detectors (*see* **Note 17**):

Sequence 1 Alexa 633: (HyD 643–776 nM).

Sequence 2 Alexa 568 (PMT 569–632 nM).

Sequence 3 DAPI (PMT 415–480 nM).

Sequential scan mode: between frames.

3. From the middle part of the LAS X software, select:

(a) Objective lens (see **Note 21**):

20× dry lens (HC PL APO CS2 20×/0.75).

(b) Light path (*see* **Note 18, 19**):

Select lasers to use (405, 448, 488, 552, and/or 638 nM) and % of the laser power according to your experiment.

Select the type of detector to use for each laser (PMT or HyD), e.g.:

PMT 1 – 405.

PMT 2–552.

HyD 1 – 638.

(c) Next to each detector (PMT or HyD), the excitation spectrum will appear. Click on it to choose the specific dye to use, and either drag the bars right below the spectrum to set the most convenient gating for your imaging or double-click on the gate to manually define it.

4. Click the "Live" bottom to visualize and focus the sample. Click again the same bottom to "Stop" visualizing (and bleaching) the sample.

5. Click "Capture Image" to acquire images of one selected channel or click "Start" to acquire images from all desired channels in XYZ.

6. Save the Project in an appropriate folder and extension by clicking onto the name of the Project (*see* **Note 22**).

3.3.2 Lightning Mode (Fig. 2)

Confocal imaging with Leica lightning mode is based on adaptive image reconstruction [34, 35, 37, 38]. Lightning enables near real-time and fully automated deconvolution and, unlike traditional technologies, optimizes images for each probed location separately. Lightning enables to resolve structures down to ~120 nM in XY and ~ 300 nM in Z resolution, in contrast to conventional confocal microscope whose best resolution is about 200 nM lateral and 500–600 nM axial direction [38]. Proceed to lightning mode as follows:

1. From the top left part of the LAS X software, select "Lightning" mode instead of TCS SP8.
2. From the "Acquisition" settings on the left side of the LAS X software, select the same settings for acquisition, sequential scan, and Z-stack that for the conventional confocal mode, as described in the 3.3.1. Subheading, **step 2**, and notes alongside (*see* also Fig. 2).
3. Acquire images with the default adaptive deconvolution mode [38].
4. Acquire images, save them, and export them in an appropriate folder, as described in the 3.3.1. Subheading, **steps 5** and **6**, and notes alongside.

4 Notes

1. A FBS stock bottle of 500 mL could take about 4–5 h to thaw at 37 °C. It is possible to keep the bottle overnight at 4 °C and proceed for using it day after. Once thawed, add the required amount for cell growth directly to the DMEM culture bottle. The rest of FBS could be aliquoted in 50 mL falcons and stored at −20 °C for some months.
2. Generally, 10% FBS is used for cells to grow, but that depends on cell line and experiment. In this protocol, 10% FBS has been added to the DMEM containing sodium pyruvate, glutamate, and antibiotics to grow, maintain, and differentiate Caco-2 cells using Transwell membranes/filters. However, asymmetric serum medium can be used to induce differentiation. In this case, DMEM containing sodium pyruvate, glutamate, and antibiotics but without FBS is added to the upper side of the filters and DMEM supplemented with 10% of FBS (in addition to the above supplements) to the bottom side of the filters, which closer resembles physiological conditions to differentiate and reduces the ethical questions regarding the use of FBS [26].
3. It is possible to use as fixation buffer 2% PFA diluted in PBS for 20 minutes. Alternatively, cells could be fixed with cold ethanol

for 5 min. However, ethanol can remove lipids, dehydrate cells, and precipitate some proteins altering structures such as actin, among others.

4. Triton from 0.1% to 0.5% could be added to the permeabilization buffer without perturbing the structure of the cell monolayer.
5. High background staining can be reduced by adding up to 0.5% Tween-20 in the washing buffer and/or allowing longer time to the washes and/or increasing the number of washes.
6. Blocking buffer could contain up to 5% BSA to reduce background in the images.
7. DAPI or Hoechst which intercalate into DNA could be used to visualize the nucleus. Some mounting media already contain DAPI. Of note, those DNA dyes enter into the nucleus even without permeabilization. Then, it is important to avoid direct skin contact with the dyes when weighting powder to make stocks and/or adding DNA dye solution to the samples.
8. Various mounting media could be used such as Aqua-Mount (which dries in a couple of hours at room temperature), Mowiol, or ProLong Gold (which are recommended to dry overnight at 4 °C). It is important to consider the objective lens requirements (water, air, oil immersion) to select the most convenient mounting medium to use in order to get the best imaging since some mounting media increase the refractive index. Some manufacturers offer mounting media with additives such as DABCO, an anti-fading agent which protects the sample from photobleaching, as well as with DNA dyes (therefore, the samples do not need to be stained with another DAPI source).
9. Sealing reagents such as nail polish could be used to avoid dehydration of samples for a few days at 4 °C and/or room temperature protected from the light. For longer storage or permanent preparations, harder sealing reagents such as Fixogum rubber cement could be used. Before sealing, it is wise to check under the microscope if the samples exhibit the expected fluorescence signal.
10. Any light microscope could be used to monitor and capture cells in phase contrast.
11. Note that Leica TCS SP8 microscopes could be purchased with the basic features for conventional confocal imaging or with accessories to image beyond the "conventional" confocal resolution. To develop this protocol, a SP8 microscope coupled to a lightning detection module was used. Such a microscope was integrated with five laser lines and detectors either

photomultiplier tubes (PMT) with standard sensitivity or hybrid detection (HyD) with high sensitivity.

12. An alternative procedure to generate Caco-2 cells at different stages of differentiation from pre-confluent subcultures could be found in [39].

13. It exists filters with different size of porous. Keep in mind that the use of large porous could result in cells migrating to the bottom part of the filters.

14. Incubation with trypsin-EDTA could be done at room temperature, but that may need longer incubation time for cells to detach.

15. Inserts which fit in 6-, 12-, or 24-well plates could be used. Be aware that the number of cells to be seeded may need to scale up/down depending on the size of the inserts to be used for the experiment.

16. To fix cells, the plate with inserts could be incubated with the fixation reagent either under the hood without agitation or outside the hood under gentle agitation keeping the lid close.

17. To start with a particular imaging setup, it is recommended to use the "Dye assistant" in order to select the appropriate light paths and channel setup for the specific dyes used in your experiment. The final choice depends on the time you wish to invest in acquiring images and the quality of those images. Individual light paths (separate channels per fluorophore) will always give you the cleanest images with no bleed through, but imaging will take long time. In this specific protocol, three independent sequences were acquired, one per wavelength, to completely avoid any bleed through. However, fluorophores significantly separated (e.g., or DAPI and Texas Red or even GFP and Far Red) could be collected on the same channel simultaneously to speed up the acquisition time. In such a case, cross talk could be reduced or eliminated by adjusting the gating of each fluorophore within the Acquire tab.

18. HyD detectors are more sensitive than PMT detectors; therefore HyD detectors are preferred for dim fluorescence. However, a 405 laser with the ability to take "a pulse" for FLIM experiments only works with a PMT detector. Defining the gate allows to control the potential cross talk when two fluorophores emit too closely. A narrow gate will eliminate noise but too narrow will only acquire the fluorescence at the peak of emission. A wide gate could be useful if the signal of the sample is too weak.

19. Settings for image acquisition is a balance between acquisition time and image quality. "Format" refers to number of pixels to capture. Higher pixels, higher resolution, but longer time

acquisition and larger file. Increase of acquisition "Speed" will reduce acquisition time as well as the exposure of the fluorophore to the laser but increase noise. Either 400 or 600 Hz are a good speed to start tests. In addition, higher number of line and frame average will enhance image quality but require longer acquisition time. "Pinhole" is the size of the opening in which the laser passes through. For confocal, choose Airy 1.00 AU as the optimum pinhole diameter. However, if the signal is weak, open the pinhole to allow imaging a thicker section with a greater signal.

20. Z-Size and number of steps are based on specimen thickness. Either fix Z size and number of steps or fine-tune the focus knob to both top and bottom of the specimen to determine the Z-stack range. Of note, a Z-stack could be taken only if the Z has been activated in the acquisition mode. If a Z-stack is not required, then activate XY.
21. Other magnifications could be used. For example, 63× or 100× objective lens to zoom in a specific area (i.e., image smaller area with bigger resolution) or 10× objective lens to image a larger area of the sample with less resolution. Check the requirements of your objective lens (dry air, water, or oil immersion) before exchanging from one to another during the experiment.
22. When clicking on "Save as," a window will open asking the desired name and type of the file to save. Always save the file as "Leica Image File" (.lif) in order to save all the raw data, including metadata which is critical for setups, reusing imaging conditions, and writing acquisition data properties in the next experiment and/or method section in a future paper. However, to facilitate opening and analysis of the imaging files in the future, it is recommended to export images as .tiff or .jpeg format.

Acknowledgments

Caco-2 cells were kindly provided by Dr. Paloma Ordoñez-Morán from the University of Nottingham, UK. This work was supported by grants from the Academy of Medical Science/the Wellcome Trust/the Government Department of Business, Energy and Industrial Strategy/the British Heart Foundation/Diabetes UK AMS Springboard Award [SBF006\1070], and the CIDEGENT Excellence Research Program from the Valencian regional gover ment CIDEGENT/2021/026 to M.A.J.

References

1. Kayama H, Okumura R, Takeda K (2020) Interaction between the microbiota, epithelia, and immune cells in the intestine. Annu Rev Immunol 38:23–48. https://doi.org/10.1146/annurev-immunol-070119-115104
2. Peterson LW, Artis D (2014) Intestinal epithelial cells: regulators of barrier function and immune homeostasis. Nat Rev Immunol 14:141–153. https://doi.org/10.1038/nri3608
3. Conigrave AD, Young JA (1996) Function of the intestine. In: Greger R, Windhorst U (eds) Comprehensive human physiology. Springer, Berlin/Heidelberg, pp 1259–1287
4. Ali A, Tan H, Kaiko GE (2020) Role of the intestinal epithelium and its interaction with the microbiota in food allergy. Front Immunol 11:604054. https://doi.org/10.3389/fimmu.2020.604054
5. Kong S, Zhang YH, Zhang W (2018) Regulation of intestinal epithelial cells properties and functions by amino acids. Biomed Res Int 2018:2819154. https://doi.org/10.1155/2018/2819154
6. van der Flier LG, Clevers H (2009) Stem cells, self-renewal, and differentiation in the intestinal epithelium. Annu Rev Physiol 71:241–260. https://doi.org/10.1146/annurev.physiol.010908.163145
7. Spit M, Koo B-K, Maurice MM (2018) Tales from the crypt: intestinal niche signals in tissue renewal, plasticity and cancer. Open Biol 8. https://doi.org/10.1098/rsob.180120
8. Umar S (2010) Intestinal stem cells. Curr Gastroenterol Rep 12:340–348. https://doi.org/10.1007/s11894-010-0130-3
9. Bonis V, Rossell C, Gehart H (2021) The intestinal epithelium – fluid fate and rigid structure from crypt bottom to villus tip. Front Cell Dev Biol 9:661931. https://doi.org/10.3389/fcell.2021.661931
10. Juanes MA (2020) Cytoskeletal control and Wnt signaling-APC's dual contributions in stem cell division and colorectal cancer. Cancers (Basel) 12. https://doi.org/10.3390/cancers12123811
11. McCarthy N, Kraiczy J, Shivdasani RA (2020) Cellular and molecular architecture of the intestinal stem cell niche. Nat Cell Biol 22:1033–1041. https://doi.org/10.1038/s41556-020-0567-z
12. Clevers H (2013) The intestinal crypt, a prototype stem cell compartment. Cell 154:274–284. https://doi.org/10.1016/j.cell.2013.07.004
13. Ouladan S, Gregorieff A (2021) Taking a step back: insights into the mechanisms regulating gut epithelial dedifferentiation. Int J Mol Sci 22. https://doi.org/10.3390/ijms22137043
14. Seishima R, Barker N (2019) A contemporary snapshot of intestinal stem cells and their regulation. Differentiation 108:3–7. https://doi.org/10.1016/j.diff.2019.01.004
15. Beumer J, Clevers H (2021) Cell fate specification and differentiation in the adult mammalian intestine. Nat Rev Mol Cell Biol 22:39–53. https://doi.org/10.1038/s41580-020-0278-0
16. Klunder LJ, Faber KN, Dijkstra G, van Ijzendoorn SCD (2017) Mechanisms of cell polarity-controlled epithelial homeostasis and immunity in the intestine. Cold Spring Harb Perspect Biol 9. https://doi.org/10.1101/cshperspect.a027888
17. Crawley SW, Mooseker MS, Tyska MJ (2014) Shaping the intestinal brush border. J Cell Biol 207:441–451. https://doi.org/10.1083/jcb.201407015
18. Ding X, Hu X, Chen Y et al (2021) Differentiated Caco-2 cell models in food-intestine interaction study: current applications and future trends. Trends Food Sci Technol 107:455–465. https://doi.org/10.1016/j.tifs.2020.11.015
19. Chantret I, Barbat A, Dussaulx E et al (1988) Epithelial polarity, villin expression, and enterocytic differentiation of cultured human colon carcinoma cells: a survey of twenty cell lines. Cancer Res 48:1936–1942
20. Simon-Assmann P, Turck N, Sidhoum-Jenny M et al (2007) In vitro models of intestinal epithelial cell differentiation. Cell Biol Toxicol 23:241–256. https://doi.org/10.1007/s10565-006-0175-0
21. van Klinken BJ, Oussoren E, Weenink JJ et al (1996) The human intestinal cell lines Caco-2 and LS174T as models to study cell-type specific mucin expression. Glycoconj J 13:757–768
22. Natoli M, Leoni BD, D'Agnano I et al (2012) Good Caco-2 cell culture practices. Toxicol In Vitro 26:1243–1246. https://doi.org/10.1016/j.tiv.2012.03.009
23. Fatmawati NND, Goto K, Mayura IPB et al (2020) Caco-2 cells monolayer as an in-vitro model for probiotic strain translocation. Bali Med J 9:137. https://doi.org/10.15562/bmj.v9i1.1633

24. Hidalgo IJ, Raub TJ, Borchardt RT (1989) Characterization of the human colon carcinoma cell line (Caco-2) as a model system for intestinal epithelial permeability. Gastroenterology 96:736–749. https://doi.org/10.1016/0016-5085(89)90897-4
25. Kenny B, Dean P (2013) Do Caco-2 subclones provide more appropriate in vitro models for understanding how human enteric pathogens cause disease? Future Microbiol 8:701–703. https://doi.org/10.2217/fmb.13.51
26. Ferruzza S, Rossi C, Scarino ML, Sambuy Y (2012) A protocol for differentiation of human intestinal Caco-2 cells in asymmetric serum-containing medium. Toxicol In Vitro 26: 1252–1255. https://doi.org/10.1016/j.tiv.2012.01.008
27. Ferruzza S, Scacchi M, Scarino ML, Sambuy Y (2002) Iron and copper alter tight junction permeability in human intestinal Caco-2 cells by distinct mechanisms. Toxicol In Vitro 16: 399–404. https://doi.org/10.1016/S0887-2333(02)00020-6
28. Noben M, Vanhove W, Arnauts K et al (2017) Human intestinal epithelium in a dish: current models for research into gastrointestinal pathophysiology. United European Gastroenterol J 5:1073–1081. https://doi.org/10.1177/2050640617722903
29. Ponce de León-Rodríguez MDC, Guyot J-P, Laurent-Babot C (2019) Intestinal in vitro cell culture models and their potential to study the effect of food components on intestinal inflammation. Crit Rev Food Sci Nutr 59:3648–3666. https://doi.org/10.1080/10408398.2018.1506734
30. Riedl A, Schlederer M, Pudelko K et al (2017) Comparison of cancer cells cultured in 2D vs 3D reveals differences in AKT/mTOR/S6. J Cell Sci 130(1):203–218. Epub 2016 Sep 23. PMID: 27663511. https://doi.org/10.1242/jcs.188102
31. Juanes MA, Bouguenina H, Eskin JA et al (2017) Adenomatous polyposis coli nucleates actin assembly to drive cell migration and microtubule-induced focal adhesion turnover. J Cell Biol 216:2859–2875. https://doi.org/10.1083/jcb.201702007
32. Juanes MA, Isnardon D, Badache A et al (2019) The role of APC-mediated actin assembly in microtubule capture and focal adhesion turnover. J Cell Biol 218:3415–3435. https://doi.org/10.1083/jcb.201904165
33. Juanes MA, Fees C, Hoeprich GJ et al (2020) EB1 directly regulates APC-mediated actin nucleation. Curr Biol 30:4763
34. Sanderson MJ, Smith I, Parker I, Bootman MD (2014) Fluorescence microscopy. Cold Spring Harb Protoc 2014:pdb.top071795. https://doi.org/10.1101/pdb.top071795
35. Downloads.leica-microsystems.com/LeicaTCSSP8/Brochures/SP8-Lightning-Product-Flyer-201910-EN.pdf
36. Wang YL, Grooms NWF, Civale SC, Chung SH (2021) Confocal imaging capacity on a widefield microscope using a spatial light modulator. PLoS One 16:e0244034. https://doi.org/10.1371/journal.pone.0244034
37. Wang X, Kress A, Brasselet S, Ferrand P (2013) High frame-rate fluorescence confocal angle-resolved linear dichroism microscopy. Rev Sci Instrum 84:053708. https://doi.org/10.1063/1.4807318
38. SP8 LIGHTNING Confocal Microscope | Products | Leica Microsystems. https://www.leica-microsystems.com/products/confocal-microscopes/p/leica-tcs-sp8/media/. Accessed 7 July 2020
39. Ferraretto A, Gravaghi C, Donetti E et al (2007) New methodological approach to induce a differentiation phenotype in Caco-2 cells prior to post-confluence stage. Anticancer Res 27:3919–3925

Chapter 20

Automated Quantitative Analysis of Shape Features in Human Epithelial Monolayers and Spheroids Generated from Colorectal Cancer Cells

Hannah M. Brown and M. Angeles Juanes

Abstract

Advancements in microscopy techniques permit us to acquire endless datasets of images. A major bottleneck in cell imaging is how to analyze petabytes of data in an effective, reliable, objective, and effortless way. Quantitative imaging is becoming crucial to disentangle the complexity of many biological and pathological processes. For instance, cell shape is a summary readout of a myriad of cellular processes. Changes in cell shape use to reflect changes in growth, migration mode (including speed and persistence), differentiation stage, apoptosis, or gene expression, serving to predict health or disease. However, in certain contexts, e.g., tissues or tumors, cells are tightly packed together, and measurement of individual cellular shapes can be challenging and laborious. Bioinformatics solutions like automated computational image methods provide a blind and efficient analysis of large image datasets. Here we describe a detailed and friendly step-by-step protocol to extract various cellular shape parameters quickly and accurately from colorectal cancer cells forming either monolayers or spheroids. We envision those similar settings could be extended to other cell lines, colorectal and beyond, either label/unlabeled or in 2D/3D environments.

Key words HCT116, Colon cancer cells, Cell size, Machine learning, Segmentation, Quantification, Monolayers, Spheroids

1 Introduction

In this era of quantitative imaging becomes critical to show numbers from all types of experiments [1]. Cell imaging experiments usually involve large datasets of images from each experimental set. Adding many repeats from independent clones and/or cell lines results in exhaustive analysis. Retrieving biological insights from intricate processes and/or endless data in an objective manner is some of the bottlenecks in nowadays cell biologist's lives. Cell shape can relate to fundamental functions such as cell growth, migration, or differentiation, serving as indicators of health or disease [2–8].

Paloma Ordóñez-Morán (ed.), *Intestinal Differentiated Cells: Methods and Protocols*, Methods in Molecular Biology, vol. 2650,
https://doi.org/10.1007/978-1-0716-3076-1_20,

The starting point to analyze cell shape is image segmentation [9], in other words transforming cell shape into a form that can be computationally processed. Multiple machine learning models for segmentation have been developed which provide satisfactory results. However, little can be used in a friendly way and/or allow users to quickly adapt and/or customize pipelines to their specific images and user needs. Customized analysis of large datasets has been facilitated by many collaborative efforts between experts from different fields of research [10].

Recent strategies have been used to develop a user-friendly pipeline for training data while maintaining the state-of-the-art segmentation performance. One example is Cellpose 2.0 package [11]. Cellpose 2.0 uses an ensemble of pretrained models with different segmentation styles, allowing to test a wide range of cell shapes [11]. In addition, it is possible to retrain the package for users' needs and datasets. This tool, combined with the traditional ImageJ/Fiji software [12, 13], allows obtaining accurate contour for cell shape, which usually takes information from the cell membrane. A series of cell shape parameters (e.g., size, roundness, solidity) could be extracted from large datasets of images.

In this chapter we describe the protocols to efficiently analyze cell shape parameters from a collection of cells forming (i) 2D monolayers using an established model (Subheading 3.1) or (ii) 3D spheroids using a customized model trained using artificial intelligence (Subheading 3.2). We provide some representative images that can be taken as examples for those analyses (doi links available in Subheadings 2 and 3). At the end of the chapter, a section named "Notes" has been added with the purpose to help researchers from any field to improve and/or troubleshoot setups when analyzing this and/or other cell lines.

2 Materials/Equipment

1. Computer with at least 8GB of RAM—larger capacity for larger datasets—and CUDA-enabled NVIDIA GPU with the appropriate drivers (*see* **Note 1**). This protocol uses Windows 10 with GPU NVIDIA GeFORCE GTX 750 Ti and GeForce Game Ready Driver version 512.59 (GeForce Game Ready Driver | 512.59 | Windows 10 64-bit, Windows 11 | NVIDIA).
2. Anaconda Distribution with conda version 4.12.0 (Anaconda | Anaconda Distribution).
3. ImageJ/FIJI version 1.8.0_172 (Downloads (imagej.net);) [12].
4. jython ImageJ plugin (GitHub – jython/jython: Python for the Java Platform) .

5. imagej_roi_converter.py ImageJ plugin (cellpose/imagej_roi_converter.py at main · MouseLand/cellpose · GitHub)
6. Cell images to analyze (own images or provided examples available in Mendeley repository, whose doi are: 10.17632/xmm5vfdmr2.1 and 10.17632/8f4gkfgh86.1)

3 Methods

3.1 Quantitative Analysis of Cell Shape Features of HCT116 Colorectal Cancer Cells Forming 2D Monolayers

1. Setup of the Cellpose2.0 software by opening the Anaconda Navigator and clicking on the CMD.exe prompt. This will open a black window which has options to type commands.
2. Write the following commands:

   conda create --name cellpose python = 3.8, then type y.

   conda activate cellpose,

   python –m pip install cellpose[all], then type y (*see* **Note** 2).

   pip uninstall torch, then type y,

   conda install pytorch cudatoolkit = 11.3 -c pytorch, then press y,

   python –m cellpose. This will bring up the GUI. Go to the CMD.exe, and check,

   for "TORCH CUDA version installed and working" (*see* **Notes 3** and **4**).
3. Open ImageJ/Fiji.
4. Drag your images to ImageJ/Fiji. Otherwise go to "Files," click on "Open," and look for the location of the images. If you do not have your own images, you could use the provided image available in this link (doi: 10.17632/xmm5vfdmr2.1).
5. In ImageJ/Fiji, click "Analyze" and then "Tool," followed by "ROI manager." Keep this box open until the average cell diameter is determined. Estimate the cell diameter of your cells of interest. For that, use the "straight" tool from ImageJ/Fiji. Click on it to draw a line from side to side of a cell. Click on "Add" from the ROI manager window or use the shortcut "Ctrl + T."
6. Repeat this step for 5-6 cells.
7. Highlight all labels from ROI manager. Click 'Measure' on ROI manager. A window with the results will open, then save this as a .csv file. Average the results to get the cell diameter for that cell line.
8. Create a folder (on the desktop for convenience) with the image(s) you wish to analyze.
9. Drag an independent image to analyze into the Cellpose 2.0 GUI. Alternatively, go to the top left bar of the Cellpose 2.0

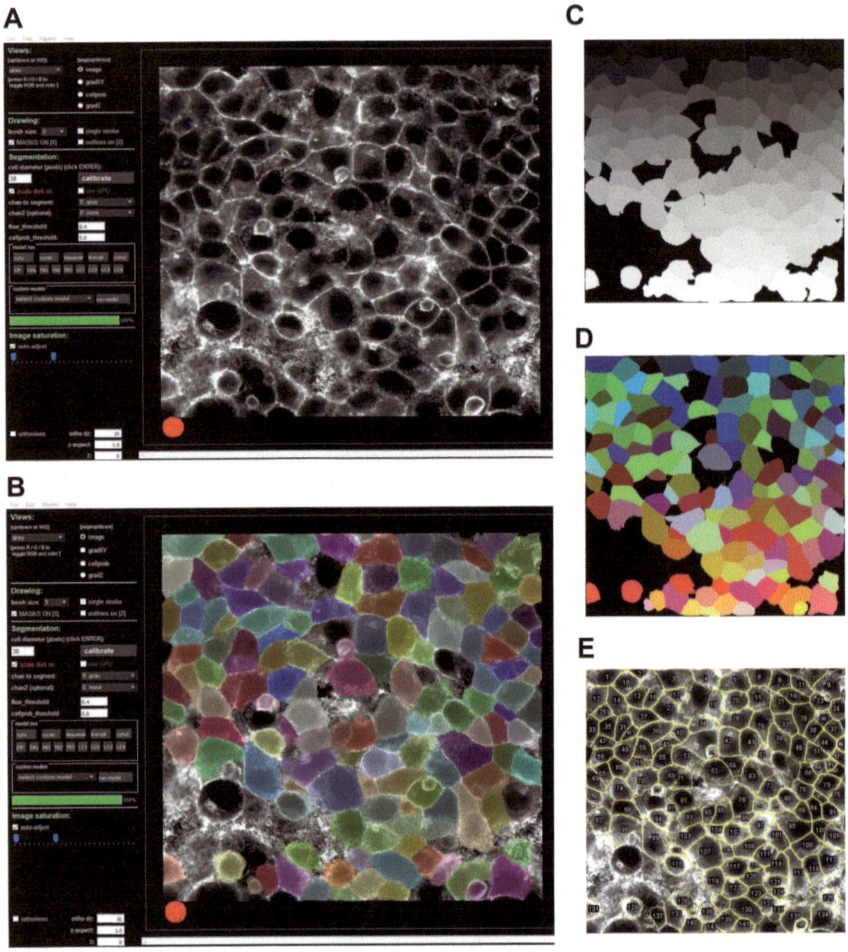

Fig. 1 Shows the interface displayed in the Cellpose 2.0 package for HCT116 monolayers after one .tiff image has been uploaded as described in the protocol using the CP model, Subheading 3.1 (**a**); the respective color corrected segmented image of the independent cells (**b**); .tiff files obtained in gray scale (upper image), RGB color (middle image), and with the ROI labels for each cell (bottom image) (**c**)

and click on "File" and then on "Load image." At this point, the image should pop-in into the right side of the Cellpose 2.0 (*see* Fig.1).

10. Under the top bar, "Views" tab, click on RGB to view all channels at once. Individual channels can be seen in their respective color that they have been acquired. If your image contains only one channel, ignore this step.
11. The tab below, "Drawing," can be ignored for 2D analysis of HCT116 cells. This could be used for training your own artificial intelligence model (AI) (*see* Subheading 3.2).
12. In the "Segmentation" tab, change the cell diameter from default to the average you have previously calculated. Press

Enter. Keep the "scale disk on" to visually compare the red dot showing cell diameter on Cellpose versus cells on your images (*see* **Note 5**).

13. Ensure that "use GPU" option is clicked (*see* **Note 4**).
14. To analyze one channel, as in the example, keep channel to segment in gray or in the color that the image has been acquired. For multichannel images, follow GitHub - MouseLand/cellpose: a generalist algorithm for cellular segmentation with human-in-the-loop capabilities.
15. Thresholds below can be kept as default (*see* **Note 6**).
16. Choose the appropriate model for your images. In particular, for the HCT116 monolayers provided, the most accurate model is CP. Click on "CP" to run the model. Allow some time for the model to segment the cells.
17. Ignore the other models and leave image saturation as default since it will auto adjust.
18. Once the run is completed, the image in the GUI will show a mask overlay (*see* Fig. 1. Save this image pressing Control + N (this saves the masks in gray as .png; despite initially appearing in color; masks can be changed to RGB colors as seen in **step 23**) and then Control + O (this saves the outlines as .txt file). These will be saved in the same location as the image (e.g., Desktop folder).
19. To start analysis in ImageJ/Fiji, first set the measurements. Under "Analyze," click on "Set measurements," e.g., area, shape descriptors (roundness, circularity, solidity), and display labels.
20. Open one raw .tiff file and run the macro named imagej_roi_-converter.py. For that, go to Plugins > Macros > Run > look for the .py file. A window will pop in asking the location of the outlines (.txt file). Click on it, and then on Run. Another ROI manager window will pop up with "Show all" and "Labels" displayed at the bottom. Click on those boxes and then on the image which will show the segmented cells with the respective ROI labels.
21. Highlight (select) all labels from the ROI manager window, then click on "Measure" from that same window.
22. Save the results as a .csv file, as it contains the raw segmentation results from the different parameters that initially have been selected. Additionally, save the ROI labels by clicking on "More" and then Save. It is advised to save also the .tiff image with overlaid/selected and labeled cells (*see* **Note 7**).
23. Convert the gray mask output from Cellpose 2.0 to a color mask by dragging the .png with the gray masks into ImageJ/

Fiji. Go to Image > Lookup Tables and click "3-3-2 RGB." Then save this image as a .tiff file.

24. Repeat **steps 4–23** for each image to analyze keeping the same cell diameter.
25. To exit the software, close the Cellpose 2.0 GUI and type "exit" into CMD.exe.
26. To open the program at another time, open the Anaconda Navigator and type "conda activate cellpose." Then, type "python -m cellpose." This should open the software without having to reinstall the packages. Then continue **steps 3–25**.

3.2 Quantitative Analysis of Individual Cell Shape Parameters Forming 3D HCT116 Spheroids

Spheroids are 3D cellular "cluster" models generated from single cells in suspension that either self-assemble or are forced to grow in a confined compartment. Spheroids are well documented to retain intrinsic microenvironment properties and functional similarities than solid tumors, thereby mimicking patient response in the laboratory [14–18]. Genetic and/or pharmacological treatments can be tested in spheroids, providing more physiological results than 2D cellular models [16, 19–24]. However, one of the main challenges of generating spheroids is the lack of uniformity. Developments have been positively achieved in spheroid reproducibility [20, 25, 26]. Regardless, it is difficult and laborious to analyze the shapes of individual cells forming each spheroid. Here we explain how to analyze single cells fast and accurately.

For stained spheroids, the same settings explained above for the CP model could be used (Subheading 3.1). For unlabeled spheroids, it is recommended to use the "live cell" model. However, we found that, at least for HCT116 colorectal cancer unlabeled spheroids, the latter model cannot accurately identify the individual cells. Therefore, we provide a protocol for developing a customized model by training the AI:

1. If Cellpose 2.0. and ImageJ/Fiji are already open, start by dragging your image to ImageJ/Fiji, and estimate the average diameter of individual cells in the spheroid (see **steps 4–9** from Subheading 3.1). Otherwise, open both software as stated in Subheading 3.1. An image can be downloaded to use as example from this link: doi: 10.17632/8f4gkfgh86.1.
2. Keep "Views" tab as above **step 10** in Subheading 3.1.
3. In the tab below, "Drawing," the brush size can be adjusted to the specific contour of your cells. In this HCT116 example, it was changed to size 1.
4. In the "Segmentation" tab, change the cell diameter from default to the average you have previously calculated. Press Enter. Keep the "scale disk on" to visually compare the red dot showing cell diameter on Cellpose versus cells on your images (*see* **Note 5**).

5. As the provided example is from one unlabeled spheroid, the channel to segment will be gray.
6. Ensure that "use GPU" option is clicked (*see* **Notes 3** and **4**).
7. Thresholds below can be kept as default (*see* **Note 6**).
8. Choose the appropriate model for your images (livecell/ LC1–4) by clicking on the "Model" bottom, and check visually which model detects the larger number of cells in an accurate manner. In the example provided for HCT116 cells as spheroid, we recommend using LC1. Allow some time for the model to segment the cells.
9. Once the run is completed, the image in the GUI will show a mask overlay (*see* Fig. 2).
10. Remove inaccurate masks from non-cells or cells which have not been fully segmented. To do this, click together the keys "Control + Left mouse."
11. Click "Right mouse" to add masks on the cells, and release the bottom, and then draw around the entire cell. This will add color and fill up the mask around that cell. Repeat the same for each cell (*see* **Note 8**).
12. Go to "Models" in the top bar. Click on "Train new model with image + mask in folder." At this point, a window will pop up with different parameters to train the AI. Change the "chan to segment" and "initial model" to options in **steps 5** and **8** and rename the model. Keep the other settings the same (*see* Fig. 3).
13. The model will be saved automatically into the same folder as the images in another folder titled "models."
14. Repeat **steps 10–12** until you are satisfied with the cell masks obtained.
15. Save this image pressing Control + N (this saves the masks in gray as .png; despite initially appearing in color; masks can be changed to RGB colors) and then Control + O (this saves the outlines as .txt file). These will be saved in the same location as the image (e.g., Desktop folder).
16. To analyze shape features, proceed as above, and follow **steps 19–25** of Subheading 3.1.
17. To reuse the custom model in future datasets and/or experiments, this same model can be loaded by clicking on Model > Add custom torch model to GUI.

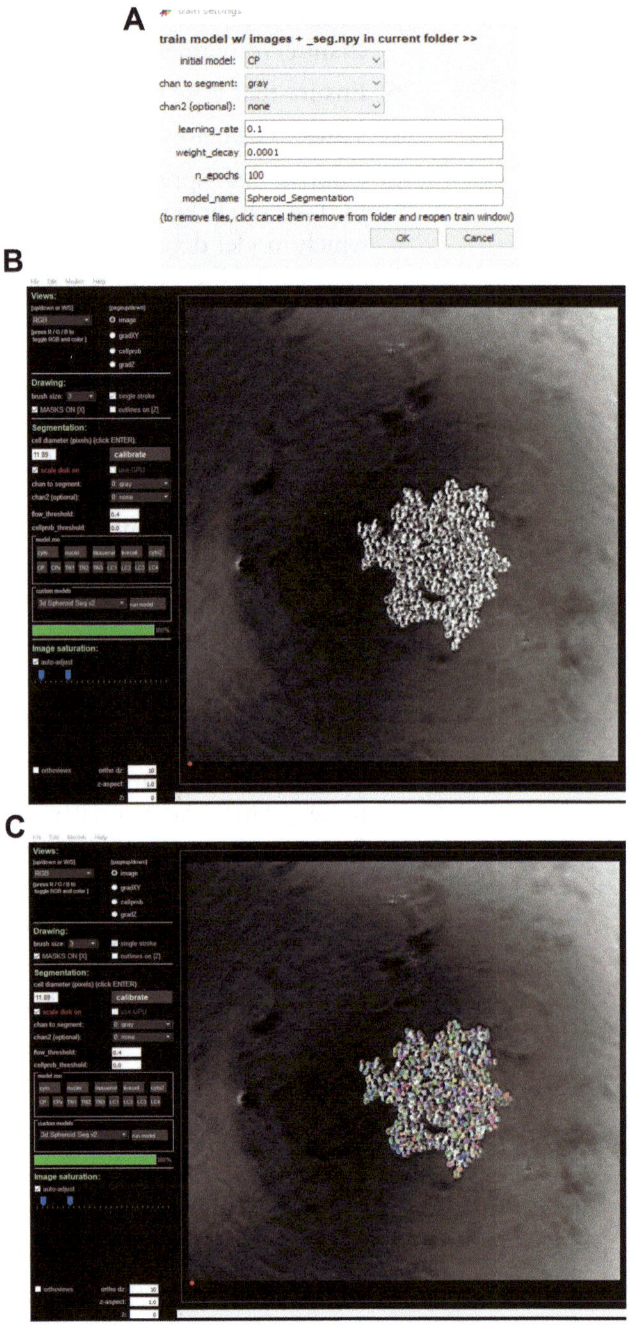

Fig. 2 Displays setups to train the model using artificial intelligence for one 3D HCT116 spheroid (**a**); the interface in the Cellpose 2.0 package once the .tiff image has been uploaded as described in the protocol for spheroids using the customized trained model, Subheading 3.2 (**b**); and segmented image showing independent cells in color (**c**)

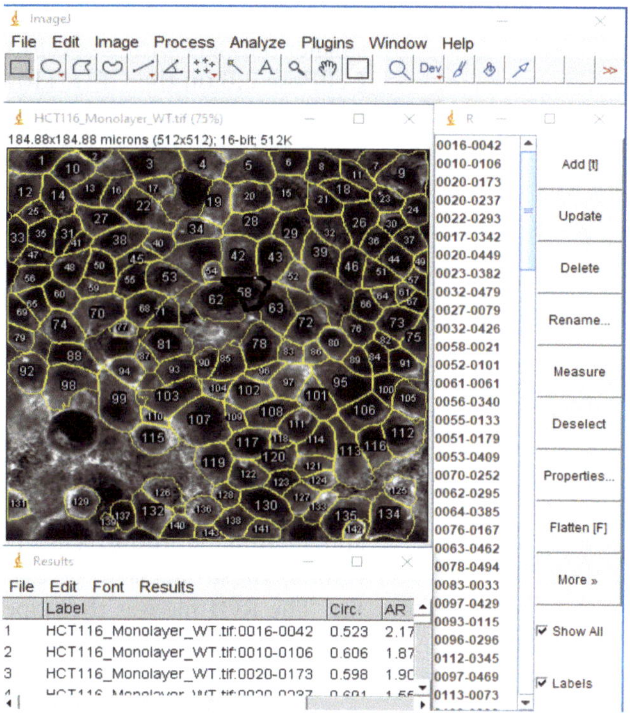

Fig. 3 Shows the interface displayed in the ImageJ/Fiji software once the .tiff image with the ROI labels for each cell has been uploaded as described in the protocol for HCT116 monolayers using the CP model, Subheading 3.1, as well as the ROI manager window (right side of the figure) and results (.csv file, bottom part of the figure)

4 Notes

1. To check which NVIDIA drivers are compatible or for help with installation, use the following links: https://docs.nvidia.com/deploy/cuda-compatibility/index.html

   https://docs.nvidia.com/cuda/cuda-installation-guide-microsoft-windows/index.html

2. If the packages installed have been blocked by an administration, it is possible to create a temporary conda environment by typing "--user" before the name of the package, e.g., "python pip install --user cellpose[all]."

3. If the TORCH CUDA is not working/installed, check which drivers the computer is using and if they are compatible with cudatoolkit version 11.3. You may have to downgrade the cudatoolkit version to 10.2, depending on GPU driver version, by typing "pip uninstall torch" and then "conda install pytorch cudatoolkit=10.2 -c pytorch," and then open the program.

4. The GPU is highly recommended for the software, but for smaller datasets with less cells to segment, it is possible to run CellPose 2.0 on the CPU. However, performance will be slow.
5. The "calibrate" button may be used to estimate average cell diameter, but we found that manually estimating the average cell diameter led to the most accurate cell segmentation.
6. The flow threshold is the maximum allowed error of the flows for each mask produced. Increasing the thresholds limits the ROIs produced. The threshold should be decreased if the masks are too ill-shaped. Lowering the cellprob threshold increases the number of pixels in each mask and vice versa. This has a range of −6 to 6.
7. We created a minimum threshold for the average area of the cell to remove inaccurate cells. This was done by calculating the lower boundary of the cell line (from previous experiments) and identifying those cells. Following the labels within the results, those labels were manually removed from the ROI file. The ROI file was then opened in ImageJ with the respective image and the new labels were measured.
8. If the mask produced by Cellpose 2.0 encompasses another cell, delete the mask, and draw around both cells as the overlays will not overlap.

Acknowledgments

Tiff images of HCT116 cell monolayers and 3D spheroids were kindly provided by Lautaro Baro and Dr. Asifa Islam, respectively, both members of our lab. Images have been deposited in Zenodo repository (doi: 10.52181/zenodo.7679086). This work was supported by grants from the Academy of Medical Science/the Wellcome Trust/the Government Department of Business, Energy and Industrial Strategy/the British Heart Foundation/Diabetes AMS Springboard Award UK [SBF006\1070], and the CIDEGENT Excellent Research Program from the Valencian regional government CIDEGENT/2021/026 to M.A.J.

References

1. Wrigley N (1986) Quantitative methods: the era of longitudinal data analysis. Prog Hum Geogr 10:84–102. https://doi.org/10.1177/030913258601000105
2. Thakuri PS, Gupta M, Plaster M, Tavana H (2019) Quantitative size-based analysis of tumor spheroids and responses to therapeutics. Assay Drug Dev Technol 17:140–149. https://doi.org/10.1089/adt.2018.895
3. Haftbaradaran Esfahani P, Knöll R (2020) Cell shape: effects on gene expression and signaling. Biophys Rev 12:895–901. https://doi.org/10.1007/s12551-020-00722-4
4. Barker CG, Petsalaki E, Giudice G et al (2022) Identification of phenotype-specific networks from paired gene expression-cell shape imaging data. Genome Res 32:750–765. https://doi.org/10.1101/gr.276059.121

5. Bodor DL, Pönisch W, Endres RG, Paluch EK (2020) Of cell shapes and motion: the physical basis of animal cell migration. Dev Cell 52: 550–562. https://doi.org/10.1016/j.devcel.2020.02.013
6. Cadart C, Zlotek-Zlotkiewicz E, Le Berre M et al (2014) Exploring the function of cell shape and size during mitosis. Dev Cell 29: 159–169. https://doi.org/10.1016/j.devcel.2014.04.009
7. Luxenburg C, Zaidel-Bar R (2019) From cell shape to cell fate via the cytoskeleton – insights from the epidermis. Exp Cell Res 378:232–237. https://doi.org/10.1016/j.yexcr.2019.03.016
8. Gupta M, Doss BL, Kocgozlu L et al (2019) Cell shape and substrate stiffness drive actin-based cell polarity. Phys Rev E 99:012412. https://doi.org/10.1103/PhysRevE.99.012412
9. Xiong Y, Iglesias PA (2010) Tools for analyzing cell shape changes during chemotaxis. Integr Biol (Camb) 2:561–567. https://doi.org/10.1039/c0ib00036a
10. Post JN, Loerakker S, Merks R, Carlier A (2022) Implementing computational modeling in tissue engineering: where disciplines meet. Tissue Eng Part A 28:542. https://doi.org/10.1089/ten.TEA.2021.0215
11. Stringer C, Wang T, Michaelos M, Pachitariu M (2021) Cellpose: a generalist algorithm for cellular segmentation. Nat Methods 18:100–106. https://doi.org/10.1038/s41592-020-01018-x
12. Schindelin J, Arganda-Carreras I, Frise E et al (2012) Fiji: an open-source platform for biological-image analysis. Nat Methods 9:676–682. https://doi.org/10.1038/nmeth.2019
13. Schneider CA, Rasband WS, Eliceiri KW (2012) NIH image to ImageJ: 25 years of image analysis. Nat Methods 9:671–675. https://doi.org/10.1038/nmeth.2089
14. Jeppesen M, Hagel G, Glenthoj A et al (2017) Short-term spheroid culture of primary colorectal cancer cells as an in vitro model for personalizing cancer medicine. PLoS One 12: e0183074. https://doi.org/10.1371/journal.pone.0183074
15. Fennema E, Rivron N, Rouwkema J et al (2013) Spheroid culture as a tool for creating 3D complex tissues. Trends Biotechnol 31: 108–115. https://doi.org/10.1016/j.tibtech.2012.12.003
16. Zanoni M, Piccinini F, Arienti C et al (2016) 3D tumor spheroid models for in vitro therapeutic screening: a systematic approach to enhance the biological relevance of data obtained. Sci Rep 6:19103. https://doi.org/10.1038/srep19103
17. Ro J, Kim J, Cho Y-K (2022) Recent advances in spheroid-based microfluidic models to mimic the tumour microenvironment. Analyst 147:2023. https://doi.org/10.1039/d2an00172a
18. Sogomonyan AS, Shipunova VO, Soloviev VD et al (2022) 3D models of cellular spheroids as a universal tool for studying the cytotoxic properties of anticancer compounds in vitro. Acta Nat 14:92–100. https://doi.org/10.32607/actanaturae.11603
19. Castro F, Leite Pereira C, Helena Macedo M et al (2021) Advances on colorectal cancer 3D models: the needed translational technology for nanomedicine screening. Adv Drug Deliv Rev 175:113824. https://doi.org/10.1016/j.addr.2021.06.001
20. Azizipour N, Avazpour R, Weber MH et al (2022) Uniform tumor spheroids on surface-optimized microfluidic biochips for reproducible drug screening and personalized medicine. Micromachines (Basel) 13. https://doi.org/10.3390/mi13040587
21. Riedl A, Schlederer M, Pudelko K et al (2017) Comparison of cancer cells in 2D vs 3D culture reveals differences in AKT-mTOR-S6K signaling and drug responses. J Cell Sci 130:203–218. https://doi.org/10.1242/jcs.188102
22. Kunz-Schughart LA, Freyer JP, Hofstaedter F, Ebner R (2004) The use of 3-D cultures for high-throughput screening: the multicellular spheroid model. J Biomol Screen 9:273–285. https://doi.org/10.1177/1087057104265040
23. Sant S, Johnston PA (2017) The production of 3D tumor spheroids for cancer drug discovery. Drug Discov Today Technol 23:27–36. https://doi.org/10.1016/j.ddtec.2017.03.002
24. Duval K, Grover H, Han L-H et al (2017) Modeling physiological events in 2D vs. 3D cell culture. Physiology (Bethesda) 32:266–277. https://doi.org/10.1152/physiol.00036.2016
25. Kelm JM, Timmins NE, Brown CJ et al (2003) Method for generation of homogeneous multicellular tumor spheroids applicable to a wide variety of cell types. Biotechnol Bioeng 83: 173–180. https://doi.org/10.1002/bit.10655
26. Cisneros Castillo LR, Oancea A-D, Stüllein C, Régnier-Vigouroux A (2016) Evaluation of consistency in spheroid invasion assays. Sci Rep 6:28375. https://doi.org/10.1038/srep28375

Index

Paloma Ordóñez-Morán (ed.), *Intestinal Differentiated Cells: Methods and Protocols*, Methods in Molecular Biology, vol. 2650, https://doi.org/10.1007/978-1-0716-3076-1,

T

V

W

MIX
Papier aus verantwortungsvollen Quellen
Paper from responsible sources
FSC® C105338

If you have any concerns about our products,
you can contact us on
ProductSafety@springernature.com

In case Publisher is established outside the EU,
the EU authorized representative is:
Springer Nature Customer Service Center GmbH
Europaplatz 3, 69115 Heidelberg, Germany

Printed by Libri Plureos GmbH
in Hamburg, Germany